Molecular Realizations
of Quantum Computing
2007

Kinki University Series on Quantum Computing

Editor-in-Chief: Mikio Nakahara *(Kinki University, Japan)*

ISSN: 1793-7299

Published

Vol. 1 Mathematical Aspects of Quantum Computing 2007
*edited by Mikio Nakahara, Robabeh Rahimi (Kinki Univ., Japan) &
Akira SaiToh (Osaka Univ., Japan)*

Vol. 2 Molecular Realizations of Quantum Computing 2007
*edited by Mikio Nakahara, Yukihiro Ota, Robabeh Rahimi, Yasushi Kondo &
Masahito Tada-Umezaki (Kinki Univ., Japan)*

Kinki University Series on Quantum Computing – Vol. 2

editors

Mikio Nakahara
Yukihiro Ota
Robabeh Rahimi
Yasushi Kondo
Masahito Tada-Umezaki

Kinki University, Japan

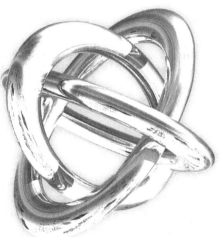

Molecular Realizations of Quantum Computing 2007

World Scientific

NEW JERSEY · LONDON · SINGAPORE · BEIJING · SHANGHAI · HONG KONG · TAIPEI · CHENNAI

Published by

World Scientific Publishing Co. Pte. Ltd.
5 Toh Tuck Link, Singapore 596224
USA office: 27 Warren Street, Suite 401-402, Hackensack, NJ 07601
UK office: 57 Shelton Street, Covent Garden, London WC2H 9HE

British Library Cataloguing-in-Publication Data
A catalogue record for this book is available from the British Library.

MOLECULAR REALIZATIONS OF QUANTUM COMPUTING 2007
Kinki University Series on Quantum Computing — Vol. 2

Copyright © 2009 by World Scientific Publishing Co. Pte. Ltd.

All rights reserved. This book, or parts thereof, may not be reproduced in any form or by any means, electronic or mechanical, including photocopying, recording or any information storage and retrieval system now known or to be invented, without written permission from the Publisher.

For photocopying of material in this volume, please pay a copying fee through the Copyright Clearance Center, Inc., 222 Rosewood Drive, Danvers, MA 01923, USA. In this case permission to photocopy is not required from the publisher.

ISBN-13 978-981-283-867-4
ISBN-10 981-283-867-8

Printed in Singapore.

PREFACE

This is the second volume of the "Kinki University Series on Quantum Computing". It contains lecture notes presented at the symposium on *Molecular Realizations of Quantum Computing 2007*, held from October 31st to November 2nd, 2007 at Yutorito Hall, Higashi-Osaka, Japan.

The main purpose of this symposium has been presenting the state of the art for quantum computing; particularly the technology related to molecular spin systems. For this symposium, we aimed at guiding the researchers and students for working toward a feasible realization of a quantum computer. Invited speakers to the symposium were asked to prepare their lecture notes in a self-contained manner and therefore we expect that each contribution will be useful for students and researchers even with less background to the topic.

Editors of this volume have carefully read and examined the lecture notes. The lecture notes being written by physicists and chemists, however, some style differences naturally inhere in them, which have been left unchanged by editors to respect each lecturer's writing style.

This symposium was supported by "Open Research Center" Project for Private Universities: matching fund subsidy from MEXT (Ministry of Education, Culture, Sports, Science and Technology).

We yearn for continued outstanding success in this field and wish to expand conferences, workshops and summer schools further in the field of quantum computing based at Kinki University.

Finally, we would like to thank Ms Zhang Ji of World Scientific for her excellent editorial work.

Osaka, March 2009

Mikio Nakahara
Yukihiro Ota
Robabeh Rahimi
Yasushi Kondo
Masato Tada

Symposium on Molecular Realizations of Quantum Computing 2007

Yutorito Hall Higashi-Osaka, Japan

31 October - 2 November 2007

31 October

Yasushi Kondo (Kinki University, Japan)
 Quantum computation using nuclear magnetic resonance: as a testbench for realistic study of quantum algorithms

Antti Niskanen (VTT, Finland)
 Flux qubits, tunable coupling and beyond

1 November

Mika Sillanpää (Helsinki University of Technology, Finland)
 Josephson phase qubits, and quantum communication via a resonant cavity

Takeji Takui (Osaka City University, Japan)
 Quantum computing using electron-nuclear double resonance

Yukihiro Ota (Kinki University, Japan)
 Errors in a plausible scheme of quantum gates in Kane's model

Akira SaiToh (Osaka Univerity, Japan)
 Conservation of nonclassical correlation useful for playing a game: a numerical study

Toshiki Ide (Okayama Institute for Quantum Physics, Japan)
 Accidental cloning of a single-photon qubit in two-channel

continuous-variable quantum teleportation

2 November

Tomonari Wakabayashi (Kinki University, Japan)
   Fullerene-C_{60}: A possible molecular quantum computer

Takayoshi Kuroda (Kinki University, Japan)
   Molecular magnets for quantum computation

List of PARTICIPANTS

Bando, Masamitsu	Kinki University, Japan
Ebrahimi Bakhtavar, Vahideh	Kinki University, Japan
Goto, Yoshito	Kinki University, Japan
Ide, Toshiki	Okayama Institute, Japan
Iwai, Toshihiro	Kyoto University, Japan
Kikuta, Toshiyuki	Kinki University, Japan
Koge, Satoru	Kinki University, Japan
Kondo, Yasushi	Kinki University, Japan
Kuroda, Takayoshi	Kinki University, Japan
Matsumoto, Hidenobu	Kinki University, Japan
Mimori, Takahiro	Kyoto University, Japan
Minami, Kaori	Kinki University, Japan
Mori, Fumiya	Kinki University, Japan
Nakahara, Mikio	Kinki University, Japan
Nakazawa, Shigeaki	Osaka City University, Japan
Niskanen, Antti	VTT, Finland
Ohmi, Tetsuo	Kinki University, Japan
Ootsuka, Takayoshi	Kinki University, Japan
Ota, Yukihiro	Kinki University, Japan
Rahimi Darabad, Robabeh	Kinki University, Japan
SaiToh, Akira	Osaka University, Japan
Sakuragi, Daizo	Kinki University, Japan
Sillanpää, Mika	Helsinki University of Technology, Finland
Takui, Takeji	Osaka City University, Japan
Tanimura, Shogo	Kyoto University, Japan
Tomita, Hiroyuki	Kyoto University, Japan
Wakabayashi, Tomonari	Kinki University, Japan
Yasuda, Taiga	Kinki University, Japan
Yoshino, Tomohiro	Kinki University, Japan
	Osaka City University, Japan

CONTENTS

Preface   v

Liquid-state NMR Quantum Computer: Working Principle
and Some Examples   1
    Y. Kondo

Flux qubits, Tunable Coupling and Beyond   53
    A. O. Niskanen

Josephson Phase Qubits, and Quantum Communication via
a Resonant Cavity   56
    M. A. Sillanpää

Quantum Computing Using Pulse-based Electron-nuclear
Double Resonance (ENDOR): Molecular Spin-qubits   58
    K. Sato, S. Nakazawa, R. D. Rahimi, S. Nishida, T. Ise,
    D. Shimoi, K. Toyota, Y. Morita, M. Kitagawa,
    P. Carl, P. Höfner, T. Takui

Fullerene C_{60}: A Possible Molecular Quantum Computer   163
    T. Wakabayashi

Molecular Magnets for Quantum Computation   193
    T. Kuroda

Errors in a Plausible Scheme of Quantum Gates in Kane's
Model   207
    Y. Ota

Yet Another Formulation for Quantum Simultaneous
Noncooperative Bimatrix Games   223
   A. SaiToh, R. Rahimi, M. Nakahara

Continuous-variable Teleportation of Single-photon
States and an Accidental Cloning of a Photonic Qubit in
Two-channel Teleportation   243
   T. Ide

LIQUID-STATE NMR QUANTUM COMPUTER: WORKING PRINCIPLE AND SOME EXAMPLES

Yasushi Kondo

Department of Physics, Kinki University
Higashi-Osaka 577-8502, Japan
E-mail: kondo@phys.kindai.ac.jp

We review the working principle of a liquid-state Nuclear Magnetic Resonance (NMR) quantum computer where spins of nuclei in isolated molecules dissolved in a solvent work as qubits and they are controlled by radio frequency magnetic field pulses. Then, we show some of our recent experimental results, "generation and suppression of artificial decoherence" and "quantum teleportation without irreversible detection".

Keywords: Quantum Computer; NMR; Teleportation; Decoherence

1. Introduction

Quantum information processing is now a very active field because of its potential power.[1] If it is realized, it may be used for factoring large numbers[2] and for breaking the encryption system widely used in INTERNET business. Among various physical systems proposed for a realization of a quantum computer,[3] a liquid-state NMR (nuclear magnetic resonance) is considered most successful thanks to its well established techniques,[4-8] which can be seen in a beautiful demonstration of Shor's factorization algorithm,[9] for example.

The idea of quantum information processing is as follows. Suppose we would like to implement a quantum algorithm, whose unitary operator representation is U_{alg}. Our task is to find the control parameters $\gamma(t)$ in the Hamiltonian $\mathcal{H}$ such that the time development operator

$$U_{\text{alg}}[\mathcal{H}, t] = \mathcal{T} \exp\left[-\frac{i}{\hbar} \int_0^t \mathcal{H}(\gamma(t'))dt'\right] \quad (1)$$

is equal to U_{alg}, where $\mathcal{T}$ stands for the time-ordered product and $\hbar$ is Planck's constant. In the quantum information processing, one can take advantages of quantum mechanical properties, such as superposition of states and entanglement, which essentially differ from those of classical mechanics.

This contribution is organized as follows. In Sec. 2 the principle of NMR is discussed. In Sec. 3 we explain the idea of a liquid-state NMR quantum computer, hereafter referred to as an NMR-QC. We show how to realize an NMR-QC and discuss its fundamental limitations.[a] Sec. 4 is devoted to discussions on "generation and suppression of artificial decoherence"[11] and "quantum teleportation without irreversible detection"[12] that are our recent experimental results.

Other aspects of NMR-QC are available in excellent reviews.[3,6,13–15]

2. Principle of NMR

Here we describe the principle of NMR for isolated spin-$\frac{1}{2}$ nuclei. Subsec. 2.1 provides an intuitive view of the NMR principle with the *vector model*. Then, we re-examine the principle from quantum mechanical viewpoint in Subsec. 2.2.

2.1. *Vector model*

The so-called *vector model* provides an intuitive view of NMR spectroscopy.

2.1.1. *Magnetic moment*

A macroscopic sample placed in a static magnetic field $\vec{H}_0$ have a bulk magnetic moment $\vec{M}_0$ which is parallel to $\vec{H}_0$. Note that we assume the linear response of the isotropic sample to the magnetic field. We set the z'-axis along the direction of $\vec{H}_0$, and thus $\vec{H}_0 = (0, 0, H_0)$ by definition. The x'- and y'-axes are arbitrarily fixed in space. We call this coordinate system the *laboratory frame*.

2.1.2. *Precession*

The dynamics of $\vec{M}$ in $\vec{H}_0$ is governed by

$$\frac{d\vec{M}}{dt} = \gamma \vec{M} \times \vec{H}_0, \qquad (2)$$

where γ is the gyromagnetic ratio which depends on the sample material. We assume $\gamma > 0$ in this contribution for the sake of illustration. When

---

[a]Braunstein *et al.* pointed out that no entanglement appears in the physical states at any stage of NMR experiments so far.[10]

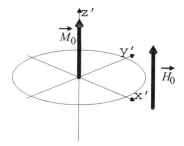

Fig. 1. Static magnetic field $\vec{H}_0$ and induced magnetic moment $\vec{M}_0$.

$\vec{M}$ is somehow tilted from $\vec{H}_0$, $\vec{M}$ starts to *precess* (rotate) about the z'-axis according to Eq. (2). The rotating frequency $\omega_0 = \gamma H_0$ is called the *Larmor* (resonance) frequency.[b] Typical values of the Larmor frequency are 42.59 MHz and 10.71 MHz for a proton and a carbon-13, respectively, under the field of flux density 1 T. The magnetization $\vec{M}$ is at rest when this precession motion is viewed in a *rotating frame*, which rotates around the z'-axis with $\vec{M}$, Therefore, an effective magnetic field for $\vec{M}$ in the rotating frame can be considered as zero. The z-axis is taken along the direction of $\vec{H}_0$ and the x- and the y-axes co-rotate with $\vec{M}$.

More generally, $\vec{M}$ in the (clockwise) rotating frame with angular fre-

---

[b]The Larmor frequency is often defined as $-\gamma H_0$ including the sense of rotation. However we explicitly indicate the sense of rotation by *clockwise* or *counterclockwise* and we always take $\omega_0 > 0$.

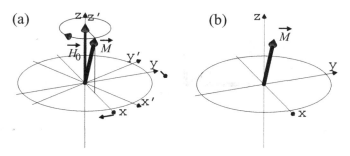

Fig. 2. Rotating frame. (a) View in the laboratory frame. The rotating frame and $\vec{M}$ rotate clockwise at the Larmor frequency $\omega_0 = \gamma H_0$. (b) View in the rotating frame. $\vec{M}$ is at rest in this rotating frame and hence the effective magnetic field vanishes.

quency ω precesses at $\omega_0 - \omega$ and thus the effective magnetic field for $\vec{M}$ is $(0, 0, H_0 - \omega/\gamma)$.

2.1.3. Rotating magnetic field

Let us apply a clockwise rotating magnetic field

$$\vec{H}'_1 = -H_1 \left(\cos\left(\omega_{\rm rf} t - \phi\right), -\sin\left(\omega_{\rm rf} t - \phi\right), 0 \right) \tag{3}$$

in addition to the static magnetic field $\vec{H}_0 = (0, 0, H_0)$. Note that " ′ " indicates that the rotating magnetic field is viewed in the laboratory frame. In the laboratory frame, the sum of $\vec{H}_0$ and $\vec{H}'_1$ moves rapidly and thus it is not easy to solve Eq. (2). However the dynamics of the system becomes simpler, when we view the system in the rotating frame whose frequency is $\omega_{\rm rf}$. The rotating magnetic field is considered as static in this rotating frame, and thus the effective magnetic field is simplified to $\vec{H}_1 = (-H_1 \cos\phi, -H_1 \sin\phi, H_0 - \omega_{\rm rf}/\gamma)$. $\vec{M}$ precesses about this effective field in the rotating frame and its precession frequency is $\gamma\sqrt{H_1^2 + (H_0 - \omega_{\rm rf}/\gamma)^2}$.

Let us consider the case in which the rotating magnetic field with frequency $\omega_{\rm rf}(=\omega_0)$ is applied. In the rotating frame with frequency $\omega_{\rm rf}$, $\vec{M}$ precesses about the axis $-(\cos\phi, \sin\phi, 0)$ at $\omega_1 = \gamma H_1$ as long as the rotating field is on. When the rotating field is turned off after $t_{\rm p}$, $\vec{M}$ with the initial condition $(0, 0, M)$ is flipped by the angle $\beta = \omega_1 t_{\rm p}$. When $\beta = \pi/2$, the rotating field is called a $\pi/2$-pulse (90°-pulse) and the initial vector $(0, 0, M)$ turns to $(M\sin\phi, -M\cos\phi, 0)$ in the xy-plane (rotating frame). Therefore, this pulse is often written as $90°_\phi$. Especially, $90°_0, 90°_{\pi/2}, 90°_\pi$, and

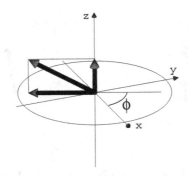

Fig. 3. Rotating magnetic field in the rotating frame. $\vec{H}_1 = (-H_1 \cos\phi, -H_1 \sin\phi, H_0 - \omega_{\rm rf}/\gamma)$ is shown.

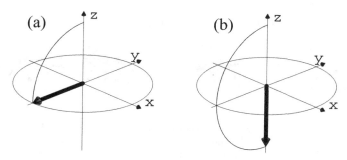

Fig. 4. RF pulses. (a) A $90°_x$-pulse turns the magnetization to the y-axis. (b) A $180°_x$-pulse flips it.

$90°_{3\pi/2}$ are written as $90°_x, 90°_y, 90°_{-x}$, and $90°_{-y}$, respectively. When $\beta = \pi$, the rotating field is called a π-pulse (or, 180°-pulse) and $\vec{M}$ turns to $-\vec{M}$.

We also note that an oscillating magnetic field

$$-2H_1 \left(\cos\left(\omega_{\rm rf}t - \phi\right), 0, 0\right)$$

is usually employed instead of a rotating magnetic field. Since

$$-2H_1 \left(\cos\left(\omega_{\rm rf}t - \phi\right), 0, 0\right) = -H_1 \left(\cos\left(\omega_{\rm rf}t - \phi\right), \sin\left(\omega_{\rm rf}t - \phi\right), 0\right) \\ -H_1 \left(\cos\left(\omega_{\rm rf}t - \phi\right), -\sin\left(\omega_{\rm rf}t - \phi\right), 0\right),$$

the oscillating magnetic filed can be considered as a superposition of a clockwise and a counterclockwise rotating magnetic filed at $\omega_{\rm rf}$. In the clockwise rotating frame with frequency $\omega_{\rm rf}$, the clockwise component is regarded as at rest while the counterclockwise component rotates at $2\omega_{\rm rf}$. The effect of the counterclockwise rotating field is averaged to vanish in the rotating frame since $t_{\rm p} \sim 1/(\gamma H_1) \gg 1/(\gamma H_0) \sim 1/\omega_{\rm rf}$ in usual NMR experiments. Therefore, we can employ an oscillating magnetic field instead of a rotating magnetic field.

2.1.4. Bloch equation

The magnetization $\vec{M}$ is $\vec{M}_0 = (0, 0, M_0)$ when the system is in thermal equilibrium. So far we have ignored relaxation processes. We now introduce

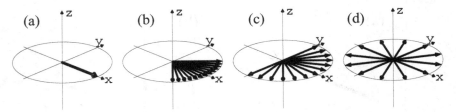

Fig. 5. T_2^* relaxation. (a) $\vec{M}$ aligned along the x-axis at $t = 0$. (b ~ d) $\vec{M}$ is distributed in the xy-plane because of the Larmor frequency difference. Since we only measure the total magnetic moment, we observe vanishing signal in (d).

them phenomenologically with the Bloch equation

$$\frac{d\vec{M}}{dt} = \gamma \vec{M} \times \vec{H}_0 - \Gamma(\vec{M} - \vec{M}_0),$$ (4)

$$\Gamma = \begin{pmatrix} 1/T_2 & 0 & 0 \\ 0 & 1/T_2 & 0 \\ 0 & 0 & 1/T_1 \end{pmatrix}$$

where T_1 and T_2 are called the *longitudinal* (spin-lattice) and the *transverse* (spin-spin) relaxation time, respectively. The second term $-\Gamma(\vec{M} - \vec{M}_0)$ in the right hand side acts as a driving force so that $\vec{M}$ eventually comes back to its thermal equilibrium value $\vec{M}_0$.

The condition $T_2 \ll T_1$ is almost always satisfied in NMR. Let us consider the dynamics of the magnetic moment. We assume that $\vec{M}(0) = M_0(\cos\chi, \sin\chi, 0)$ at $t = 0$. Then, $\vec{M}(t)$ decays with the transverse relaxation time constant T_2 and becomes $\vec{M}(t) = \vec{0}$ at $T_1 \gg t \gg T_2$. After this decay (relaxation) in the xy-plane, the longitudinal relaxation toward its thermal equilibrium $\vec{M}_0$ takes place.

Besides T_1 and T_2, T_2^*, which characterizes the signal decay caused by inhomogeneous environment, is important experimentally. Let us consider the simple case where inhomogeneous environment is caused by a small static field inhomogeneity. Then the Larmor frequency is slightly different from one place to another. $\vec{M}$ aligned along the x-axis at $t = 0$ is going to distribute in the xy-plane because of the Larmor frequency difference, as shown in Fig. 5. Since we only measure the total magnetic moment, we observe vanishing signal when $\vec{M}$ distributes uniformly in the xy-plane.

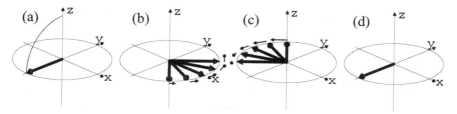

Fig. 6. Principle of spin echo. (a) All $\vec{M}_i$'s align to the $-y$-axis by a $90°_x$ pulse. (b) $\vec{M}_i$'s after τ distribute in the xy-plane because of different Larmor frequencies. Small arrows near the magnetic moment vectors indicate their velocity in the rotating frame. (c) Then a $180°_y$-pulse flips $\vec{M}_i$'s to their mirror image positions on the other side of the y-axis. (d) After another period of τ, all $\vec{M}_i$'s converge along the $-y$-axis.

2.1.5. Spin echo

Spin echo is one of the most important NMR phenomena and can be understood within the vector model. Suppose that there are several magnetic moments $\vec{M}_i$ whose Larmor frequencies ω_0^i are different from each other. This Larmor frequency difference may be caused by many reasons, such as, static field inhomogeneity, as we discussed before. By applying a strong $90°_x$-pulse which allows us to ignore $H_0^i - \omega_{\rm rf}/\gamma$ compared with H_1, all $\vec{M}_i$ align to the $-y$-axis in good approximation. Free precession of $\vec{M}_i$ for a period τ leads to the distribution of $\vec{M}_i$ in the xy-plane because of different Larmor frequencies. Then a strong $180°_y$-pulse flips $\vec{M}_i$ to their mirror image positions on the other side of the y-axis. After another free precession for an identical period τ, all $\vec{M}_i$ converge along the $-y$-axis, irrespective of the Larmor frequencies of $\vec{M}_i$'s. This $180°$-pulse is said to *refocus* $\vec{M}_i$ and called as a *refocusing* pulse.

2.1.6. NMR equipment and signal detection

The block diagram of a simplified NMR equipment is shown in Fig. 7. The radio frequency wave is carved by the mixer according to the output of the pulse generator and becomes an rf-pulse. These rf-pulses are amplified and fed into the tank circuit. Then, oscillating (as we discussed before, equivalent to rotating) magnetic fields are generated in the coil and control the sample magnetic moment in the test tube. The motion of the sample magnetic moment causes a signal in the tank circuit via electromagnetic induction. This signal is amplified and then detected. We employ the tank circuit for both generating a strong oscillating magnetic fields and detecting

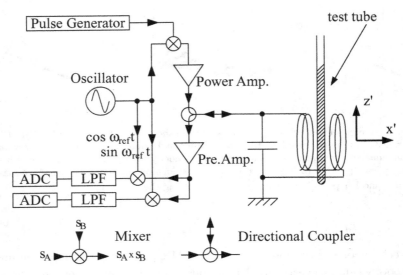

Fig. 7. Block diagram of NMR equipment. Rf-pulses are generated by the oscillator and pulse generator. Then, these are amplified and fed into the tank circuit. Oscillating magnetic fields are generated in the coil and control the sample magnetic moment in the test tube. The motion of the sample magnetic moment causes a signal in the tank circuit via electromagnetic induction. This signal is amplified and then detected. LPF and ADC denote a low pass filter and an analog to Digital converter, respectively. Note that a directional coupler controls the signal flow as indicated by arrows and that a mixer outputs the product of two inputs.

a large signal.

We now discuss details of signal detection. If there is no relaxation mechanism exists, the magnetization $\vec{M} = (M_x, M_y, 0) = M(\cos \chi, \sin \chi, 0)$ in the xy-plane (rotating frame) is constant. However, $\vec{M}(t)$ decays as

$$\vec{M}(t) = M(\cos \chi, \sin \chi, 0) \exp(-t/T_2),$$

because of the transverse relaxation. Here we assume $T_2 \ll T_1$ and thus the longitudinal relaxation may be ignored. Note that the condition $T_2 \ll T_1$ is almost always satisfied in NMR. When we see the magnetic moment in the laboratory frame, $\vec{M}'(t)$ (note the " $'$ " symbol) is

$$\vec{M}'(t) = M(\cos(\omega_0 t - \chi), -\sin(\omega_0 t - \chi), 0) \exp(-t/T_2).$$

Note that ω_0 is the Larmor frequency and that the rotation is clockwise. We assume here that the x-component $M \cos(\omega_0 t - \chi) \exp(-t/T_2)$ can be measured.[c] We call this signal as the *Free Induction Decay* (= FID) signal. The FID signal is multiplied by $\cos\omega_{\text{ref}}t$

$$M\cos(\omega_0 t - \chi)\exp(-t/T_2) \times \cos\omega_{\text{ref}}t$$
$$= \frac{1}{2}M\left(\cos(\Delta\omega\, t - \chi) + \cos((\Delta\omega + 2\omega_{\text{ref}})t - \chi)\right)\exp(-t/T_2),$$

where $\omega_{\text{ref}} > 0$ and $\Delta\omega = \omega_0 - \omega_{\text{ref}}$. Then by dropping the high frequency term with $(\Delta\omega + 2\omega_{\text{ref}})$, we obtain

$$\frac{1}{2}M\cos(\Delta\omega\, t - \chi)\exp(-t/T_2).$$

Note that the high frequency term may be dropped by passing the signal through a low pass filter whose cut-off frequency is much smaller than $2\omega_{\text{ref}}$. Similarly we obtain

$$\frac{1}{2}M\sin(\Delta\omega\, t - \chi)\exp(-t/T_2),$$

by multiplying by $\sin\omega_{\text{ref}}t$. We note here the typical frequency scales: $\omega_{\text{ref}} \sim \omega_0 \sim 100$ MHz, $\Delta\omega \sim 10$ kHz, and $1/T_2 \sim 1$ Hz. We construct a complex time domain signal as follows,

$$s(t) = M\left(\cos(\Delta\omega t - \chi) + i\sin(\Delta\omega t - \chi)\right)\exp(-t/T_2)$$
$$= M\exp(-i\chi)\exp(i\Delta\omega t)\exp(-t/T_2)$$

at $t \geq 0$ and $s(t) = 0$ at $t < 0$. We convert $s(t)$ into the frequency-domain function (*spectrum*) $S(\omega)$ by the Fourier Transformation

$$S(\omega) = \int_{-\infty}^{\infty} s(t)\exp(-i\omega t)dt$$
$$= M\exp(-i\chi)\int_0^{\infty} \exp(i\Delta\omega t)\exp(-t/T_2)\exp(-i\omega t)dt$$
$$= M\exp(-i\chi)\frac{1/T_2 - i(\omega - \Delta\omega)}{(1/T_2)^2 + (\omega - \Delta\omega)^2}.$$

Therefore, we can measure M, χ, and ω_0 from the spectrum $S(\omega)$.

---

[c]The signal in a coil is generated by electromagnetic induction. Therefore, the signal in the cylindrical coil, with its axis along the x'-axis, is proportional to

$$\frac{dM'_x}{dt} = -M\omega_0\sin(\omega_0 t - \chi)\exp(-t/T_2),$$

where we ignore the time derivative of $\exp(-t/T_2)$ since $\omega_0 \gg 1/T_2$. We know ω_0 and can shift the time origin so that $M\cos(\omega_0 t - \chi)\exp(-t/T_2)$ is recovered.

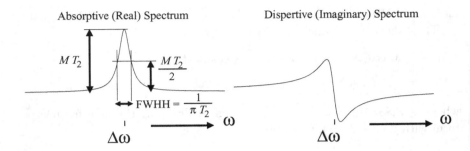

Fig. 8. Absorptive and dispersive spectra are shown. The peak frequency is at $\omega = \Delta\omega = \omega_0 - \omega_{\rm ref}$, while M and T_2 can be measured from the peak height and FWHH.

When $\chi = 0$, the real part of $S(\omega)$ is an *absorptive* Lorentzian curve centered on $\Delta\omega$,

$$\Re(S(\omega)) = \frac{M/T_2}{(1/T_2)^2 + (\omega - \Delta\omega)^2}.$$

The height at $\omega = \Delta\omega$ gives MT_2 while a full width at half-height (FWHH), or the frequency range where $\Re(S(\omega)) > MT_2/2$, gives $1/\pi T_2$. Therefore, T_2 and M can be determined from the spectrum. On the other hand, the imaginary part of $S(\omega)$ is a *dispersive* Lorentzian curve,

$$\Im(S(\omega)) = -\frac{M(\omega - \Delta\omega)}{(1/T_2)^2 + (\omega - \Delta\omega)^2}.$$

When $\chi \neq 0$, the real and imaginary spectra are linear combinations of the absorptive and dispersive spectra.

2.2. Quantum mechanical view

We re-examine the principle of NMR from quantum mechanical viewpoint here.

2.2.1. Magnetic moment

The magnetic moment operator $\vec{\mu}$ is associated with the spin operator $\vec{I}$ as

$$\vec{\mu} = \gamma \hbar \vec{I}.$$

In the case of spin 1/2 nucleus, the spin operator $\vec{I}$ (matrix representation) is

$$\vec{I} = \frac{1}{2}\left(\begin{pmatrix} 0 & 1 \\ 1 & 0 \end{pmatrix}, \begin{pmatrix} 0 & -i \\ i & 0 \end{pmatrix}, \begin{pmatrix} 1 & 0 \\ 0 & -1 \end{pmatrix} \right) = \frac{1}{2}(\sigma_x, \sigma_y, \sigma_z),$$

where σ_k are standard Pauli matrices.

The Hamiltonian in a magnetic field $\vec{H}_0 = (0, 0, H_0)$ is

$$\mathcal{H} = -\vec{\mu} \cdot \vec{H}_0 = -\hbar\omega_0 I_z, \tag{5}$$

where I_k is the kth component of the spin operator. Therefore, the eigenstates of the Hamiltonian (5) are

$$|0\rangle = \begin{pmatrix} 1 \\ 0 \end{pmatrix}, |1\rangle = \begin{pmatrix} 0 \\ 1 \end{pmatrix},$$

and their eigenvalues are $\mp\hbar\omega_0/2$, respectively. Note that $\langle 0| = (1, 0)$ and $\langle 1| = (0, 1)$.

The state of ensemble of nuclei of spin 1/2 is described by a density matrix

$$\rho = \sum_{i=0}^{1} \sum_{j=0}^{1} \rho_{ij} |i\rangle\langle j|. \tag{6}$$

Note that there are only 3 independent parameters in ρ_{ij} and we set $\rho_{00} + \rho_{11} = 1$ for normalization. The ensemble average of an observable O is

$$\text{Tr}(\rho O) = \sum_{i,j=0}^{1} \rho_{ij} \langle j|O|i\rangle.$$

The density matrix $\rho(T)$ of the thermal equilibrium state is generally

$$\rho(T) = \frac{e^{-H/k_B T}}{Z(T)},$$

where T is the temperature, $Z(T) = \text{Tr}(\exp(-H/k_B T))$, and k_B is Boltzmann's constant. In the case of spin 1/2,

$$\rho(T) = \frac{e^{\hbar\omega_0 I_z/k_B T}}{\text{Tr}(e^{\hbar\omega_0 I_z/k_B T})}$$

$$= I_0 + \frac{\hbar\omega_0}{2k_B T} I_z + O\left(\frac{\hbar\omega_0}{2k_B T}\right)^2,$$

where $I_0 = I/2$ and I is the identity matrix of dimension 2. The third term is usually negligible since $\hbar\omega_0/k_B T \sim 10^{-5}$ at room temperature. The ensemble average of $\vec{\mu}$ in the thermal state is

$$\text{Tr}\left[\vec{\mu}\rho(T)\right] = \text{Tr}\left[\left(0, 0, \gamma\hbar I_z \frac{\hbar\omega_0}{2k_B T} I_z\right)\right]$$

$$= \left(0, 0, \frac{\hbar^2 \gamma^2 H_0}{4k_B T}\right), \tag{7}$$

where $\text{Tr}(I_z^2) = 1/2$ is taken into account. The magnetic moment vector $\vec{M}_0$ appeared in the previous subsection can be obtained from Eq. (7) by multiplying the number of molecules in the sample.

The density matrix (6) can be expanded as

$$\rho(M, n_x, n_y, n_z) = I_0 + \frac{2M}{\gamma\hbar} \sum_{k=x,y,z} n_k I_k, \qquad (8)$$

where $n_x^2 + n_y^2 + n_z^2 = 1$ and thus there are 3 free parameters. I_0 is added for normalization. The meaning of M and the vector $\vec{n} = (n_x, n_y, n_z)$ becomes clear when $\text{Tr}(\vec{\mu}\rho(M, n_x, n_y, n_z))$ is calculated.

$$\text{Tr}(\vec{\mu}\,\rho(M, n_x, n_y, n_z)) = \text{Tr}\left[\gamma\hbar\vec{I}(I_0 + \frac{2M}{\gamma\hbar} \sum_{k=x,y,z} n_k I_k)\right]$$
$$= M\,(n_x, n_y, n_z),$$

where $\text{Tr}(I_k) = 0$ and $\text{Tr}(I_k^2) = 1/2$ for $k = x, y, z$ are taken into account. Therefore, $\rho(M, n_x, n_y, n_z)$ represents the bulk magnetic moment with the amplitude M and the direction (n_x, n_y, n_z).

2.2.2. *Precession*

The dynamics of the density matrix ρ is governed by the Liouville equation

$$i\hbar\frac{d\rho}{dt} = [\mathcal{H}, \rho]. \qquad (9)$$

Let us solve the dynamics in a static magnetic field $\vec{H}_0 = (0, 0, H_0)$ in the Hamiltonian (5). ρ is expanded as Eq. (8). Then the Liouville equation is

$$i\hbar\frac{d}{dt}\left(I_0 + \frac{2M}{\gamma\hbar}\sum_{k=x,y,z} n_k I_k\right) = \left[-\hbar\omega_0 I_z, I_0 + \frac{2M}{\gamma\hbar}\sum_{k=x,y,z} n_k I_k\right]$$
$$\downarrow$$
$$\frac{dn_x}{dt}I_x + \frac{dn_y}{dt}I_y = -\omega_0 n_x I_y + \omega_0 n_y I_x, \quad \frac{dn_z}{dt}I_z = 0.$$

And thus,

$$\rho(t) = I_0 + \frac{2M}{\gamma\hbar}\left[\sqrt{1-n_z^2}(\cos\omega_0 t\, I_x - \sin\omega_0 t\, I_y) + n_z I_z\right],$$

where we omit an arbitrary phase. Therefore, $\text{Tr}(\vec{\mu}\rho(t))$ gives the clockwise rotating magnetic moment as in the case of the vector model.

We introduce a time-dependent unitary operator $U_R(t)$ and define the unitary transformation of the density matrix as

$$\tilde{\rho} = U_R \rho U_R^\dagger.$$

Then Liouville equation (9) becomes

$$i\hbar \frac{d\tilde{\rho}}{dt} = [\tilde{\mathcal{H}}, \tilde{\rho}]$$

with the transformed Hamiltonian

$$\tilde{\mathcal{H}} = U_R \mathcal{H} U_R^\dagger - i\hbar U_R \frac{dU_R^\dagger}{dt}.$$

When we take $U_R = \exp(-i\omega t I_z)$ and the Hamiltonian (5), we obtain

$$\tilde{\mathcal{H}} = -\hbar(\omega_0 - \omega) I_z$$

which is the Hamiltonian for the magnetic moment (spin 1/2) in an effective magnetic field $(0, 0, H_0 - \omega/\gamma)$. In other words, this unitary transformation gives the density matrix and the Hamiltonian in the rotating frame. Especially, we obtain a vanishing Hamiltonian $\tilde{\mathcal{H}} = 0$ when $\omega_0 = \omega$. Therefore, there is no motion in $\tilde{\rho}$.

2.2.3. Rotating magnetic field

Let us apply a clockwise rotating magnetic field $\vec{H}_1'$ in addition to the static magnetic field $\vec{H}_0$. The corresponding Hamiltonian is

$$\begin{aligned}\mathcal{H} &= -\vec{\mu} \cdot (\vec{H}_0 + \vec{H}_1') \\ &= -\hbar\omega_0 I_z + \hbar\omega_1 \left(\cos(\omega_{\text{rf}} t - \phi) I_x - \sin(\omega_{\text{rf}} t - \phi) I_y\right).\end{aligned}$$

It is not easy to calculate its time development operator because it involves time-ordered product. However the Hamiltonian becomes simpler when it is transformed by $U_R = \exp(-i\omega_{\text{rf}} t I_z)$. We obtain

$$\tilde{\mathcal{H}} = -\hbar(\omega_0 - \omega_{\text{rf}}) I_z + \hbar\omega_1 (\cos\phi\, I_x + \sin\phi\, I_y) \tag{10}$$

Note that $\tilde{\mathcal{H}} = -\vec{\mu} \cdot (-H_1\cos\phi, -H_1\sin\phi, H_0 - \omega_{\text{rf}}/\gamma)$. The time development operator U_{rf}, when the rotating magnetic field is applied for the duration t_p, is easy to calculate as

$$U_{\text{rf}}(\omega_1, t_p, \phi) = e^{i(\omega_0 - \omega_{\text{rf}}) t_p I_z - i\omega_1 t_p (\cos\phi\, I_x + \sin\phi\, I_y)} \tag{11}$$

since the Hamiltonian (10) is independent of time.

We define the unitary operator

$$\begin{aligned}
R(\beta, \phi) &= e^{-i\beta(\cos\phi I_x + \sin\phi I_y)} \\
&= \cos\frac{\beta}{2} I - 2i \sin\frac{\beta}{2}(\cos\phi I_x + \sin\phi I_y) \\
&= \begin{pmatrix} \cos\beta/2 & -ie^{-i\phi}\sin\beta/2 \\ -ie^{i\phi}\sin\beta/2 & \cos\beta/2 \end{pmatrix},
\end{aligned} \qquad (12)$$

which can be obtained from Eq. (11) by substituting $\omega_0 = \omega_{\rm rf}$ and $\beta = \omega_1 t_{\rm p}$. We call β as the tipping angle. The density matrix $\rho(M, 0, 0, 1)$ (see Eq. (8)) is transformed by $R(\beta, \phi)$ as

$$\begin{aligned}
&{\rm Ad}(R(\beta, \phi))\rho(M, 0, 0, 1) \\
&= R(\beta, \phi)\rho(M, 0, 0, 1)R^\dagger(\beta, \phi) \\
&= \begin{pmatrix} \frac{1}{2} + \frac{M}{\gamma\hbar}\cos\beta & \frac{M}{\gamma\hbar}\sin\beta(i\cos\phi + \sin\phi) \\ \frac{M}{\gamma\hbar}\sin\beta(-i\cos\phi + \sin\phi) & \frac{1}{2} - \frac{M}{\gamma\hbar}\cos\beta \end{pmatrix} \\
&= \rho(M, \sin\phi\sin\beta, -\cos\phi\sin\beta, \cos\beta).
\end{aligned}$$

Here we introduced the adjoint action of U on ρ as ${\rm Ad}(U)\rho \equiv U\rho U^\dagger$ to simplify equations. When $\beta = \pi/2$, it is reduced to $\rho(M, \sin\phi, -\cos\phi, 0)$. Therefore, $R(\pi/2, \phi)$ corresponds to a $90°_\phi$-pulse in the vector model. Similarly, $R(\pi, \phi)$ corresponds to a $180°_\phi$-pulse.

An oscillating magnetic field

$$(-2H_1\cos(\omega_{\rm rf} t - \phi), 0, H_0)$$

leads to the Hamiltonian

$$-\hbar\omega_0 I_z + 2\hbar\omega_1 \cos(\omega_{\rm rf} t - \phi)I_x.$$

When this Hamiltonian is transformed by $U_R = \exp(-i\omega_{\rm rf} t I_z)$, we get

$$\begin{aligned}
\tilde{\mathcal{H}} &= -\hbar(\omega_0 - \omega_{\rm rf})I_z + \hbar\omega_1(\cos\phi I_x + \sin\phi I_y) \\
&\quad + \hbar\omega_1(\cos(\phi - 2\omega_{\rm rf}t)I_x - \sin(\phi - 2\omega_{\rm rf}t))I_y) \\
&\approx -\hbar(\omega_0 - \omega_{\rm rf})I_z + \hbar\omega_1(\cos\phi I_x + \sin\phi I_y),
\end{aligned}$$

since the effect of the terms with frequency $2\omega_{\rm rf}$ is small compared with the others. Note that $2\omega_{\rm rf} \gg \omega_1, (\omega_0 - \omega_{\rm rf})$. This approximation is called the rotating wave approximation. As in the vector model, we can employ an oscillating field instead of a rotating field.

2.2.4. Bloch equation

The Hamiltonian for a spin-1/2 nucleus is generally described as

$$\mathcal{H} = \hbar \sum_k \omega_k I_k,$$

and thus this Hamiltonian is traceless; $\mathrm{Tr}(\mathcal{H}) = 0$. Therefore the time development operator $\exp(-i\mathcal{H}t/\hbar)$ generated by $\mathcal{H}$ is an element of SU(2). In other words, we can consider no relaxation and there is no simple correspondence with the Bloch equation (5) in quantum mechanics.

The signal decay characterized with T_2^* can be reproduced since the "sum" is not a unitary operation. We discuss the signal decay in inhomogeneous environment with the phenomena "spin echo" next.

2.2.5. Spin echo

The phenomenon of the *spin echo* can be understood as follows. Suppose that there are several magnetic moment $\vec{M}_i$ whose Larmor frequencies ω_0^i's ($0 \leq i \leq N$) are different from each other. The Hamiltonian for each magnetic moment in the clockwise rotating frame at $\omega_0 = \omega_0^0$ is

$$\tilde{\mathcal{H}}^i = -\hbar \Delta^i I_z,$$

where $\Delta^i = \omega_0^i - \omega_0^0$.

We operate $R(\pi/2, 0)$ to ρ_i (corresponding to M_i) which is equivalent to applying a strong 90_x°-pulse to M_i. Free precession of $\vec{M}_i$ during τ corresponds to applying

$$U_{\mathrm{fp}}^i(\tau) = e^{-i\tilde{\mathcal{H}}^i \tau/\hbar} = e^{i\Delta^i \tau I_z}$$

to ρ_i. We assume that the pulse width is negligible compared with τ. The total time development operator is

$$U_{\mathrm{fp}}^i(\tau) R(\pi/2, 0) = \frac{1}{\sqrt{2}} \begin{pmatrix} e^{i\Delta^i \tau/2} & -i e^{i\Delta^i \tau/2} \\ -i e^{-i\Delta^i \tau/2} & e^{-i\Delta^i \tau/2} \end{pmatrix}$$

When the density matrix at $t = 0$ is $\rho(M, 0, 0, 1)$, that at $t = \tau$ is

$$\mathrm{Ad}(U_{\mathrm{fp}}^i(\tau) R(\pi/2, 0)) \rho(M, 0, 0, 1) = \rho(M, -\sin \Delta^i \tau, -\cos \Delta^i \tau, 0).$$

When $\tau \gg 2\pi/\Delta^i$, $\Delta^i \tau \bmod (2\pi)$ can be approximated with random variables and thus $\sum_i \rho(M, -\sin \Delta^i \tau, -\cos \Delta^i \tau, 0) = \vec{0}$ is obtained. In other words, we lose the signal because of the inhomogeneity in environment.

We consider the spin echo phenomena from the following operation

$$U_{\text{fp}}^i(\tau)\, R(\pi,\pi/2)\, U_{\text{fp}}^i(\tau) = \begin{pmatrix} 0 & -1 \\ 1 & 0 \end{pmatrix}.$$

This operation corresponds to the sequence of (1) free-precession for τ (2) $180°_y$-pulse (3) free-precession for τ. Therefore the density matrix at $t = 2\tau$ is

$$\text{Ad}\left(U_{\text{fp}}^i\, R(\pi,\pi/2)\, U_{\text{fp}}^i(\tau) R(\pi/2, 0)\right) \rho(M, 0, 0, 1) = \rho(M, 0, -1, 0),$$

regardless of the index i. We interpret this result as $R(\pi, \pi/2)$ *refocused* ρ.
We note

$$R(-\pi, \phi)\, U_{\text{fp}}^i(\tau)\, R(\pi, \phi)\, U_{\text{fp}}^i(\tau) = I$$

which means that a pair of 180°-pulses can null the effect of the Larmor frequency differences.

2.2.6. NMR equipment and signal detection

Since NMR measurements are based on ensemble ones, there is nothing to be added to the section 2.1.6 here. We only need to remember $\vec{M} = \text{Tr}(\vec{\mu}\rho)$, and thus

$$M_x = \gamma\hbar\, \text{Tr}(I_x \rho) = \frac{\gamma\hbar}{2}\, \text{Tr}(\sigma_x \rho)$$

which is observed as an absorptive signal in NMR measurements. M_y corresponds to a dispersive signal.

3. Principle of Liquid-State NMR Quantum Computer

The density matrix for n-spin molecules in NMR can be generally written as

$$\rho = I_0^{\otimes n} + \Delta\rho$$

where "$\otimes n$" denotes the n-th tensor power. The first term $I_0^{\otimes n}$ represents an isotropic maximally mixed ensemble, in which all spin states appear with equal probability. The second term $\Delta\rho$ represents a deviation from the uniform ensemble. A unitary transformation U acts on the density matrix as

$$\text{Ad}(U)\rho = U\rho U^\dagger = U\left[I_0^{\otimes n} + \Delta\rho\right] U^\dagger = I_0^{\otimes n} + U\Delta\rho U^\dagger. \tag{13}$$

This implies that only the deviation from the isotropic ensemble has relevance in time development of the system.

Therefore, we can treat the following mixed state

$$\rho_{\text{pps}} = I_0^{\otimes n} + \alpha \operatorname{diag}(1, \underbrace{0, \ldots, 0}_{2^n - 1}), \quad (14)$$

as a pure state $|00\ldots 0\rangle$ in terms of its dynamics and we call it a *pseudopure state*.

We discuss here the liquid-state NMR quantum computation with a molecule that contains two spins (two qubits), which is the smallest molecule to implement quantum algorithms. We discuss its Hamiltonian, show a way how to construct unitary and non-unitary gates, and discuss about the separability of its density matrix. Note that we employ the natural unit $\hbar = 1$ hereafter.

3.1. *Two-qubit Hamiltonian*

We consider a minimum Hamiltonian for performing a quantum computation. Molecules in a solution are moving and rotating rapidly and randomly. Therefore the intermolecular and intramolecular dipole-dipole interactions are averaged out to vanish, while the intramolecular scalar-couplings remain nonvanishing. The Hamiltonian without an rf magnetic field is

$$\mathcal{H}_{\text{drift}} = -\omega_{0,1} I_z \otimes I - \omega_{0,2} I \otimes I_z + J \sum_{k=x,y,z} I_k \otimes I_k, \quad (15)$$

where $\omega_{0,i} = \gamma_i H_0$ is the Larmor frequency of the i-th spin in a static field $(0, 0, H_0)$ and J characterizes the strength of the scalar coupling.[5d] Note that γ_i is the gyromagnetic ratio of the i'th spin, which includes the effect of chemical shift. The symbol $\otimes$ denotes the tensor product.

The spins are manipulated by applying a series of rf pulses in order to realize a desired quantum algorithm. The Hamiltonian corresponding to the rf pulse, in which the amplitude $2H_1$, the frequency ω_{rf}, and the phase $-\phi_{\text{rf}}$ are controllable parameters, is

$$\mathcal{H}_{\text{rf}}(t) = 2\cos(\omega_{\text{rf}} t - \phi_{\text{rf}}) \left(\omega_{1,1} I_x \otimes I + \omega_{1,2} I \otimes I_x \right), \quad (16)$$

where $\omega_{1,i} = \gamma_i H_1$. Note that the coil axis to generate rf pulses is assumed to be along the x-axis.

---

[d] $2\pi J$ instead of J in $\mathcal{H}_{\text{drift}}$ is often used in literature.

3.2. *One-qubit gates*

We show how to construct a one-qubit gate here. We assume

- $\omega_{\mathrm{rf}} = \omega_{0,1}$
- $|\omega_{0,1} - \omega_{0,2}| \gg |\omega_{1,i}|$
- $J \ll \omega_{1,i}$.

We transform the Hamiltonian $\mathcal{H}_{\mathrm{tot}} = \mathcal{H}_{\mathrm{drift}} + \mathcal{H}_{\mathrm{rf}}(t)$ with the time-dependent unitary operator

$$U_c(t) = e^{-i\omega_{\mathrm{rf}} t I_z} \otimes e^{-i\omega_{\mathrm{rf}} t I_z}$$

to obtain

$$\begin{aligned}\tilde{\mathcal{H}}_{\mathrm{tot}} &= \bigl(\omega_{1,1}(\cos\phi_{\mathrm{rf}} I_x + \sin\phi_{\mathrm{rf}} I_y)\bigr) \otimes \\ &\quad \bigl(-(\omega_{0,2} - \omega_{0,1})I_z + \omega_{1,2}(\cos\phi_{\mathrm{rf}} I_x + \sin\phi_{\mathrm{rf}} I_y)\bigr) \\ &= e^{-i\omega_{1,1}\,\vec{n}_1 \cdot \vec{I}} \otimes e^{-i\Delta\sqrt{1+\epsilon^2}\,\vec{n}_2 \cdot \vec{I}},\end{aligned} \quad (17)$$

by applying the rotating wave approximation and by ignoring the scalar coupling term from the third assumption. Here,

$$\Delta = \omega_{0,1} - \omega_{0,2}$$
$$\vec{n}_1 = (\cos\phi, \sin\phi, 0)$$
$$\vec{n}_2 = (\epsilon\cos\phi, \epsilon\sin\phi, 1)/\sqrt{1+\epsilon^2}$$

where $\epsilon = \omega_{1,2}/\Delta$. Note that $\vec{n}_2 \approx (0,0,1)$ from the assumption.

The Hamiltonian (17) implies that the pulse, whose duration t_{p} satisfies $\beta = \omega_{1,1} t_{\mathrm{p}}$, rotates qubit 1 by the angle β around the axis $(\cos\phi, \sin\phi, 0)$ in the rotating frame with frequency $\omega_{0,1} = \omega_{\mathrm{rf}}$. On the other hand, qubit 2 is not flipped since the corresponding time development operator is

$$e^{-it_{\mathrm{p}}\Delta\sqrt{1+\epsilon^2}\,\vec{n}_2 \cdot \vec{I}} \approx e^{-it_{\mathrm{p}}\Delta\sqrt{1+\epsilon^2} I_z}.$$

If there is no rf field applied to qubit 2, its time development operator during the period t_{p} is $e^{-it_{\mathrm{p}}\Delta I_z}$ in this rotating frame. The difference between $e^{-it_{\mathrm{p}}\Delta I_z}$ and $e^{-it_{\mathrm{p}}\Delta\sqrt{1+\epsilon^2} I_z}$ is called Bloch-Siegert effect.[16,17] This effect can be ignored under our assumptions, since

$$t_{\mathrm{p}}\Delta\sqrt{1+\epsilon^2} - t_{\mathrm{p}}\Delta \approx \frac{t_{\mathrm{p}}\omega_{1,2}}{2}\epsilon \sim \epsilon \ll 1.$$

Note that $t_{\mathrm{p}}\omega_{1,2} \sim 1$ since $t_{\mathrm{p}}\omega_{1,1}$ is typically $\pi/2$ or π and since $\gamma_1 \sim \gamma_2$ for both homo- and hetero-nucleus molecules. Therefore, qubit 1 can be selectively manipulated without making any influence on qubit 2.[e] In other words, we can implement the operation $R(\beta, \phi) \otimes I$. The operation $I \otimes R(\beta, \phi)$ can be trivially implemented by taking $\omega_{\rm rf} = \omega_{0,2}$.

The operation $Z(\alpha)$ at $t = t_0$, which rotates a qubit around the z-axis by an angle α, can be implemented without rf pulses, simply by taking the unitary transformation by $e^{-i(\omega_{0,i}-\alpha)tI_z}$ at $t > t_0$ instead of $e^{-i\omega_{0,i}tI_z}$ at $t < t_0$.

Any one qubit operation can be constructed with the operator set

$$R_1(\beta, \phi) \equiv R(\beta, \phi) \otimes I,$$
$$R_2(\beta, \phi) \equiv I \otimes R(\beta, \phi),$$
$$Z_1(\alpha) \equiv Z(\alpha) \otimes I,$$
$$Z_2(\alpha) \equiv I \otimes Z(\alpha). \tag{18}$$

We also define

$$R_i(\beta, \phi) \equiv I \otimes I \otimes \cdots \otimes \overbrace{R(\beta, \phi)}^{i\text{-th}} \otimes \cdots \otimes I$$

$$Z_i(\alpha) \equiv I \otimes I \otimes \cdots \otimes \overbrace{Z(\alpha)}^{i\text{-th}} \otimes \cdots \otimes I$$

for treating more spins as qubits.

3.3. *Two-qubit gates*

We need controlled-NOT gates (CNOT's) in addition to one-qubit gates in order to obtain a universal set of unitary operations for quantum computation. The $\text{CNOT}_{21} = I \otimes |0\rangle\langle 0| + \sigma_x \otimes |1\rangle\langle 1|$, where qubit 1 is a target qubit and qubit 2 is a control one, can be constructed as

$$\text{CNOT}_{21} = e^{i\pi/4} Z_1(-\pi/2) Z_2(\pi/2) R_1(\pi/2, 0) U_{\rm E} R_1(\pi/2, \pi/2), \tag{19}$$

where $U_{\rm E} = e^{-i\pi I_z \otimes I_z}$. Since we have already shown how to construct any one-qubit operations, we need to show how to implement $U_{\rm E}$.

---

[e] Bloch-Siegert effect is often not negligible, and thus the phase of the second spin is influenced by the rf pulse which controls the first spin. Let us estimate the effects in both hetero- and homo-nucleus molecules. A chloroform molecule in the magnetic field of $B = 11.4$ T is taken as an example of heteronucleus molecules. In this case, $\omega_{0,1}/2\pi = 125$ MHz for ^{13}C, $\omega_{0,2}/2\pi = 500$ MHz for proton, $t_{\rm p} \sim 20$ μs, $\omega_{1,1}/2\pi \sim 25$ kHz, and $\omega_{1,2}/2\pi \sim 100$ kHz are typical figures for a π pulse on ^{13}C. Then, $t_{\rm p}\omega_{1,2}\epsilon/2 \sim 2 \times 10^{-3}$. A cytosine molecule in the same magnetic filed is taken as an example of homonucleus molecules with two proton spins. In this case, $\Delta/2\pi \sim 700$ Hz, $t_{\rm p} \sim 6$ ms, and $\omega_{1,2}/2\pi \sim 80$ Hz are typical figures for a π-pulse on one of the protons. Then, $t_{\rm p}\omega_{1,2}\epsilon/2 \sim 0.2$.

We transform $\mathcal{H}_{\text{drift}}$ by the unitary operator $U_{\text{i}} = e^{-i\omega_{0,1} I_z t} \otimes e^{-i\omega_{0,2} I_z t}$ and obtain

$$\tilde{\mathcal{H}}_{\text{i}} = U_{\text{i}} \mathcal{H}_{\text{drift}} U_{\text{i}}^\dagger - iU_{\text{i}} \frac{d}{dt} U_{\text{i}}^\dagger = JI_z \otimes I_z + \frac{J}{2} M(\Delta t), \qquad (20)$$

where

$$M(x) = \begin{pmatrix} 0 & 0 & 0 & 0 \\ 0 & 0 & e^{-ix} & 0 \\ 0 & e^{ix} & 0 & 0 \\ 0 & 0 & 0 & 0 \end{pmatrix}.$$

Here, qubit 1 is viewed in the rotating frame with frequency $\omega_{0,1}$ while qubit 2 in the other frame with frequency $\omega_{0,2}$. We call this frame an individual rotating frame.

In the case of *weak coupling limit*, i.e. $|J/\Delta| \ll 1$, the second term $M(\Delta t)$ in Eq. (20) is time-averaged to vanish in the time scale $t \gtrsim \pi/J$, and thus we obtain

$$\tilde{\mathcal{H}}_{\text{i}} \approx \tilde{\mathcal{H}}_{\text{i,w}} = JI_z \otimes I_z. \qquad (21)$$

Therefore, we obtain the operator U_{E} by letting the system develop freely without rf pulses for the period $t_{\text{E}} = \pi/J$ as

$$U_{\text{E}} = U[\tilde{\mathcal{H}}_{\text{i,w}}, \pi/J],$$

where the definition of $U[\cdot, \cdot]$ is given in Eq. (1).

Although the individual rotating frame is convenient for understanding the principle, it is not very compatible with numerical calculations since the transformed Hamiltonian $\tilde{\mathcal{H}}_{\text{i}}$ is time-dependent thus we have to take into account time ordered product. When we transform $\mathcal{H}_{\text{drift}}$ by $U_{\text{c}} = e^{-i\omega_{0,1} I_z t} \otimes e^{-i\omega_{0,1} I_z t}$, we obtain

$$\tilde{\mathcal{H}}_{\text{c}} = U_{\text{c}} \mathcal{H}_{\text{drift}} U_{\text{c}}^\dagger - iU_{\text{c}}(d/dt) U_{\text{c}}^\dagger = \Delta I \otimes I_z + JI_z \otimes I_z + \frac{J}{2} M(0). \qquad (22)$$

Here, both qubit 1 and 2 are viewed in the rotating frame with frequency $\omega_{0,1}$. We call this frame a common rotating frame.

$U[\tilde{\mathcal{H}}_c, t]$ can be factored as

$$U[\tilde{\mathcal{H}}_c, t] = \begin{pmatrix} 1 & 0 & 0 & 0 \\ 0 & f(t) & -g(t)^* & 0 \\ 0 & g(t) & f(t)^* & 0 \\ 0 & 0 & 0 & 1 \end{pmatrix} \cdot U[\tilde{\mathcal{H}}_{c,w}, t], \qquad (23)$$

$$\tilde{\mathcal{H}}_{c,w} = \Delta I \otimes I_z + J I_z \otimes I_z,$$

$$f(t) = e^{-i\frac{1}{2}\Delta t}\left(\cos\frac{\Omega t}{2} + i\frac{\Delta}{\Omega}\sin\frac{\Omega t}{2}\right),$$

$$g(t) = -i\, e^{-i\frac{1}{2}\Delta t}\frac{J}{\Omega}\sin\frac{\Omega t}{2},$$

where $\Omega = \sqrt{\Delta^2 + J^2}$. We note that $f(t) \approx 1$ and $g(t) \approx 0$ when $t \gtrsim \pi/J$ and $|J/\Delta| \ll 1$. This leads to the approximation $U[\tilde{\mathcal{H}}_c, t] \approx U[\tilde{\mathcal{H}}_{c,w}, t]$. Therefore, $\tilde{\mathcal{H}}_{c,w}$ can be employed as an approximate Hamiltonian in the common rotating frame.[f]

In order to construct U_E with $\tilde{\mathcal{H}}_{c,w}$, we need to compensate the effect of $\Delta I \otimes I_z$ in $\tilde{\mathcal{H}}_{c,w}$ by employing a spin echo method. For example,

$$U_E = \big(R_1(-\pi, 0)R_2(-\pi, 0)\big)U\Big[\tilde{\mathcal{H}}_{c,w}, \frac{1}{2}t_E\Big]$$
$$\big(R_1(\pi, 0)R_2(\pi, 0)\big)U\Big[\tilde{\mathcal{H}}_{c,w}, \frac{1}{2}t_E\Big].$$

One qubit operations (Eqs. (18)) and the CNOT gates discussed here form a universal gate set for arbitrary quantum gates.

3.4. Non-unitary operations

For a quantum computation the system should be initialized to be in a pure state, or at least a pseudopure state. However, no unitary transformation $\rho \mapsto U\rho U^\dagger$ changes the rank of ρ, see Eq. (13). Therefore, it is impossible to get the (pseudo)pure state from a thermal state by a unitary time development. Therefore non-unitary transformations are required to prepare (pseudo)pure states.

---

[f]Let us consider an extreme common rotating frame with frequency 0. $\tilde{\mathcal{H}}_c$ in Eq. (22) is $\mathcal{H}_{\text{drift}}$ in Eq. (15), and thus $\tilde{\mathcal{H}}_{c,w}$ in Eq. (23) is replaced by

$$H_{\text{drift,w}} = -\omega_{0,1} I_z \otimes I - \omega_{0,2} I \otimes I_z + J I_z \otimes I_z.$$

This is another way to understand a secular approximation in the case of weak coupling limit.

A naive method to produce a pure state would be to cool the system under consideration. When the thermal energy $k_B T$ becomes much smaller than the energy difference between the ground state and the first excited state, the system is definitely in the ground state. However, this method is not applicable to liquid-state NMR, since solvent may freeze when it is cooled.

Non-unitary transformations to produce a pseudopure state are classified into three categories: temporal averaging, spatial averaging, and logical labeling.[g] Temporal and spatial averagings are based on the linearity of quantum mechanics. Suppose that there are N initial density matrices $\rho_{I,i}$. Let the same unitary operator U act on them. Then it yields N output density matrices $\rho_{O,i}$. Quantum mechanical linearity guarantees that

$$\sum_i \rho_{O,i} = U \left(\sum_i \rho_{I,i} \right) U^\dagger. \tag{24}$$

If $\sum_i \rho_{I,i}$ is considered as a pseudopure state ρ_{pps}, then $\sum_i \rho_{O,i}$ is a desired density matrix $U \rho_{\text{pps}} U^\dagger$.

3.4.1. *Temporal averaging*

In the case of homonucleus molecules, we can take temporal average as follows. Its thermal density matrix is

$$\rho_{\text{th}} = I_0^{\otimes 2} + \epsilon \left(I_z \otimes I_0 + I_0 \otimes I_z \right), \tag{25}$$

where we ignore the scalar coupling term since $\omega_{0,i} \gg J$. Equation (25) implies that the thermal state is regarded as a mixture of two states. From this observation, we find that ρ_{pps} is constructed as follows. Let

$$\rho_{\text{th}} \xrightarrow{\text{CNOT}_{12}} \rho_1 = I_0^{\otimes 2} + \epsilon \left(I_z \otimes I_0 + I_z \otimes I_z \right)$$

$$\rho_{\text{th}} \xrightarrow{\text{CNOT}_{21}} \rho_2 = I_0^{\otimes 2} + \epsilon \left(I_z \otimes I_z + I_0 \otimes I_z \right)$$

Therefore,

$$\rho_{\text{pps}} = \frac{1}{3} \left(\rho_{\text{th}} + \rho_1 + \rho_2 \right) = \frac{2\epsilon}{3} \text{diag}(1,0,0,0)$$

where we omitted the irrelevant term proportional to $I_0^{\otimes 2}$. This is a pseudopure state. The above construction is called the product operator approach.[15]

---

[g]The logical labeling is not discussed in this contribution.

The general thermal density matrix of two qubit molecules, including heteronucleus ones, is given as

$$\rho_{\text{th}} = I_0^{\otimes 2} + \text{diag}(\rho_{11}, \rho_{22}, \rho_{33}, \rho_{44})$$

and two CNOT operations permute the diagonal elements as

$$\rho_{\text{th}} \xrightarrow{\text{CNOT}_{12}\text{CNOT}_{21}} \rho_1 = I_0^{\otimes 2} + \text{diag}(\rho_{11}, \rho_{44}, \rho_{22}, \rho_{33}),$$

$$\rho_{\text{th}} \xrightarrow{\text{CNOT}_{21}\text{CNOT}_{12}} \rho_2 = I_0^{\otimes 2} + \text{diag}(\rho_{11}, \rho_{33}, \rho_{44}, \rho_{22}).$$

Note that the element ρ_{11} is left unchanged under these transformations. We obtain, after averaging over the density matrices,

$$\rho_{\text{pps}} = \frac{1}{3}(\rho_{\text{th}} + \rho_1 + \rho_2)$$
$$= \frac{1}{3}(3\rho_{11} - \rho_{22} - \rho_{33} - \rho_{44})\,\text{diag}(1,0,0,0).$$

When the number of qubits is more than two, the product operator approach is more advantageous. For example, three initial states are enough for the product operator approach even for three-qubit molecules, while the cyclic permutation approach requires $2^3 - 1 = 7$ initial states.

3.4.2. *Spatial averaging*

Spatial averaging approach employs pulsed field gradients to prepare $\rho_{I,i}$ in Eq. (24).

Here we discuss effects of field gradient on a two-spin system. Field gradient modifies the system Hamiltonian (15) to

$$\mathcal{H}_{\text{drift}}(z) = -\omega_{0,1}(1+\nabla z)I_z \otimes I - \omega_{0,2}(1+\nabla z)I \otimes I_z + J\sum_i I_i \otimes I_i,$$

where ∇ is the normalized strength of the field gradient along the z-direction. We assume that the sample extends over $-L_0 < z < L_0$ along the z-axis. In the individual rotating frame, the relevant Hamiltonian is

$$\tilde{\mathcal{H}}_{\text{drift}}(z) = -\omega_{0,1}\nabla z\, I_z \otimes I - \omega_{0,2}\nabla z\, I \otimes I_z + J\sum_i I_i \otimes I_i.$$

After application of the field gradient for τ, the state is unitarily transformed by the operator

$$U_G(z) = \exp[-i(-\omega_{0,1}\nabla z I_z \otimes I - \omega_{0,2}\nabla z I \otimes I_z + J I_z \otimes I_z)\tau].$$

The density matrix $\tilde{\rho} = (\rho_{ij})$ in the rotating frame is a 4×4 Hermitian matrix. After application of the field gradient, the density matrix has z-dependence as

$$\tilde{\rho}_G(z) = U_G(z)\tilde{\rho}U_G^\dagger(z) = \begin{pmatrix} \rho_{11} & * & * & * \\ * & \rho_{22} & e^{i\Delta\tau\nabla z}\rho_{23} & * \\ * & e^{-i\Delta\tau\nabla z}\rho_{32} & \rho_{33} & * \\ * & * & * & \rho_{44} \end{pmatrix}, \quad (26)$$

where $*$ are terms containing $e^{\pm i 2\omega_{0,i}\nabla z\tau}$ or $e^{\pm i(\omega_{0,1}+\omega_{0,2})\nabla z\tau}$. The parameters ∇ and τ are chosen so that these oscillating terms vanish after taking average over the sample coordinate z, which is a non-unitary operation. Note that the coil detects the sum of the signals from the each parts of the sample.

In the case of heteronucleus molecules, $|\Delta| = |\omega_{0,1} - \omega_{0,2}|$ and $\omega_{0,i}$ are on the same order of magnitude. Therefore, the application of a pulsed field gradient is equivalent to applying the following non-unitary operation on the density matrix,

$$\rho \xrightarrow{D_G} \rho_G = \begin{pmatrix} \rho_{11} & 0 & 0 & 0 \\ 0 & \rho_{22} & 0 & 0 \\ 0 & 0 & \rho_{33} & 0 \\ 0 & 0 & 0 & \rho_{44} \end{pmatrix}. \quad (27)$$

On the other hand, ρ_G for homonucleus molecules takes an approximate form

$$\rho_G = \begin{pmatrix} \rho_{11} & 0 & 0 & 0 \\ 0 & \rho_{22} & \rho_{23} & 0 \\ 0 & \rho_{32} & \rho_{33} & 0 \\ 0 & 0 & 0 & \rho_{44} \end{pmatrix}. \quad (28)$$

Some of the off-diagonal components are left unchanged in the latter case since $|\Delta|L\nabla\tau \ll 1$ for a typical choice of parameters.

We consider here homonucleus molecules first. The thermal state is transformed by $D_{G1}R_2(\pi/3, \pi/2)$ as,

$$\rho_{th} \longrightarrow \epsilon\left(I_z \otimes I_0 + \frac{1}{2}I_0 \otimes I_z\right).$$

We omit irrelevant terms that are proportional to $I_0^{\otimes 2}$ hereafter. Subsequent operations $D_{G2}R_1(\pi/4, 3\pi/2)U_E(\pi)R_1(\pi/4, 0)$ yield

$$\longrightarrow \epsilon\left(I_z \otimes I_z + I_z \otimes I_0 + I_0 \otimes I_z\right) = \epsilon\,\mathrm{diag}(1, 0, 0, 0).$$

If heteronucleus molecules, such as ^{13}C-labeled chloroform, are employed, the first operation $D_{G1}R_2(\pi/3, \pi/2)$ is replaced by $D_{G1}R_1(\eta, \pi/2)$ where η satisfies $\omega_{0,1}\cos\eta = 2\omega_{0,2}$. Note that $\omega_{0,1}/\omega_{0,2} \approx 4$ for ^{13}C-labeled chloroform when spin 1 and 2 are proton and ^{13}C, respectively. Then, the thermal state is transformed as,

$$\epsilon\left(I_z \otimes I_0 + \frac{\omega_{0,2}}{\omega_{0,1}} I_0 \otimes I_z\right) \longrightarrow 2\epsilon\frac{\omega_{0,2}}{\omega_{0,1}}\left(I_z \otimes I_0 + \frac{1}{2}I_0 \otimes I_z\right).$$

The same second operation is applicable to generate a pseudopure state. We can apply the operation after equalizing populations of two spins.[18]

General procedures for arbitrary number of qubits are proposed by several groups.[19,20] The method discussed here is a two-qubit version of Sakaguchi, Ozawa and Fukumi[19] which does not require an ancilla spin. In contrast, a method proposed by Sharf, Havel and Cory[20] requires an ancilla spin.

3.5. *Separability*

Braunstein *et al.* pointed out that no entanglement appears in the physical states at any stage of NMR experiments so far.[10] We follow their discussions.

Let us consider the density matrix for N-qubits

$$\rho_\epsilon = (1 - \epsilon)I_0^{\otimes N} + \epsilon\rho_1.$$

If ρ_ϵ is rewritten as

$$\rho_\epsilon = \sum_{ij...k} p_{ij...k} O_{1i} \otimes O_{2j} \otimes \cdots \otimes O_{Nk}$$

with non-negative $p_{ij...k}$, then $p_{ij...k}$ can be interpreted as the probabilities realizing the state described by the pure direct product density matrix $O_{1i} \otimes O_{2j} \otimes \cdots \otimes O_{Nk}$ without any entangled density matrices. In other words, ρ_ϵ can be considered as a classical density matrix and is called *separable*.

In the case of 2-qubit molecule, its density matrix can be expanded as

$$\rho_1 = \sum_{i,j=0,x,y,z} c_{ij} I_i \otimes I_j.$$

Note that $c_{00} = 1$ from the normalization requirement and $-1 \leq c_{ij} \leq 1$. On the other hand, ρ_1 can be rewritten as,

$$\rho_1 = \sum_{\alpha,\beta=\pm 1} \sum_{i,j=x,y,z} p_{i\alpha j\beta} \, P_{i,\alpha} \otimes P_{j,\beta},$$

where $P_{k,\pm 1} = I_0 \pm I_k$ and $p_{i\alpha j\beta} = (1 + 3\alpha c_{i0} + 3\beta c_{0j} + 9\alpha\beta c_{ij})/36$.

The maximally mixed density matrix $I_0^{\otimes 2}$ corresponds to the case that $c_{00} = 1$ and $c_{ij} = 0$ for $(i,j) \neq (0,0)$. And thus, $I_0^{\otimes 2}$ can be rewritten as $\frac{1}{36}\sum_{i\alpha j\beta} P_{i,\alpha} \otimes P_{j,\beta}$. Therefore, the thermal state density matrix ρ_{th}, which is in a small neighborhood of the maximally mixed density matrix, is given as,

$$\rho_{\text{th}} = (1-\epsilon)I_0^{\otimes 2} + \epsilon\rho_1 = \sum_{i\alpha j\beta}(\frac{1-\epsilon}{36} + \epsilon p_{i\alpha j\beta})P_{i,\alpha} \otimes P_{j,\beta}.$$

The coefficients $(1-\epsilon)/36 + \epsilon p_{i\alpha j\beta}$ is non-negative when $\epsilon \sim 10^{-5}$. And thus, ρ_{th} is separable.

The bound $\epsilon_{\max}$ that ρ_ϵ is separable for any ρ_1 is obtained as follows. The minimum value of $p_{i\alpha j\beta}$ is $(1 - 3 - 3 - 9)/36 = -7/18$ since $|c_{ij}| \leq 1$. Then, $(1-\epsilon_{\max})/36 - 7\epsilon_{\max}/18 \geq (1-\epsilon)/36 + \epsilon p_{i\alpha j\beta} \geq 0$ gives $\epsilon_{\max} = 1/15$. Note that $\epsilon_{\max}$ is far larger than ϵ in usual liquid-state NMR.[h]

Braunstein et al. discussed a general N-qubits case.[10] Here we refer only their conclusion. ρ_ϵ is separable if

$$\epsilon \leq \frac{1}{1+2^{2N-1}} \overset{N\to\infty}{\sim} \frac{2}{4^N}.$$

See also discussions by Schack and Caves.[22] They provided a better bound

$$\epsilon \leq \frac{1}{1 \pm 2^N + 2^{2N-2}},$$

$+(-)$ applying when N is even (odd).

4. Quantum Computation

We discuss here on "generation and suppression of artificial decoherence"[11] and "quantum teleportation without irreversible detection"[12] that are our recent experimental results. For understanding these, we first discuss quantum channel and related topics.

4.1. *Quantum channel*

4.1.1. *Quantum channel and entanglement fidelity*

A general state change in quantum mechanics is described by a map,

$$\rho \mapsto \frac{\mathcal{E}(\rho)}{\text{Tr}[\mathcal{E}(\rho)]}, \qquad (29)$$

---

[h]Anwar et al. prepared a two-spin molecule in an almost pure state by using parahydrogen.[21]

which connects an input state ρ and an output state $\mathscr{E}(\rho_s)/\mathrm{Tr}[\mathscr{E}(\rho_s)]$. The general form for $\mathscr{E}$, which satisfies (1) linearity, (2) trace decreasing, (3) preserving positivity, and (4) complete positivity, is

$$\mathscr{E}(\rho) = \sum_i A_i \rho A_i^\dagger, \qquad (30)$$

where the system operators A_i must satisfy $\sum_i A_i^\dagger A_i \leq I$. The above equation is called an operator sum representation of the channel $\mathscr{E}$.[1,23–26] The entanglement fidelity is defined by

$$F_e(\rho, \mathscr{E}) = \frac{\sum_i |\mathrm{Tr}[A_i \rho]|^2}{\mathrm{Tr}[\mathscr{E}(\rho)]}. \qquad (31)$$

In this contribution, we only consider the case of

$$\sum_i A_i^\dagger A_i = I. \qquad (32)$$

Equation (32) implies that $\mathrm{Tr}_s \mathscr{E}(\rho) = \mathrm{Tr}\,\rho$ and hence it is called the trace-preserving condition. Hereafter, we employ E_i instead of A_i in order to emphasize that we consider a special case. Then, Eqs. (29) and (31) reduce to

$$\rho \to \mathscr{E}(\rho),$$
$$F_e(\rho, \mathscr{E}) = \sum_i |\mathrm{Tr}[E_i \rho]|^2.$$

4.1.2. *Phase flip channel*

Here we introduce the phase flip channel, which is a typical example of a channel,

$$\mathscr{E}(\rho) = E_0 \rho E_0^\dagger + E_1 \rho E_1^\dagger \qquad (33)$$

with

$$E_0 = \sqrt{p}\, I, \quad E_1 = \sqrt{1-p}\, \sigma_z.$$

The channel (33) transforms ρ as

$$\rho(M, n_x, n_y, n_z) \mapsto p\rho(M, n_x, n_y, n_z) + (1-p)\rho(M, -n_x, -n_y, n_z). \qquad (34)$$

The right hand side shows that the azimuthal angle (phase) of the Bloch vector is left unchanged with probability p while it is flipped with probability $1 - p$. Hence it is natural to call this a phase flip channel. In particular, when $p = \frac{1}{2}$,

$$\rho(M, n_x, n_y, n_z) \mapsto \rho(M, 0, 0, n_z).$$

This means that the components of the Bloch vector in the xy-plane ($M(n_x, n_y, 0)$) vanish after the channel is applied and the information about the phase is completely lost. However, n_z, which represent the population difference between the states $|0\rangle$ and $|1\rangle$ remains unchanged. Due to these properties, decoherence generated via the phase flip channel is called the phase decoherence.

The entanglement fidelity of the phase flip channel is calculated as

$$F_e(I_0, \mathscr{E}) = p. \tag{35}$$

4.1.3. Quantum process tomography

A quantum channel is considered to be known when all E_i's are known. Knowing or measuring E_i's is called *quantum process tomography*.

The idea of quantum process tomography is as follows. Let us assume that ρ is a $d \times d$ matrix, then the map $\rho \to \mathscr{E}(\rho)$ should generally be described with d^4 independent real parameters. When the map is trace-preserving, this gives d^2 additional constraints. Therefore, a trace-preserving $\mathscr{E}$ is described with $d^4 - d^2$ independent real parameters. When we prepare d^2 independent input states and measure all output states, the linearity of quantum mechanics guarantees to determine all parameters. We also note that knowing a quantum state is called *quantum state tomography*.

We, now, consider the case of $d = 2$.[i] To determine E_i from measurements, E_i is parameterized as

$$E_i = \sum_m e_{im} \tilde{E}_m,$$

with real numbers, e_{im}, where

$$\tilde{E}_1 = I, \quad \tilde{E}_2 = \sigma_x, \quad \tilde{E}_3 = -i\sigma_y, \quad \tilde{E}_4 = \sigma_z$$

are selected for convenience. Then, $\mathscr{E}$ is rewritten as

$$\mathscr{E}(\rho) = \sum_{mn} \chi_{mn} \tilde{E}_m \rho \tilde{E}_n^\dagger, \tag{36}$$

where $\chi_{mn} = \sum_i e_{im} e_{in}^*$. We can accomplish our goal by determining χ from measurements.

Four ($= d^2$) independent ρ_i

$$\rho_1 = \begin{pmatrix} 1 & 0 \\ 0 & 0 \end{pmatrix}, \rho_2 = \begin{pmatrix} 0 & 1 \\ 0 & 0 \end{pmatrix}, \rho_3 = \begin{pmatrix} 0 & 0 \\ 1 & 0 \end{pmatrix}, \rho_4 = \begin{pmatrix} 0 & 0 \\ 0 & 1 \end{pmatrix}$$

---

[i]See, also Chap. 8 in Ref.1

are selected. Then, $\mathscr{E}(\rho_j)$ is parameterized as

$$\mathscr{E}(\rho_j) = \sum_k \lambda_{jk}\, \rho_k = \begin{pmatrix} \lambda_{j,1} & \lambda_{j,2} \\ \lambda_{j,3} & \lambda_{j,4} \end{pmatrix}, \qquad (37)$$

where λ_{jk} can be determined from measurements. Note that there are only 3 independent λ_{jk} for each j, because of the trace preserving condition. We introduce β_{jk}^{mn} which is defined by

$$\tilde{E}_m \rho_j \tilde{E}_n^\dagger = \sum_k \beta_{jk}^{mn}\, \rho_k. \qquad (38)$$

Combining Eqs. (36), (37) and (38), we obtain

$$\sum_k \sum_{mn} \chi_{mn} \beta_{jk}^{mn}\, \rho_k = \sum_k \lambda_{jk} \rho_k, \quad \text{or} \quad \sum_{mn} \chi_{mn} \beta_{jk}^{mn} = \lambda_{jk}, \qquad (39)$$

and then χ by solving Eq. (39).[j] We can calculate χ more conveniently as

$$\chi = \Lambda \begin{pmatrix} \mathscr{E}(\rho_1) & \mathscr{E}(\rho_2) \\ \mathscr{E}(\rho_3) & \mathscr{E}(\rho_4) \end{pmatrix} \Lambda = \Lambda \begin{pmatrix} \lambda_{11} & \lambda_{12} & \lambda_{21} & \lambda_{22} \\ \lambda_{13} & \lambda_{14} & \lambda_{23} & \lambda_{24} \\ \lambda_{31} & \lambda_{32} & \lambda_{41} & \lambda_{42} \\ \lambda_{33} & \lambda_{34} & \lambda_{43} & \lambda_{44} \end{pmatrix} \Lambda,$$

where

$$\Lambda = \frac{1}{2} \begin{pmatrix} \tilde{E}_1 & \tilde{E}_2 \\ \tilde{E}_2 & -\tilde{E}_1 \end{pmatrix},$$

in terms of block matrices.

Finally, E_i is obtained as follows. Let the unitary matrix $U^\dagger$ diagonalize χ, or

$$U^\dagger \chi U = \mathrm{diag}(d_1, d_2, d_3, d_4), \quad \text{or} \quad \chi_{mn} = \sum_{ik} U_{mi} d_i \delta_{ik} U_{kn}^*.$$

Substitution of the above definition into Eq. (36) yields

$$\mathscr{E}(\rho) = \sum_{mnik} U_{mi} d_i \delta_{ik} U_{kn}^* \tilde{E}_m \rho \tilde{E}_n^\dagger = \sum_{mni} \sqrt{d_i}\, U_{mi} \tilde{E}_m \rho \sqrt{d_i}\, U_{in}^* \tilde{E}_n^\dagger.$$

---

[j] In our particular case, Eq. (39) is

$$\begin{pmatrix} \chi_{11} + \chi_{14} + \chi_{41} + \chi_{44} & \chi_{12} + \chi_{13} + \chi_{42} + \chi_{43} & \chi_{21} + \chi_{24} + \chi_{31} + \chi_{34} & \chi_{22} + \chi_{23} + \chi_{32} + \chi_{33} \\ \chi_{12} - \chi_{13} + \chi_{42} - \chi_{43} & \chi_{11} - \chi_{14} + \chi_{41} - \chi_{44} & \chi_{22} - \chi_{23} + \chi_{32} - \chi_{33} & \chi_{21} - \chi_{24} + \chi_{31} - \chi_{34} \\ \chi_{21} + \chi_{24} - \chi_{31} - \chi_{34} & \chi_{22} + \chi_{23} - \chi_{32} - \chi_{33} & \chi_{11} + \chi_{14} - \chi_{41} - \chi_{44} & \chi_{12} + \chi_{13} - \chi_{42} - \chi_{43} \\ \chi_{22} - \chi_{23} - \chi_{32} + \chi_{33} & \chi_{21} - \chi_{24} - \chi_{31} + \chi_{34} & \chi_{12} - \chi_{13} - \chi_{42} + \chi_{43} & \chi_{11} - \chi_{14} - \chi_{41} + \chi_{44} \end{pmatrix}$$

$$= \begin{pmatrix} \lambda_{11} & \lambda_{12} & \lambda_{13} & \lambda_{14} \\ \lambda_{21} & \lambda_{22} & \lambda_{23} & \lambda_{24} \\ \lambda_{31} & \lambda_{32} & \lambda_{33} & \lambda_{34} \\ \lambda_{41} & \lambda_{42} & \lambda_{43} & \lambda_{44} \end{pmatrix}.$$

Then
$$E_i = \sqrt{d_i} \sum_m U_{mi} \tilde{E}_m$$
is obtained.

4.2. *Generation and suppression of artificial decoherence*

In this section we show how to understand decoherence in quantum mechanics and how to simulate decoherence phenomena in NMR-QC.[11]

4.2.1. *Decoherence*

Decoherence is an irreversible change of a state of a quantum system which has quantum correlation with its environment. Decoherence is irreversible due to our lack of knowledge about the state of the environment. Let $\mathcal{H}_s$ and $\mathcal{H}_e$ be the Hilbert spaces of the system and the environment, respectively. The initial state of the system is represented by the density matrix ρ_s while that of the environment by ρ_e. The state of the whole system changes following the time-evolution law,

$$\rho_s \otimes \rho_e \mapsto U \rho_s \otimes \rho_e U^\dagger. \tag{40}$$

Here U is a unitary operator acting on the Hilbert space of the composite system $\mathcal{H}_s \otimes \mathcal{H}_e$. We consider the case in which the initial state is an uncorrelated state $\rho = \rho_s \otimes \rho_e$. The states of the system and the environment are correlated via the transformation (40). If we are interested only in the state of the system, the measurement outcomes are completely described by the reduced density matrix

$$\rho_s' = \mathcal{E}(\rho_s) = \mathrm{Tr}_e\left(U \rho_s \otimes \rho_e U^\dagger\right), \tag{41}$$

where the symbol Tr_e denotes the partial trace over $\mathcal{H}_e$. The partial trace operation is non-invertible and the associated loss of information is interpreted as decoherence. Equation (41) defines a quantum channel representing decoherence phenomenon.

There is another approach to defining channels without resort to partial trace over the Hilbert space of environment. Assume that we have a set of unitary operators $\{U_k\}$, which acts on $\mathcal{H}_s$, and that we have a set of real numbers $\{p_k\}$ such that $0 \leq p_k \leq 1$ and $\sum_k p_k = 1$. We then define a transformation of the system density matrix ρ_s by

$$\rho_s \mapsto \mathcal{M}(\rho_s) = \sum_k p_k U_k \rho_s U_k^\dagger. \tag{42}$$

They satisfy the condition (32) if we put $E_k = \sqrt{p_k}\, U_k$. This argument tells us that if we apply a set of time-evolution unitary operators $\{U_k\}$ on the system with a probability distribution $\{p_k\}$, we will observe a decoherence-like phenomenon after taking an average of the measured data over k. We call the transformation $\mathcal{M}$ a mixing process.[k]

4.2.2. *Phase flip channel*

We formulate the phase flip channel through the interaction between the system and environment, as follows. We take a one-qubit system and a one-qubit environment. Assume that the initial state of the environment is

$$|\psi_e\rangle = \sqrt{p}\,|0\rangle + \sqrt{1-p}\,|1\rangle$$

with a real number p ($0 \leq p \leq 1$). We take a unitary operator

$$U = I \otimes |0\rangle\langle 0| + \sigma_z \otimes |1\rangle\langle 1| = \begin{pmatrix} 1 & 0 & 0 & 0 \\ 0 & 1 & 0 & 0 \\ 0 & 0 & 1 & 0 \\ 0 & 0 & 0 & -1 \end{pmatrix}$$

which acts on $\mathcal{H}_s \otimes \mathcal{H}_e = \mathbb{C}^2 \otimes \mathbb{C}^2$. By substituting them into Eq. (41), we obtain the phase flip channel (33).

We can also obtain a phase flip channel by introducing the mixing process defined previously. Let us consider a single spin Hamiltonian

$$\mathcal{H}_{\text{random}} = -\omega(t) I_z, \tag{45}$$

where $\omega(t)$ is a random fluctuating field. The time-evolution operator associated with the Hamiltonian (45) is the phase shift gate

$$S(\theta) = e^{i\theta I_z} \tag{46}$$

with

$$\theta = \int_0^t \omega(\tau)\, d\tau.$$

---

[k]We can construct a virtual environment Hilbert space $\mathcal{H}_e = \{\sum_k c_k |k\rangle\}$ by demanding formally that $\{|k\rangle\}$ is a complete orthonormal set. Moreover, we define an environment density matrix

$$\rho_e = \sum_k p_k |k\rangle\langle k| \tag{43}$$

and define a unitary operator $U = \sum_k U_k \otimes |k\rangle\langle k|$ that acts on $\mathcal{H}_s \otimes \mathcal{H}_e$ as

$$U\big(|\psi_s\rangle \otimes |k\rangle\big) = \big(U_k|\psi_s\rangle\big) \otimes |k\rangle. \tag{44}$$

By substituting them into Eq. (41), we obtain the mixing process (42).

The phase θ integrates the effect of $\omega(\tau)$ in the interval $[0, t]$. The phase shift gate acts on the density matrix as

$$S(\theta)\rho_s S^\dagger(\theta) = \begin{pmatrix} \rho_{00} & e^{i\theta}\rho_{01} \\ e^{-i\theta}\rho_{10} & \rho_{11} \end{pmatrix}.$$

Given a probability distribution $p(\theta)$ which characterizes the random fluctuating field, the mixing process is evaluated as[27]

$$\mathcal{M}(\rho_s) = \int_{-\infty}^{\infty} p(\theta) S(\theta) \rho_s S^\dagger(\theta) d\theta = \begin{pmatrix} \rho_{00} & \langle e^{i\theta}\rangle \rho_{01} \\ \langle e^{-i\theta}\rangle \rho_{10} & \rho_{11} \end{pmatrix}.$$

For any probability distribution $p(\theta)$,

$$|\langle e^{-i\theta}\rangle| = \left| \int_{-\infty}^{\infty} p(\theta) e^{-i\theta} d\theta \right| \leq \int_{-\infty}^{\infty} \left| p(\theta) e^{-i\theta} \right| d\theta = 1.$$

Therefore the absolute value of the off-diagonal elements of $\mathcal{M}(\rho_s)$ is smaller than those of ρ_s. When the average $\langle e^{-i\theta}\rangle$ is a real number, the map $\mathcal{M}(\rho_s)$ reproduces the phase flip channel (34). This applies when $p(\theta) = p(-\theta)$ for example.

We can further simplify the model without losing the essence of the random fluctuating field model. Suppose that $\omega(t)$ takes only two values, $\omega_0 \pm \delta\omega$, with the corresponding probabilities $p(\pm\delta\omega)$. We simulate phase decoherence phenomena according to this simplified random fluctuating field model.

4.2.3. Suppressing decoherence by the bang-bang control

Several groups have proposed and analyzed a useful technique to suppress decoherence, which is called a quantum bang-bang control.[28,29] We briefly explain the principle of the bang-bang control.

We assume that $\omega(t)$ in the Hamiltonian (45) is constant, so that Eq. (46) becomes

$$S(\omega t) = e^{i\omega t I_z}.$$

If, however, time-evolution of the qubit could be reversed for the next duration t by some methods, the total effect of the environment for $2t$ is nulled as

$$S(-\omega t) S(\omega t) = I.$$

A time-reversal operation for the phase shift can be simply implemented with a pair of π-pulses, as

$$R(-\pi, 0) S(\omega t) R(\pi, 0) = S(-\omega t) = S(\omega t)^{-1},$$

where R is defined in Eq. (12). Note that $R(\pi,0) = -i\sigma_x$ and $R(-\pi,0) = i\sigma_x$.

In general circumstances, $\omega(t)$ is not constant. Then $S(\omega t)$ is replaced with

$$U(t;t_0) = S\left(\int_{t_0}^{t} \omega(\tau)d\tau\right).$$

We, here, introduce a characteristic time τ_c and assume that the autocorrelation function $\langle \omega(t_0+t)\omega(t_0)\rangle/\langle \omega^2(t_0)\rangle \approx 1$ for $t \lesssim \tau_c$, where $\langle \cdot \rangle$ denotes the average over t_0. Then, $U(t;t_0)$ dose not vary rapidly and satisfies

$$U(t_0 + 2t_b; t_0 + t_b) \approx U(t_0 + t_b; t_0)$$

for $t_b (\ll \tau_c)$. Then, it is legitimate to use an approximation

$$R(-\pi,0)U(t_0 + 2t_b; t_0 + t_b)R(\pi,0)U(t_0 + t_b; t_0) \approx I.$$

The above equation implies that the phase shift of the system will be mostly canceled by inserting pairs of π-pulses with a short interval $t_b (\ll \tau_c)$. Therefore the associated decoherence of the system will be suppressed.

4.2.4. Artificial environment

Any system in an environment has a Hamiltonian of the form

$$\mathcal{H}_t = \mathcal{H}_s + \mathcal{H}_e + \mathcal{H}_{se},$$

where $\mathcal{H}_s$ and $\mathcal{H}_e$ govern intrinsic behaviors of the system and the environment, respectively, while $\mathcal{H}_{se}$ represents interaction between them. The relation between the system and the environment is schematically depicted in Fig. 9.

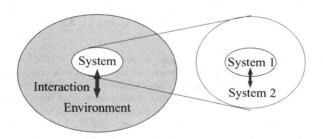

Fig. 9. System and its environment. The system consists of two subsystems, 1 and 2.

Suppose the interaction $\mathcal{H}_{se}$ is so weak that its effect on the system qubits is negligible compared with that of $\mathcal{H}_s$ for a certain time scale τ. Assume further that the system consists of two subsystems, which are referred to as subsystems 1 and 2. Then the system Hamiltonian $\mathcal{H}_s$ is decomposed as

$$\mathcal{H}_s = \mathcal{H}_1 + \mathcal{H}_2 + \mathcal{H}_{12}.$$

Here $\mathcal{H}_1$ and $\mathcal{H}_2$ govern intrinsic behaviors of the subsystems, while $\mathcal{H}_{12}$ describes interaction between them. Under this decomposition, we may regard the subsystem 1 as a new system and the subsystem 2 as an artificial environment. The subsystem 1 will exhibit a decoherence-like behavior if the subsystem 2 simulates an environment that has many degrees of freedom.[1]

Teklemariam et al.[32] proposed to apply a stochastic classical field[33] on the artificial environment to generate artificial decoherence. Although the artificial environment has only a few degrees of freedom, it can simulate an open environment if it is randomized by an external stochastic field. With this idea, they observed decoherence-like behavior by using NMR. Although we closely follow Teklemariam et al., we simplify the principle of generating artificial phase decoherence phenomena to the limit.

4.2.5. *Phase decoherence in a quantum memory*

Suppose a qubit sits in a quantum memory device (quantum register). The qubit is exposed to a noisy environment and loses its phase coherence. The noise is described by the random fluctuating field in the Hamiltonian (45). We again assume that $\langle \omega(t_0 + t)\omega(t_0)\rangle/\langle\omega^2(t_0)\rangle/ \approx 1$ for $t \lesssim \tau_c$. This can be also interpreted as a model of the T_2 relaxation process in NMR.

We can construct a model which realizes the above situation with a two-qubit system as follows. Two qubits are referred to as qubit 1 and qubit 2, respectively. Qubit 1 is regarded as a quantum register (system) while qubit 2 as an environment coupled to qubit 1. We take the Hamilto-

---

[1]Zhang et al.[30] experimentally studied the spin dynamics of ^{13}C-labeled trichloroethane, which has three spins in a molecule. They regarded three spins as a composite of a two-qubit system and a one-qubit environment. They claimed that they observed decoherence in the two-qubit system. However, the artificial environment must have a large number of degrees of freedom to introduce irreversible decoherence-like behavior in the system while the one-qubit environment employed in their experiments had not enough degrees of freedom. Hence their system exhibited a periodic behavior and failed to introduce an irreversible change in the system. Ryan et al.[31] overcame this difficulty by employing a seven-qubit molecule for simulating complex dynamics of an environment.

nian

$$\mathcal{H} = J I_z \otimes I_z. \tag{47}$$

We assume $J > 0$ without loss of generality. When the Hamiltonian (47) acts on states $|\psi\rangle \otimes |0\rangle$ and $|\psi\rangle \otimes |1\rangle$, it yields

$$\begin{aligned}\mathcal{H}|\psi\rangle \otimes |0\rangle &= \tfrac{J}{2} I_z |\psi\rangle \otimes |0\rangle, \\ \mathcal{H}|\psi\rangle \otimes |1\rangle &= -\tfrac{J}{2} I_z |\psi\rangle \otimes |1\rangle,\end{aligned} \tag{48}$$

respectively. Thus the Hamiltonian $\mathcal{H}$ describes an effective magnetic field acting on qubit 1. The effective magnetic field is $J/2$ when qubit 2 is in the state $|1\rangle$ and it is $-J/2$ when qubit 2 is in $|0\rangle$. Hence, by flipping qubit 2 randomly, we can realize a random fluctuating field acting on qubit 1.

Set the initial state of qubit 2 to $|0\rangle$ at time $t_0 = 0$, turn it to $|1\rangle$ at t_1, turn it back to $|0\rangle$ at t_2, and repeat flipping qubit 2 at t_3, t_4 and so on. Under this manipulation qubit 2 effectively works as a noisy environment acting on qubit 1. The time interval between consecutive flippings is denoted as

$$\Delta_j = t_{j+1} - t_j = \bar{\Delta}(1 + \alpha \xi_j) \quad (j = 0, 1, 2, \cdots). \tag{49}$$

We emphasize that $\bar{\Delta}$ in the above equation corresponds to τ_c, the time scale with which the environment retains its memory. Here $\{\xi_j\}$ are independent random variables obeying the probability distribution function

$$p(\xi_j) = \frac{1}{\sqrt{2\pi}} e^{-\xi_j^2/2}, \tag{50}$$

which is selected in order to mimic the phase decoherence due to the fluctuating environment.[34] The parameter α ($0 \leq \alpha \leq 1/4$) characterizes variance of the time intervals. To ensure that Δ_j in (49) is positive, the range of ξ_j should be $-\frac{1}{\alpha} < \xi_j$. However, if α is not too large, the probability of having negative Δ_j is negligible. Hence, it is legitimate to extend the range of ξ_j-integration to $-\infty < \xi_j < \infty$ when we take an average. The time evolution operator for qubit 1 at time t_{2n} is

$$U_\xi = S\left(\frac{J}{2}(-\Delta_1 + \Delta_2 + \cdots - \Delta_{2n-1} + \Delta_{2n})\right). \tag{51}$$

In this case, the mixing process (42) yields

$$\mathcal{M}(\rho_s) = \int_{-\infty}^{\infty} d\xi_1 \cdots \int_{-\infty}^{\infty} d\xi_{2n} \; p(\xi_1) \cdots p(\xi_{2n}) \, U_\xi \rho_s U_\xi^\dagger. \tag{52}$$

$\mathcal{M}(\rho_s)$ is calculated, by noticing that ξ_i are independent of each other, as

$$\mathcal{M}(\rho_s) = \begin{pmatrix} \rho_{00} & \lambda^n \rho_{01} \\ \lambda^n \rho_{10} & \rho_{11} \end{pmatrix},$$

at $\overline{t_{2n}} = 2n\bar{\Delta}$, the average value of time t_{2n}. Here,

$$\lambda = \int_{-\infty}^{\infty} d\xi_1 \, p(\xi_1) \int_{-\infty}^{\infty} d\xi_2 \, p(\xi_2) e^{-i(J/2)\bar{\Delta}(1+\alpha\xi_1)} e^{i(J/2)\bar{\Delta}(1+\alpha\xi_2)}$$
$$= e^{-\frac{1}{4}(J\bar{\Delta}\alpha)^2}.$$

Then λ^n is found as

$$\lambda^n = \exp\left\{-\frac{1}{4}(J\bar{\Delta}\alpha)^2 \frac{\overline{t_{2n}}}{2\bar{\Delta}}\right\}$$
$$= \exp\left(-\frac{1}{8}J^2\alpha^2\bar{\Delta}\,\overline{t_{2n}}\right) = e^{-\overline{t_{2n}}/T_2^*}. \quad (53)$$

We see that phase decoherence in the presence of a random fluctuating field is characterized by the decay constant

$$T_2^* = \frac{8}{J^2\alpha^2\bar{\Delta}}. \quad (54)$$

It is clear that T_2^* becomes smaller for a larger variance α of fluctuation of the effective magnetic field. We call T_2^* the effective transverse relaxation time since $\mathscr{M}(\rho_s)$ is calculated only at $\overline{t_{2n}}$. We note that $\bar{\Delta}$ is regarded as the correlation time of the artificial environment and that the phase decoherence is non-exponential for $t \lesssim \bar{\Delta}$.

The reader may refer to our publication[11] for the detailed analysis when the bang-bang control is applied.

4.2.6. Simulation of phase decoherence in a quantum memory

We experimentally demonstrate generation and suppression of phase decoherence in a quantum memory. A 0.6 ml, 200 mM sample of ^{13}C-labeled chloroform (Cambridge Isotope) in d-6 acetone is employed as a two-qubit molecule. The spin of carbon nucleus in chloroform is referred to as spin 1 (qubit 1), while the spin of hydrogen nucleus is referred to as spin 2 (qubit 2). Data is taken at room temperature with a JEOL ECA-500 NMR spectrometer,[35] whose hydrogen Larmor frequency is approximately 500 MHz.[m]

Both spins are initially set to the up-state $|0\rangle$. Spin 1 is turned to the x-axis by a $\pi/2$-pulse at $t = t_0 = 0$ while spin 2 is flipped from $|0\rangle$ to $|1\rangle$ by

---

[m]The measured spin-spin coupling constant is $J/2\pi = 215.5$ Hz. The transverse relaxation times in a natural environment are $T_2 \sim 0.30$ s for the carbon nucleus and $T_2 \sim 7.5$ s for the hydrogen nucleus. The longitudinal relaxation times are measured to be $T_1 \sim 20$ s for both nuclei. The duration of a π-pulse for both nuclei is set to 50 μs throughout our experiments. Precision of pulse duration control is 100 ns.

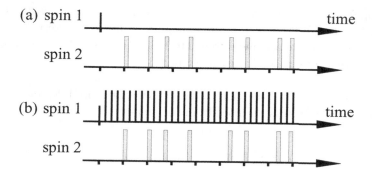

Fig. 10. Pulse sequences to implement artificial phase decoherence in a quantum memory. Both spins are initially in the up-states. A short bar is a $\pi/2$-pulse while a long bar is a π-pulse. (a) A series of pulses with random intervals is applied on spin 2. The variance α of pulse intervals is adjustable. (b) The bang-bang control pulses are applied on spin 1 while random pulses are applied on spin 2.

a π-pulse at $t = t_{2k-1}$ and is flipped back from $|1\rangle$ to $|0\rangle$ by a subsequent π-pulse at $t = t_{2k}$ ($k = 1, 2, 3, \cdots$).[n] The time intervals $\{\Delta_j = t_{j+1} - t_j\}$ between adjacent π-pulses distribute according to the Gaussian distribution (49) and (50). See, Fig. 10 (a). The spin 1 evolves with the Hamiltonians (48).

In the first run, we put $\alpha = 0$ and hence the time interval between π-pulses is constant, $\Delta_j = \bar{\Delta} = 2.0$ ms. In this case a regular alternating field acts on spin 1. If the π-pulses and the spin dynamics were perfect, there would be no decoherence. However, in reality, it is impossible to avoid pulse imperfection, measurement errors and intrinsic decoherence. And thus, the measurement with $\alpha = 0$ is necessary as a reference. M's in $\rho(M, n_x, n_y, n_z)$ under the pulses with $\alpha = 0$ are plotted as filled squares ■ in Fig. 11. The measured decoherence factor for $\alpha = 0$ is

$$e^{-T/T_2^*} = 0.45 \quad \text{at } T = 100 \text{ ms}. \tag{56}$$

Next, we modulate the time interval of spin 2 flippings randomly with nonvanishing $\alpha = 0.25$. A series of random variables $\Xi = (\xi_0, \xi_1, \cdots, \xi_r)$ is prepared according to the Gaussian distribution (50). Then a series of inter-

---

[n]The rotation axes of these π-pulses are cyclically permutated as

$$(x, -x, y, -y, -x, x, -y, y) \tag{55}$$

to reduce undesired influence of imperfections in the π-pulses. Thus the number of π-pulses are set to a multiple of 8.

vals $(\Delta_0, \Delta_1, \cdots, \Delta_r)$ is defined via (49) and then M's of spin 1 is measured at time $t = T$. We prepare 128 different pulse sequences corresponding to $\{\Xi_1, \Xi_2, \cdots, \Xi_{128}\}$, repeat measurements 128 times and take an average of 128 measured amplitudes for each set of values of α and T. The magnitude of the averaged amplitude is plotted as open squares $\square$ ($\alpha = 0.25$ and $\bar{\Delta} = 2.0$ ms) in Fig. 11. The plotted data show exponential decrease in M. The decoherence factors are read from the slopes of the graphs in Fig. 11 as

$$e^{-T/T_2^*} = 0.05 \quad \text{at } T = 100 \text{ ms.} \tag{57}$$

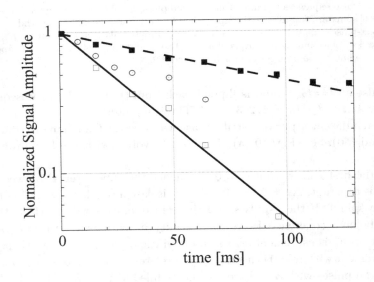

Fig. 11. Time evolution of M's in $\rho(M, n_x, n_y, n_z)$ of spin 1. Decrease in these magnitude indicates phase decoherence. The correspondence between the symbol and the variance α of the pulse interval distribution is ($\blacksquare$: $\alpha = 0$), and ($\square$: $\alpha = 0.25$). The broken line is the least square fit for $\{\blacksquare\}$ while the solid line is that for $\{\square\}$. Magnitudes M's of spin 1 with the bang-bang control ($t_b = 0.5$ ms) applied are shown with the symbol $\{\circ\}$.

Let us introduce a dimensionless quantity

$$R(\alpha) = -\frac{1}{\alpha^2} \ln \left\{ \frac{e^{-T/T_2^*}|_{\alpha \neq 0}}{e^{-T/T_2^*}|_{\alpha = 0}} \right\} \tag{58}$$

to compare the measured data with the theoretical estimation. The theoretical prediction Eq. (53) yields

$$R_{\text{theory}} = \frac{1}{8} J^2 \bar{\Delta} T = 46 \qquad (59)$$

for $J = 2\pi \times 215.5$ Hz, $\bar{\Delta} = 2$ ms, $T = 100$ ms, independently of α. On the other hand, by substituting the measured values Eqs. (56) and (57) into Eq. (58), we obtain 35 in good agreement with R_{theory}. For other $\alpha = 0.1, 0.15$, and 0.20, we obtain $R(\alpha) = 37, 36$, and 38, respectively. We observe that the value $R(\alpha)$ is almost independent of α, implying that the decoherence rate $(T_2^*)^{-1}$ is proportional to α^2 as predicted in Eq. (54).

Next, we apply the bang-bang control to spin 1. The pulse sequence to incorporate the bang-bang control is shown in Fig. 10 (b). A regular sequence of π-pulses with interval $t_b = 0.5$ ms is applied on spin 1.° During this run, a sequence of π-pulses whose interval fluctuates with variance $\alpha = 0.25$ is also applied on spin 2. We finally measure M of spin 1 at $t = T$. We repeat the measurement by preparing 128 different sequences corresponding to $\{\Xi_1, \Xi_2, \cdots, \Xi_{128}\}$. The averaged M's are plotted as a function of T with the symbol (○) in Fig. 11. Comparing the data points {○} with {□}, we observe that decoherence is suppressed by the bang-bang control.

4.3. *Quantum teleportation*

The quantum teleportation[36] of an unknown quantum state is a surprising demonstration,[37,38] since it may be intuitively thought to violate the uncertainty principle of quantum mechanics, which forbids extracting all the information in an quantum object. By employing a reversible detector, Braunstein theoretically clarified how the quantum information is hidden within the correlations between the system and the environment and how it recovers from them.[39] More general discussions were made by Nielsen and Caves.[40]

The purpose of this section is (i) to realize Braunstein's reversible detector in a quantum teleportation algorithm and confirm his discussions with a concrete example and (ii) to show technical details of NMR-QC experiments.

°The rotation axes of these π-pulses are cyclically permutated as given in Eq. (55)

4.3.1. Theory

In order to make our discussions as simple as possible, we restrict ourselves to the case of two-state system teleportation.

The original quantum teleportation scheme[36] is as follows. Qubit 1 in an unknown state $|\psi\rangle$ is jointly measured with qubit 2, which forms an entangled pair $|\beta_{00}\rangle$ with qubit 3, and then qubit 3 is unitary transformed according to the result of the joint measurement and becomes $|\psi\rangle$. The mathematics behind the quantum teleportation is summarized by the decomposition of the initial state $|\psi\rangle \otimes |\beta_{00}\rangle$,

$$|\psi\rangle \otimes |\beta_{00}\rangle = \frac{1}{2} \sum_{i,j=0}^{1} |\beta_{ij}\rangle \otimes U_{ij}^\dagger |\psi\rangle, \qquad (60)$$

where

$$|\beta_{00}\rangle = \frac{|00\rangle + |11\rangle}{\sqrt{2}}, |\beta_{01}\rangle = \frac{|01\rangle + |10\rangle}{\sqrt{2}},$$
$$|\beta_{10}\rangle = \frac{|00\rangle - |11\rangle}{\sqrt{2}}, |\beta_{11}\rangle = \frac{|01\rangle - |10\rangle}{\sqrt{2}},$$

and

$$U_{00}^\dagger = I, U_{01}^\dagger = \sigma_x, U_{10}^\dagger = \sigma_z, U_{11}^\dagger = \sigma_z \sigma_x.$$

After a joint measurement, the system becomes one of the four states, for example, $|\beta_{01}\rangle \otimes \sigma_x |\psi\rangle$ and the information "01" is obtained. Then, operating σ_x, which corresponds to the information "01", on qubit 3 leads the state $|\psi\rangle$.

Braunstein introduced a detector, that is labeled ij and a 2^2-states system.[39] Now, the total initial state is

$$|00\rangle \otimes |\psi\rangle \otimes |\beta_{00}\rangle,$$

including the detector state $|00\rangle$. The joint measurement in the quantum teleportation operation is described with a unitary operator U_{meas},

$$U_{\text{meas}} = \sum_{i,j=0}^{1} \Sigma_{ij} \otimes |\beta_{ij}\rangle\langle\beta_{ij}| \otimes I,$$

where Σ_{ij} operates only on the detector and

$$\Sigma_{ij} = \begin{cases} I & i=0, j=0 \\ |01\rangle\langle 00| + |00\rangle\langle 01| + |10\rangle\langle 10| + |11\rangle\langle 11| & i=0, j=1 \\ |10\rangle\langle 00| + |01\rangle\langle 01| + |00\rangle\langle 10| + |11\rangle\langle 11| & i=1, j=0 \\ |11\rangle\langle 00| + |01\rangle\langle 01| + |10\rangle\langle 10| + |00\rangle\langle 11| & i=1, j=1 \end{cases}.$$

After the system interacts with the detector, or after the "measurement", the total state becomes

$$\frac{1}{2}\sum_{i,j=0}^{1}|ij\rangle\otimes|\beta_{ij}\rangle\otimes U_{ij}^{\dagger}|\psi\rangle.$$

When the detector decoheres to a certain state $|ij\rangle$, it is the same as the conventional quantum teleportation scheme.

However, even if the detector does not decohere, applying the unitary operator $\sum_{ij}|ij\rangle\langle ij|\otimes I\otimes I\otimes U_{ij}$ gives

$$\left(\frac{1}{2}\sum_{ij}|ij\rangle\otimes|\beta_{ij}\rangle\right)\otimes|\psi\rangle.$$

The above equation implies that an unknown quantum state is teleportated from qubit 1 to qubit 3 without the irreversible amplification of an intermediate detector.

4.3.2. *Quantum circuit*

Nielsen, Knill, and Laflamme performed the first complete quantum teleportation experiment by using an NMR-QC.[38] The quantum circuit proposed by Brassard, Braunstein, and Cleve[41] was implemented and decoherence of qubit 1 and 2 was interpreted as an observation or measurement by the environment.[42]

The quantum teleportation circuit including a detector is shown in Fig. 12. The detector qubits (qubit $d1$ and $d2$) and the multi-qubit gate $\sum_{ij}$ are added to the quantum circuit proposed by Brassard, Braunstein, and Cleve.[41] Qubit 1 and 2 are entangled by $\sum_{ij}$. Therefore, qubit 1 and 2 decohere, when qubit $d1$ and $d2$ decohere. However, even when qubit 1 and 2 decohere, the output of the circuit does not change[41,43] and thus the unknown state $|\psi\rangle$ appears as the output of qubit 3.

4.3.3. *Sample and spectrometer*

We employ the same JEOL ECA-500 NMR spectrometer[35] as in Sec. 4.2.6. We employ Gaussian pulses, whose pulse widths are $t_p = 0.7$ ms, to implement quantum algorithms.

A 0.6 ml, 0.78 M sample of ^{13}C-labeled L-alanine (98% purity, Cambridge Isotope) solved in D_2O is used. The structure of L-alanine is shown

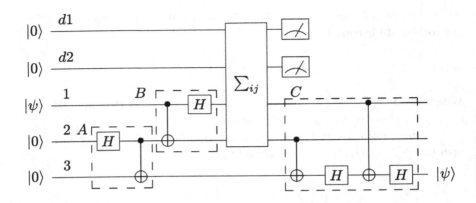

Fig. 12. Quantum teleportation circuit including a detector. The block A makes an entangled pair $|\beta_{00}\rangle$ out of $|00\rangle$. The block B converts $|\beta_{00}\rangle, |\beta_{01}\rangle, |\beta_{10}\rangle$ and $|\beta_{11}\rangle$ to $|00\rangle, |01\rangle, |10\rangle$ and $|11\rangle$, respectively. The block C is a circuit which realizes a conditional unitary gate for recovering the unknown state $|\psi\rangle$. The detector qubits (qubit $d1$ and $d2$) and the multi-qubit gate $\sum_{ij}$ are added to the quantum circuit proposed by Brassard, Braunstein, and Cleve.[41]

in Fig. 13. We label the methyl carbon qubit 1, the α carbon qubit 2, and the carboxyl carbon qubit 3. We note that rapid exchange of the two amine protons and the single carboxyl proton (see, Fig. 13) with the deuterated solvent makes them inactive as qubits (spins).[p]

4.3.4. Hamiltonian and detector

The Hamiltonian of L-alanine in the individual rotating frame,[44] when protons are decoupled, is approximated by

$$\mathcal{H} = J_{12} I_{zz}^{12} + J_{23} I_{zz}^{23}, \tag{61}$$

---

[p]The detailed parameters measured from the spectra with and without decoupling protons are as follows. The Larmor frequency differences are $(\omega_{02} - \omega_{01})/2\pi = 4.4$ kHz and $(\omega_{03} - \omega_{02})/2\pi = 15.8$ kHz. Here, ω_{0i} denotes the Larmor frequency of qubit i. The scalar couplings are $J_{12}/2\pi = 34.8$ Hz between qubit 1-2 and $J_{23}/2\pi = 53.8$ Hz between qubit 2-3. The scalar coupling between qubit 1-3, J_{13}, satisfies $J_{13} \ll J_{12}, J_{23}$. The scalar couplings, $J_{d1}/2\pi = 130.0$ Hz between qubit 1 and the three protons in the methyl group and $J_{d2}/2\pi = 145.5$ Hz between qubit 2 and the α-proton, are also measured. Relaxation times are $T_1(1) = 1.4$ s, $T_1(2) = 2.8$ s, $T_1(3) = 14$ s and $T_2(1) = 0.32$ s, $T_2(2) = 0.26$ s, $T_2(3) = 0.41$ s, where the argument labels the qubit. Qubit 2 has the shortest T_2. Qubit 3 has the longest T_1. We also measure relaxation times of the methyl and α protons. The results are $T_1(\text{methyl}) = 1.2$ s, $T_1(\alpha) = 2.1$ s, $T_2(\text{methyl}) = 1.0$ s and $T_2(\alpha) = 1.3$ s.

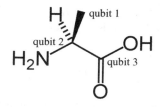

Fig. 13. Structure of L-alanine.

where

$$I_{zz}^{12} = \overbrace{I \otimes I \otimes I \otimes I}^{\text{detector}} \otimes \overbrace{I_z \otimes I_z \otimes I}^{\text{qubit}},$$
$$I_{zz}^{23} = I \otimes I \otimes I \otimes I \otimes I \otimes I_z \otimes I_z.$$

The scalar coupling between qubits 1 and 3 is ignored in Eq. (61) since $J_{13} \ll J_{12}, J_{23}$.

The essence of the "detector" discussed before is a capability of entangling the qubits with the environment (the detector). Therefore any unitary operation, which can entangle the qubits and the environment, may be employed for introducing a "detector" into the quantum teleportation circuit. Therefore, we employ the three protons in the methyl group and the α-proton, shown in Fig. 13, as a "detector" that interacts with qubit 1 and 2 thorough the scalar couplings[q].

The Hamiltonian without decoupling protons is approximated by

$$\mathcal{H}_d = \mathcal{H} + \mathcal{H}_{d1} + \mathcal{H}_{d2}, \qquad (62)$$

where

$$\mathcal{H}_{d1} = J_{d1} \overbrace{(I_z \otimes I \otimes I + I \otimes I_z \otimes I + I \otimes I \otimes I_z)}^{\text{detector}} \otimes I \otimes \overbrace{I_z \otimes I \otimes I}^{\text{qubits}},$$
$$\mathcal{H}_{d2} = J_{d2} \overbrace{I \otimes I \otimes I \otimes I_z}^{\text{detector}} \otimes \overbrace{I \otimes I_z \otimes I}^{\text{qubits}}.$$

$\mathcal{H}_{d1}$ and $\mathcal{H}_{d2}$ in Eq. (62) are the scalar couplings between qubit 1 and the three protons in the methyl group and between qubit 2 and the α-proton, respectively. We note that the "detector" can be switched on and off by (not) operating heteronucleus decoupling.

---

[q]The three protons in the methyl group are coupled with qubit 1, while the α-proton is coupled with qubit 2.

The unitary operator $\exp[-i(\mathcal{H}_{d1} + \mathcal{H}_{d2})\tau]$, which represents a "detector", is realized as follows. Here τ is a period while the "detector" is on.

$$\exp[-i(\mathcal{H}_{d1} + \mathcal{H}_{d2})\tau] \qquad (63)$$
$$= R_2(\pi, \beta + \pi)\exp(-i\mathcal{H}(\tau + \tau'))R_2(\pi, \beta)\exp(-i\mathcal{H}\tau')\exp(-i\mathcal{H}_d\tau).$$

where $R_i(\theta, \phi)$ is an selective one-qubit operator acting on qubit i as we discussed in the case of two-qubit molecules.[r] The operators $\exp(-i\mathcal{H}\tau)$ and $\exp(-i\mathcal{H}_d\tau)$ can be obtained by free time evolutions for a duration τ with and without decoupling protons, respectively. The operation $R_2(\pi, \beta + \pi)\exp(-i\mathcal{H}(\tau + \tau'))R_2(\pi, \beta)\exp(-i\mathcal{H}\tau')$ is for nulling the scalar couplings between qubits 1 and 2 and qubits 2 and 3.

4.3.5. *Pulse sequence*

We discuss an NMR pulse sequence implementing the quantum teleportation algorithm with L-alanine molecule in detail.

First, we modify the quantum circuit shown in Fig. 12 since J_{13} is too small to implement a CNOT gate between qubit 1 and 3. A standard technique to realize a CNOT gate between two qubits that have no direct interaction

$$\mathrm{CNOT}_{13} = \mathrm{SWAP}_{12}\,\mathrm{CNOT}_{23}\,\mathrm{SWAP}_{12}$$

is employed for realizing CNOT_{13}. Here,

$$\mathrm{SWAP}_{12} = \mathrm{CNOT}_{12}\,\mathrm{CNOT}_{21}\,\mathrm{CNOT}_{12}$$

swaps qubit 1 and 2. We omit the swap gate after CNOT_{23} since we can leave qubit 1 and 2 swapped. Therefore, we modified as follows.

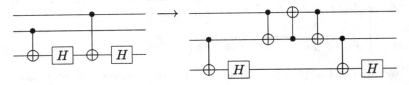

---

[r]We employed Eq. (63) instead of

$$\exp[-i(\mathcal{H}_{d1} + \mathcal{H}_{d2})\tau] = R_2(\pi, \beta + \pi)\exp(-i\mathcal{H}(\tau))R_2(\pi, \beta)\exp(-i\mathcal{H}_d\tau),$$

for an experimental convenience. We keep $\tau + \tau'$ constant for better comparison among experiments with different τ's.

Second, we simplify the pulse sequence by removing unnecessary Z-rotations. A pulse sequence realizing a Hadamard gate acting on qubit i is

$$H_i = iR_i\left(\frac{\pi}{2}, \frac{\pi}{2}\right) Z_i(\pi). \tag{64}$$

On the other hand, a pulse sequence realizing CNOT$_{ij}$ is

$$\text{CNOT}_{ij} = \exp\left(\frac{i\pi}{4}\right) Z_i\left(\frac{\pi}{2}\right) Z_j\left(-\frac{\pi}{2}\right) R_j\left(\frac{\pi}{2}, 0\right) e^{-\pi I_{zz}^{ij}} R_j\left(\frac{\pi}{2}, \frac{\pi}{2}\right), \tag{65}$$

where

$$\begin{aligned}\exp(-\pi I_{zz}^{12}) &= R_3(\pi, \beta + \pi)\exp(-i\mathcal{H}\tau_{12})R_3(\pi, \beta)\exp(-i\mathcal{H}\tau_{12}),\\ \exp(-\pi I_{zz}^{23}) &= R_1(\pi, \beta + \pi)\exp(-i\mathcal{H}\tau_{23})R_1(\pi, \beta,)\exp(-i\mathcal{H}\tau_{23}).\end{aligned} \tag{66}$$

Here, $\tau_{ij} = (2\pi/J_{ij})/4$.[s] We construct the pulse sequence realizing a quantum teleportation faithfully by substituting Eqs. (64), (65) and (66) into the quantum circuit in Fig. 12. Then, we collect all Z-rotations at the end or at the beginning of the quantum circuit[t] using the rules

$$\begin{aligned}R_i(\theta, \phi)Z_i(\psi) &= Z_i(\psi)R_i(\theta, \phi - \psi),\\ \exp(-\theta I_{zz}^{ij})Z_i(\psi) &= Z_i(\psi)\exp(-\theta I_{zz}^{ij}).\end{aligned} \tag{67}$$

The Z-rotations acting on qubits 1 and 2 at the end of operation can be omitted since we need no information on qubits 1 and 2 at the end of the algorithm. The Z-rotation acting on qubit 3 at the beginning can be also omitted since the initial state of qubit 3 is $|0\rangle$.

Third, we take into account the Bloch-Siegert effect and a finite pulse width. Because of the Bloch-Siegert effect, an experimentally realizable rf-pulse not only rotates a target qubit around the axis in the xy-plane but

---

[s]Although $\exp(-I_{zz}^{ij})$ and CNOT$_{ij}$ are independent of β in principle, we find that averaging over four measurements corresponding $\beta = 0, \pi/2, \pi, 3\pi/2$ in (66) provides better results than those without averaging.

[t]We obtain the pulse sequence

$\quad\quad\quad\quad\quad : \exp(-i\pi/4)Z_1(\pi)Z_2(5\frac{\pi}{2})$
Block C : $R_3(\frac{\pi}{2}, \frac{\pi}{2})CN_{23}(\pi)R_3(\frac{\pi}{2}, \pi)CN_{21}(-\pi)CN_{12}(-3\frac{\pi}{2})CN_{21}(-\pi)$
$\quad\quad\quad\quad\quad CN_{23}(3\frac{\pi}{2})$
$\sum_{ij}\quad\quad : R_2(\pi, \beta + \pi)\exp(-i\mathcal{H}(\tau + \tau'))R_2(\pi, \beta)\exp(-i\mathcal{H}\tau')\exp(-i\mathcal{H}_d\tau)$
Block B : $R_1(\frac{\pi}{2}, -\pi)CN_{12}(-\frac{\pi}{2})$
Block A : $CN_{23}(\pi)R_2(\frac{\pi}{2}, -\frac{\pi}{2})$
$\quad\quad\quad\quad\quad : Z_3(\frac{\pi}{2}),$
where

$$\begin{aligned}CN_{12}(\alpha) &= R_2(\frac{\pi}{2}, \alpha - \frac{\pi}{2})\exp(-\pi I_{zz}^{12})R_2(\frac{\pi}{2}, \alpha),\\ CN_{21}(\alpha) &= R_1(\frac{\pi}{2}, \alpha - \frac{\pi}{2})\exp(-\pi I_{zz}^{12})R_1(\frac{\pi}{2}, \alpha),\\ CN_{23}(\alpha) &= R_3(\frac{\pi}{2}, \alpha - \frac{\pi}{2})\exp(-\pi I_{zz}^{23})R_3(\frac{\pi}{2}, \alpha),\\ CN_{32}(\alpha) &= R_2(\frac{\pi}{2}, \alpha - \frac{\pi}{2})\exp(-\pi I_{zz}^{23})R_2(\frac{\pi}{2}, \alpha).\end{aligned}$$

also rotates the other qubits around the z-axis.[u] Let us consider the case of $CN_{32}(\alpha)$ as an example. We need to compensate Bloch-Siegert effects by adding Z-rotations as follows.

$$CN_{32}(\alpha) = \tilde{R}_2(\tfrac{\pi}{2}, \alpha - \tfrac{\pi}{2}) \overbrace{Z_1(-\phi_{21})Z_3(-\phi_{23})}^{\text{for compensation}}$$
$$\tilde{R}_1(\pi, \beta) \quad Z_2(-\phi_{12})Z_3(-\phi_{13})$$
$$\exp(-i\mathcal{H}\tau_{23})$$
$$\tilde{R}_1(\pi, \beta) \quad Z_2(-\phi_{12})Z_3(-\phi_{13})$$
$$\exp(-i\mathcal{H}\tau_{23})$$
$$\tilde{R}_2(\tfrac{\pi}{2}, \alpha) \quad Z_1(-\phi_{21})Z_3(-\phi_{23}),$$

where $\tilde{R}_i(\theta, \phi) = R_i(\theta, \phi) Z_j(\phi_{ij}) Z_k(\phi_{ik})$ $(i \neq j \neq k)$ is a unitary operator approximating a real rf pulse by taking into account the Bloch-Siegert effect. By employing Eq. (67), CN_{32} can be simplified as follows.

$$CN_{32}(\alpha)$$
$$= Z_1(-2\phi_{21})Z_2(-2\phi_{12})Z_3(-2\phi_{13} - 2\phi_{23})\tilde{R}_2(\tfrac{\pi}{2}, \alpha - \tfrac{\pi}{2} + 2\phi_{12}) \quad (68)$$
$$\tilde{R}_1(\pi, \beta + \phi_{21} + \pi) \exp(-i\mathcal{H}\tau_{23}) \tilde{R}_1(\pi, \beta + \phi_{21}) \exp(-i\mathcal{H}\tau_{23}) \tilde{R}_2(\tfrac{\pi}{2}, \alpha).$$

Since $2\phi_{12} \sim 1$ rad can easily happen, the compensation in $\tilde{R}_2$ is important.[v] When the number of qubits is increased, the method shown here is not efficient and we have to rely on numerical optimization of pulses, called strongly modulated pulses.[45]

---

[u]The following is a typical case. We consider a square rf pulse ($t_p = 0.7$ ms) which rotates qubit 1 by an angle π. Here, $\omega_1 = \pi/(0.7 \times 10^{-3}) = 4.49 \times 10^3$ rad/s. The frequency difference between qubit 2 and 1 is $\Delta = -2\pi \cdot 4.4 \times 10^3 = -2.76 \times 10^4$ rad/s, which gives $\epsilon = -0.162$ in (17). Then, an extra Z-rotation acting on qubit 2 is calculated as $\exp[-i\tau\Delta(\sqrt{1 + \epsilon^2} - 1)I_z] = \exp(i\,0.253 I_z)$. We note that Gaussian pulses employed in our experiments make about 1.7 times larger Bloch-Siegert effects (here, 0.43 rad) than square pulses do when the pulse widths are the same.[44]

[v]We measure ϕ_{12}. The thermal state ρ_{th} is described by

$$\rho_{\text{th}} = I_z \otimes I_0 \otimes I_0 + I_0 \otimes I_z \otimes I_0 + I_0 \otimes I_0 \otimes I_z,$$

where the thermal factor $1/k_B T$ and $I_0^{\otimes 3}$ are omitted. By applying CNOT$_{32}$ to ρ_{th}, we obtain

$$\text{CNOT}_{32}\, \rho_{\text{th}}\, \text{CNOT}_{32}^\dagger = I_z \otimes I_0 \otimes I_0 + I_0 \otimes I_z \otimes I_0 + I_0 \otimes I_z \otimes I_z,$$

which produces no FID signal. We apply CN_{32}'s with various ϕ_{12}'s to ρ_{th} and compare obtained FID signals with the expected one, or no FID signal. We note that τ_{23} in Eq. (68) is replaced with $(\tau_{23} - t_p)$ in experiments because of the finite pulse width of $\tilde{R}_1$. We experimentally find that the case of $2\phi_{12} = 0.80$ rad produces the smallest FID signal. The measured value of $\phi_{12} = 0.40$ rad is in good agreement with the already calculated value of 0.43 rad.

The pulse sequence for the quantum teleportation is obtained from Eq. (67) as in the case of CN_{32}. The pulse sequence consists of 25 pulses and 12 free evolution periods and takes about 95 ms. Qubits 1 , 2 and 3 are set initially in the pseudopure state $|0\rangle$ by employing a spatial averaging method.[19] All considerations for implementing the quantum teleportation algorithm are also taken into account for making the pulse sequence which creates $|000\rangle$ state. The pulse sequence consists of 26 pulses, 4 pulsed field gradients, and 12 free evolution periods and takes about 100 ms.

4.3.6. *Experiments*

We perform a quantum teleportation experiment with the pulse sequence discussed in Sec. 4.3.5. We take the following 4 initial states of qubit 1 and apply the pulse sequence to the system.

$$|0\rangle\langle 0| = I_i + I_z, \quad |+\rangle\langle +| = I_i + I_x, \\ |-\rangle\langle -| = I_i + I_y, \quad |1\rangle\langle 1| = I_i - I_z. \tag{69}$$

The states $|+\rangle\langle +|, |-\rangle\langle -|$ and $|1\rangle\langle 1|$ are prepared by applying rf pulses $R_1(\frac{\pi}{2}, \frac{\pi}{2}), R_1(\frac{\pi}{2}, \pi)$ and $R_1(\pi, 0)$ to the state $|0\rangle\langle 0|$, respectively. Qubits 2 and 3 are in $|0\rangle\langle 0|$ and the detector qubits are in the thermal state.

The states of qubit 3, ρ_{qt}, is determined via FID signals. ρ_{qt} is parameterized as

$$\rho_{\mathrm{qt}} = \rho(M, n_x, n_y, n_z),$$

where M of qubit 3 in the state $|000\rangle$ is used as unit for normalization. Mn_x (Mn_y) is measured by numerically integrating the real (imaginary) part of the qubit 3 peak in a spectrum. We note that the peak shape is neither simple absorptive nor dispersive Lorentzian curve, because of interactions with other spins. Mn_z is measured after applying $R(\pi/2, \pi/2)$, called a read pulse, which converts Mn_z to Mn_x.

The density matrix ρ_i in Eq. (37) is rewritten with $|0\rangle\langle 0|, |+\rangle\langle +|, |-\rangle\langle -|$ and $|1\rangle\langle 1|$ as follows.

$$\rho_1 = |0\rangle\langle 0|, \\ \rho_2 = |+\rangle\langle +| + i|-\rangle\langle -| - \tfrac{1+i}{2}(|0\rangle\langle 0| + |1\rangle\langle 1|), \\ \rho_3 = |+\rangle\langle +| - i|-\rangle\langle -| - \tfrac{1-i}{2}(|0\rangle\langle 0| + |1\rangle\langle 1|), \\ \rho_4 = |1\rangle\langle 1|.$$

Because of linearity in quantum mechanics, $\mathscr{E}(\rho_i)$ is described with $\mathscr{E}(|0\rangle\langle 0|), \mathscr{E}(|1\rangle\langle 1|), \mathscr{E}(|+\rangle\langle +|), \mathscr{E}(|-\rangle\langle -|)$. Therefore, we are able to determine λ_{jk} in Eq. (37) experimentally. Then, E_i and the entanglement fidelity can be calculated.

4.3.7. *Results*

We perform 6 sets of experiments with $\tau = 0, 1, 2, 3, 4, 4.9$ ms where $\tau + \tau' = 5$ ms in Eq. (63), while each set starts from 4 different initial states Eq. (69). These sets correspond to different "detector"s. Note that $\tau = 0$ ms corresponds to the case that the "detector" is always off. The spectra after quantum teleportation, when the initial states are $I_i + I_z$, are shown in Fig. 14. We observe in these spectra (1) the peak of qubit 3 is almost independent of τ, (2) the peaks of qubits 1 and 2 depend on τ, and (3) the peak height of qubit 1 decreases as τ increases and crosses zero at $\tau = 3$ ms. The above observations indicate (a) qubit 1 and 2 are "measured" and the state of qubit 1 is teleportated to qubit 3 and (b) our "detector" is reversible.

Six sets of E_i corresponding to 6 τ's are also visualized in Fig. 14 as a map from the Bloch sphere. Although those mapped surfaces are distorted, we can again see that the quantum teleportation algorithm works. The entanglement fidelities are summarized here.

$$\tau = 0, \quad 1, \quad 2, \quad 3, \quad 4, \quad 4.9 \text{ ms},$$
$$F_e(I_i, \mathscr{E}) = 0.78, \ 0.78, \ 0.79, \ 0.77, \ 0.76 \ 0.76,$$
$$\text{Tr}(\mathscr{E}) = 1.06, \ 1.09, \ 1.04, \ 1.10, \ 1.10 \ 1.02.$$

The entanglement fidelities are not unity as expected, but larger than 0.5 which is the maximum for perfect classical transmission.[26] It indicates that some quantum information is teleportated, although not perfect. Note that not only imperfections in the pulse sequence but also T_2 relaxation of qubit 3 causes decrease of the entanglement fidelity. The time required for quantum teleportation operation is 95 ms and is not negligible compared with $T_2(3) = 0.41$ s of qubit 3, and thus p in Eq. (35) is $0.5 + \exp(-t_{\text{ps}}/T_2(3))/2 \approx 0.9$. We observe that the entanglement fidelities are almost constant despite of disturbances by the "detector"s. $\text{Tr}(\mathscr{E})$s are close to unity as expected, although there are some errors.

5. Conclusions

The current NMR-QC is not considered as a true quantum computer because (1) measurements are ensemble ones instead of projection ones, as we discussed in Sec. 2.2.6 and (2) a spin polarization is too small at room temperature, as we discussed in Sec. 3.5. Despite of these limitations, the NMR-QC is considered as a model of other true quantum computers, and thus its understanding should be useful for researchers, especially for new comers who have just started their study on quantum computer. We have

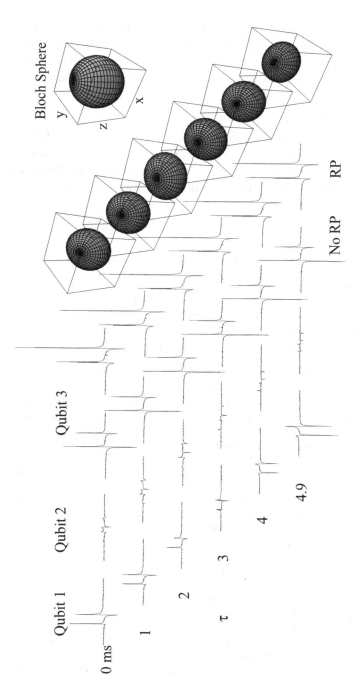

Fig. 14. Spectra after quantum teleportation when the initial state is $I_0 + I_z$. RP (No RP) denotes with (without) a read pulse ($R(\pi/2, \pi/2)$) applying to qubit 3. The six spectra correspond to $\tau = 0, 1, 2, 3, 4, 4.9$ ms. The frequency range for each peak is 2 ppm. The set of E_i's with various τ's are visualized as a map from the Bloch sphere to somewhat distorted spheres. The lengths of each sides of the cubes are 2.

been employing the NMR-QC for examining new techniques for decoherence control,[11] theoretical ideas,[12] and implementations of new quantum algorithms.[46,47] We believe that the liquid state NMR quantum computer can play an important role in the development of quantum computer.

Acknowledgments

Sections 3 and 4 are based on our publications. We would like to thank these coauthors, Mikio Nakahara, Shogo Tanimura, Sachiko Kitajima Chikako Uchiyama, and Fumiaki Shibata. We would also like to acknowledge Mikio Nakahara and Yukihiro Ota for their careful reading of the manuscript and for discussions.

References

1. M. A. Nielsen, and I. L. Chuang, *Quantum Computation and Quantum Information*, (Cambridge University Press, Cambridge, 2000).
2. A. Ekert and R. Jozsa, *Rev. Mod. Phys.* **68**, 733 (1996).
3. M. Nakahara and T. Ohmi, *Quantum Computing: From Linear Algebra to Physical Realizations*, (Taylor & Francis, New York, 2008).
4. R. R. Ernst, G. Bodenhausen, and A. Wokaun, *Principles of Nuclear Magnetic Resonance in One and Two Dimensions*, (Oxford University Press, Oxford, 1991).
5. M. H. Levitt, *Spin Dynamics* (John Wiley and Sons, New York, 2001).
6. L. M. K. Vandersypen and I. L. Chuang, *Rev. Mod. Phys.* **76**, 1037 (2004).
7. R. Freeman, *Spin Choreography* (Oxford University Press, Oxford, UK, 1998).
8. T. E. W. Claridge, *High-Resolution NMR techniques in Organic Chemistry*, (Elsevier, Amsterdam, 2004)
9. L. M. K. Vandersypen, M. Steffen, G. Breyta, C. S. Yannonl, M. H. Sherwood, and I. L. Chuang, *Nature* **414**, 883 (2001).
10. S. L. Braunstein, C. M. Caves, R. Jozsa, N. Linden, S. Popescu, and R. Schack, *Phys. Rev. Lett.* **83**, 1054 (1999).
11. Y. Kondo, M. Nakahara, S. Tanimura, S. Kitajima C. Uchiyama and F. Shibata, *J. Phys. Soc. Jpn.* **76**, 074002 (2007).
12. Y. Kondo, *J. Phys. Soc. Jpn.* **76**, 104004 (2007).
13. D. G. Cory, A. F. Fahmy, and T. F. Havel, *Proc. Natl. Acad. Sci. USA* **94**, 1634, March (1997).
14. J. A. Jones, *Prog. NMR Spectrosc.* **38**, 325 (2001).
15. L. M. K. Vandersypen, Stanford University Thesis (2001).
16. F. Bloch and A. Siegert, *Phys. Rev.* **57**, 522 (1940).
17. N. F. Ramsey, *Phys. Rev.* **100**, 1191 (1955).
18. M. A. Pravia, E. Fortunato, Y. Weinstein, M. D. Price, G. Teklemariam, R. J. Nelson, Y. Sharf, S. Somaroo, C. H. Tseng, T. F. Havel, D. G. Cory, *Concepts Magn. Res.* **11**, 225 (1999).
19. U. Sakaguchi, H. Ozawa, and T. Fukumi, *Phys. Rev. A* **61**, 042313 (2000).

20. Y. Sharf, T. F. Havel, D. G. Cory, *Phys. Rev. A* **62**, 052314 (2000).
21. M. S. Anwar, D. Blazina, H. A. Carteret, S. B. Duckett, T. K. Halstead, J. A. Jones, C. M. Kozak, and R. J. K. Taylor, *Phys. Rev. Lett.* **93**, 040501 (2004).
22. R. Schack and C. M. Caves, *J. Mod. Opt.* **47**, 387 (2000).
23. K.-E. Hellwig and K. Kraus, *Commun. Math. Phys.* **11**, 214 (1969).
24. K. Kraus, *Ann. Phys.* **64**, 311 (1971).
25. E. C. G. Sudarshan, P. M. Mathews, and J. Rau, *Phys. Rev.* **121**, 920 (1961).
26. H. Barnum, M. A. Nielsen, and B. Schumacher, *Phys. Rev. A* **57**, 4153 (1998).
27. D. Leung, L. Vandersypen, X. Zhou, M. Sherwood, C. Yannoni, M. Kubinec, and I. Chuang, *Phys. Rev. A* **60**, 1924 (1999).
28. M. Ban, *J. Mod. Opt.* **45**, 2315 (1998), L. Viola and S. Lloyd, *Phys. Rev. A* **58**, 2733 (1998), L. Duan and G. Guo, *Phys. Lett. A* **261**, 139 (1999), L. Viola, E. Knill, and S. Lloyd, *Phys. Rev. Lett.* **82**, 2417 (1999), L. Viola, S. Lloyd, and E. Knill, *Phys. Rev. Lett.* **83**, 4888 (1999), L. Tian and S. Lloyd, *Phys. Rev. A* **62**, 050301(R) (2000). See, also, H. Gutmann, F. K. Wilhelm, W. M. Kaminsky, and S. Lloyd, *Bang-Bang Refocusing of a Qubit Exposed to Telegraph Noise* in *Experimental Aspects of Quantum Computing*, edited by H. O. Everitt (Springer, New York, 2005).
29. C. Uchiyama and M. Aihara, *Phys. Rev. A* **66**, 032313 (2002), *Phys. Rev. A* **68**, 052302 (2003).
30. J. Zhang, Z. Lu, L. Shan, and Z. Deng, quant-ph/0202146, quant-ph/0204113.
31. C. A. Ryan, J. Emerson, D. Poulin, C. Negrevergne, and R. Laflamme, *Phys. Rev. Lett.* **95**, 250502 (2005).
32. G. Teklemariam, E. M. Fortunato, C. C. López, J. Emerson, J. P. Paz, T. F. Havel, and D. G. Cory, *Phys. Rev. A* **67**, 062316 (2003).
33. R. R. Ernst, *J. Chem. Phys.* **45**, 3845 (1966).
34. D. Pines and P. Slichter, *Phys. Rev.* **100**, 1014 (1955). The phase θ of the spin due to white noise evolves in time and is randomly distributed at a later time. The distribution function is $p(\theta) = \frac{1}{\sqrt{2\pi}s} e^{-\theta^2/2s^2}$, where s^2 is proportional to the evolution time t.
35. http://www.jeol.com/.
36. C. H. Bennett, G. Brassard, C. Cre'peau, R. Jozsa, A. Peres, and W. K. Wootters, *Phys. Rev. Lett.* **70**, 1895 (1993).
37. D. Bouwmeester, J. Pan, K. Mattle, M. Eibl, H. Weinfurter and A. Zeilinger, *Nature* **390**, 575 (1997), D. Boschi, S. Branca, F. De Martini, L. Hardy, and S. Popescu, *Phys. Rev. Lett.* **80**, 1121 (1998), A. Furusawa, J. L. Sørensen, S. L. Braunstein, C. A. Fuchs, H. J. Kimble, and E. S. Polzik, *Science* **282**, 706 (1998), I. Marcikic, H. de Riedmatten, W. Tittel, H. Zbinden and N. Gisin, *Nature* **421**, 509 (2003), M. Riebe, H. Häffner, C. F. Roos, W. Hänsel, J. Benhelm, G. P. T. Lancaster, T. W. Körber, C. Becher, F. Schmidt-Kaler, D. F. V. James and R. Blatt, *Nature* **429**, 734 (2004), M. D. Barrett, J. Chiaverini, T. Schaetz, J. Britton, W. M. Itano, J. D. Jost, E. Knill, C. Langer, D. Leibfried, R. Ozeri and D. J. Wineland, *Nature* **429**, 737 (2004), R. Ursin, T. Jennewein, M. Aspelmeyer, R. Kaltenbaek, M. Lindenthal, P. Walther and A. Zeilinger, *Nature* **430**, 849 (2004).

38. M. A. Nielsen, E. Knill, and R. Laflamme, *Nature* **396**, 52 (1998).
39. S. L. Braunstein, *Phys. Rev. A* **53**, 1900 (1996).
40. M. A. Nielsen and C. M. Caves, *Phys. Rev. A* **55**, 2547 (1996).
41. G. Brassard, S. L. Braunstein, and R. Cleve, *Physica D* **120**, 43 (1998).
42. W. H. Zurek, *Rev. Mod. Phys.* **75**, 715 (2003).
43. R. B. Griffiths and C. -S. Niu, *Phys. Rev. Lett.* **76**, 3228 (1996), Note that a controlled-U gate, if its control qubit is to be measured in the standard basis, leads to the same final outcome regardless whether its control qubit is measured either before or after the gate is executed.
44. Y. Kondo, M. Nakahara, and S. Tanimura, "Liquid-state NMR quantum computer: Hamiltonian Formalism and Experiments", in "Quantum Computing: Are the DiVincenzo criteria fulfilled in 2004?" (World Scientific. Singapore (2006/4)) 127.
45. E. M. Fortunato, M. A. Pravia, N. Boulant, G. Teklemariam, T. F. Havel and D. G. Cory, *J. Chem. Phys.* **116**, 7599 (2002).
46. M. Nakahara, Y. Kondo, K. Hata, and S. Tanimura, *Phys. Rev. A* **70**, 052319 (2004).
47. M. Nakahara, J. J. Vartiainen, Y. Kondo, S. Tanimura, and K. Hata, *Phys. Lett. A* **350**, 27 (2006).

FLUX QUBITS, TUNABLE COUPLING AND BEYOND

ANTTI O. NISKANEN

VTT Technical Research Centre of Finland,
Sensors, P.O. Box 1000, 02044 VTT, Finland
E-mail: Antti.Niskanen@vtt.fi

Superconducting flux qubits are attractive candidates for quantum bits because when properly operated they have quite good coherence properties. They are also flexible to fabricate. We describe how to achieve tunable coupling between such qubits in theory and experiment without sacrificing quantum coherence.

Keywords: flux qubits; coherent tunable coupling scheme

Superconducting circuits containing Josephson junctions are promising devices for the realization of quantum information processing. In particular, scalable quantum computing devices are expected to be fabricated with current lithographic techniques. For the implementation of quantum protocols with many qubits in such devices, it is inevitable to control the interactions between individual qubits without destroying quantum coherence. In this paper, we explain the scheme to accomplish the tunable coupling of superconducting flux qubits.[1] What is an important thing is that our scheme is realized at an optimal bias point, where the effect of dephasing on flux qubits is minimum. We also demonstrate several quantum protocols with our coherent tunable coupling scheme.[2]

The device for a flux qubit is a superconducting loop interrupted by several Josephson junctions. The Hamiltonian of the circuit has discrete energy levels at low temperature. The two lowest energy eigenstates correspond to a qubit. Applying microwave pluses, the single flux qubit is controlled. As for the flux qubit, one can minimize the effect of dephasing, controlling bias parameters (e.g., an external magnetic flux).

Let us consider the circuits containing several superconducting loops with Josephson junctions. The natural interaction between two flux quibts (qubits k and l) is described by the mutual inductance between two superconducting loops. We can find the coupling at the optimal bias point for the single flux qubit is very weak if $|\Delta_k - \Delta_l| > J_{kl}$, where Δ_k is the tunnelling energy of qubit k between the states and J_{kl} is the coupling constant. Thus, it is easy to decouple two flux qubits. However, it is impossible to realize a two–qubit operation (e.g., controlled–NOT gate) only with deriving the local magnetic fluxes.

Our proposal is to couple two flux qubits (qubits 1 and 2) via an additional flux qubit (qubit 3). The circuit is composed of three superconducting loops. In each loop, four Josephson junction are set. The interaction between qubits 1 (2) and 3 is induced by their mutual inductance. The external magnetic flux in qubit 3 is adiabatically changed. Thus, the quantum state of qubit 3 is always in its ground energy level. We can derive the effective coupling between qubits 1 and 2: $\tilde{J}_{12}(t)\sigma_x^1\sigma_x^2$. The coupling constant $\tilde{J}_{12}(t)$ is an nonlinear function of the instantaneous energy eigenvalue of qubit 3. Then, the application of a microwave pulse to the loop of qubit 3 allows us to control the coupling constant parametrically. In particular, when the microwave pulse is off, the coupling is also off. Note that the above coupling term appears at the optimal bias point for qubits 1 and 2.

We demonstrate the above scheme. First, we have implemented single–qubit transitions and a two–qubit operation. The single–qubit transitions for qubit 1 are $|00\rangle \leftrightarrow |10\rangle$ and $|01\rangle \leftrightarrow |11\rangle$. Similarly, as for qubit 2, they are $|00\rangle \leftrightarrow |01\rangle$ and $|10\rangle \leftrightarrow |11\rangle$. The single–qubit frequencies for qubits 1 and 2 are about 4.000 GHz and 6.889 GHz, respectively. We can find each single–qubit transition does not depend on the initial states of the other qubit. Thus, unconditional single–qubit gates are achieved. As a two–qubit operation, the transition $|00\rangle \leftrightarrow |11\rangle$ is measured. In our model, the coupling term generating the two–qubit operation is $-\Omega_{12}(t)(\sigma_x^1\sigma_x^2 - \sigma_y^1\sigma_y^2)/4$. On the other hand, when implementing the two–qubit operation, unwanted coupling term $-\kappa\sigma_z^1\sigma_z^2/2$ in the Hamiltonian at a rotating frame takes place, where $\kappa/h = 1.23$ MHz. Comparing $\Omega_{12}^{max}(= \max\Omega_{12}(t))$ to κ, the efficiency of our coupling scheme is quantified: $\Omega_{12}^{max}/\kappa \approx 19$. The coherent properties are characterized by the relaxation time T_1 and the Ramsey dephasing time T_2^{Ramsey}. Qubit 2 has $T_1 = 1.0\,\mu s$ and $T_2^{Ramsey} = 0.8\,\mu s$. On the other hand, qubit 1 has $T_1 = 0.3\,\mu s$ and $T_2^{Ramsey} = 0.2\,\mu s$. Moreover, we have demonstrated a simple quantum protocol related to quantum coin tossing.

In summary, we have proposed a coherent tunable coupling scheme of flux qubits and demonstrated several simple quantum protocols using it. Every quantum operation is implemented at the optimal bias point. All control processes of the coupling constant are accomplished by employing microwave pulses. Our future targets are the demonstration of large–scale quantum protocols and the integration with non–demolition readout of an individual qubit.

References

1. A. O. Niskanen, Y. Nakamura and J. S. Tsai, *Phys. Rev. B* **73** (2006) 094506.
2. A. O. Niskanen, K. Harrabi, F. Yoshihara, Y. Nakamura, S. Lloyd and J. S. Tsai, *Science* **316** (2007) 723.

JOSEPHSON PHASE QUBITS, AND QUANTUM COMMUNICATION VIA A RESONANT CAVITY

MIKA A. SILLANPÄÄ

National Institute of Standards and Technology, 325 Broadway,
Boulder CO 80305, USA
E-mail: masillan@cc.hut.fi

We discuss in detail the physics of single phase qubits, as well as building an elementary quantum information link between them by means of an on-chip transmission line cavity resonator. We present the experiment at NIST [Nature **449**, 438 (2007)], where we experimentally demonstrated the ability to coherently transfer quantum states between two phase qubits through such a quantum bus.

Keywords: phase qubits; circuit cavity QED; quantum bus

Transmission of quantum states is an important issue to realize quantum communication and quantum network. We try to construct quantum network in a solid–state system. Our quantum memory or local quantum computer is a flux–based Josephson phase qubit. It is necessary to prepare quantum channel to transfer quantum information. For this purpose, we use a circuit cavity QED system, which is implemented by a transmission–line resonator. We report the time–domain measurements indicating coherent interactions between two phase qubits and a cavity mode (i.e., quantum bus).[1]

For a flux-biased Josephson phase qubit, the quantum states are encoded in the phase difference across a large–area Josephson junction placed in a superconducting loop. Phase qubits have several good properties; addressability, the application of single–shot readout, and it is easy to couple together, and so on. The resonant cavity is a $\lambda/2$ transmission–line resonator, which can be regarded as a LC–circuit with $1/2\pi\sqrt{LC}(\equiv \omega_r/2\pi) \approx 8.74\,\mathrm{GHz}$. The total system is composed of two phase qubits (qubits A and B) and a cavity mode. We can find the Hamiltonian of our quantum system is equivalent to the Jaynes–Cummings Hamiltonian:

$\hat{H} = \hat{H}_r + \sum_{j=A,B} \hat{H}_j + \sum_{j=A,B} \hbar g_j \left(\hat{a}^\dagger \hat{\sigma}_-^j + \hat{a} \hat{\sigma}_+^j \right)$, where $\hat{a}$ ($\hat{a}^\dagger$) is the annihilation (creation) operator for a cavity mode, $\hat{\sigma}_-^j$ ($\hat{\sigma}_+^j$) is the lowering (raising) operator for the jth qubit, $H_r = \hbar \omega_j \left(\hat{a}^\dagger \hat{a} + \frac{1}{2} \right)$ is the free Hamiltonian for the cavity mode, and $H_j = \frac{1}{2} \hbar \omega_j \hat{\sigma}_+^j \hat{\sigma}_-^j$ is the single–qubit Hamiltonian for jth qubit. One can control the value of ω_j by the flux bias. The coupling constant g_j is chosen so that the experiment is in the strong coupling regime; $g_j > \gamma_j > \kappa$, where $\gamma_A \approx 20\,\mathrm{MHz}$ and $\gamma_B \approx 5\,\mathrm{MHz}$ are qubit decay rates, and κ is a cavity decay rate ($\kappa/2\pi \approx 1\,\mathrm{MHz}$).

We first show our system works as a strongly coupled cavity QED system. We have observed two basic phenomena in cavity QED: vacuum Rabi splittings and vacuum Rabi oscillations. Now, we implement a crucial protocol in quantum network, quantum state transfer, using the vacuum Rabi interaction of qubits A and B. We explain our scheme. First, we prepare a superposition state for qubit A by a microwave pulse. Secondly, qubit A is placed on resonance with the cavity for a time duration, applying a shift pulse. After this, we wait for a short storage time with the detuning of qubit A restored. Then, applying a shift pulse to qubit B, qubit B is placed on resonance with the cavity for a time duration. Finally, qubit B is returned to its detuned position and both of qubits are measured. We have observed the oscillating behaviors of the measured populations of qubits A and B in the time–domain. Our results are in good agreement with the theoretical calculation based on the Jaynes–Cummings Hamiltonian. Taking account of the case that the initial quantum state of qubit A is a superposition state, we have also verified quantum coherence is kept during quantum state transfer. Furthermore, the measured populations of both qubits imply that an entangled state between them can be generated.

In summary, we have successfully couple two phase qubits via a cavity mode. We have done quantum state transfer and verified quantum coherence is maintained during the protocol. In addition, our results indicate that an entangled state between qubits is created. Our demonstrations are important for realizing quantum network in solid–state systems and reveals the possibility of quantum information processing by solid–state cavity QED.

References

1. M. A. Sillanpää, J. I. Park and R. W. Simmonds, *Nature* **449** (2007) 438.

QUANTUM COMPUTING USING PULSE-BASED ELECTRON-NUCLEAR DOUBLE RESONANCE (ENDOR): MOLECULAR SPIN-QUBITS

Kazuo Sato[§], Shigeki Nakazawa, Robabeh D. Rahimi[*], Shinsuke Nishida, Tomoaki Ise, Daisuke Shimoi, Kazuo Toyota, Yasushi Morita[†], Masahiro Kitagawa[‡], Parick Carl[*], Peter Höfner[*] and Takeji Takui[¶]

Departments of Chemistry and Materials Science, Graduate School of Science, Osaka City University, Osaka 558-8585, Japan
[*]*Interdiscipilinary Graduate School of Science and Engineering, Kinki University, Higasi-Osaka 577-8502, Japan*
[†]*Department of Chemistry, Graduate School of Science, Osaka University, Osaka 560-0043, Japan*
[‡]*Department of System Innovations, Graduate School of Engineering Science, Osaka University, Osaka 560-8531, Japan*
[*]*Bruker BioSpin GmbH, Silberstreifen 4, 76287 Rheinstetten, Germany*
E-mail: [§]*sato@sci.osaka-cu.ac.jp*
[¶]*takui@sci.osaka-cu.ac.jp*

Electrons with the spin quantum number 1/2, as physical qubits, have naturally been anticipated for implementing quantum computing and information processing (QC/QIP). Recently, electron spin-qubit systems in organic molecular frames have emerged as a hybrid spin-qubit system along with a nuclear spin-1/2 qubit. Among promising candidates for QC/QIP from the materials science side, the reasons for why electron spin-qubits such as molecular spin systems, i.e., unpaired electron spins in molecular frames, have potentialities for serving for QC/QIP will be given in the lecture (Chapter), emphasizing what their advantages or disadvantages are entertained and what technical and intrinsic issues should be dealt with for the implementation of molecular-spin quantum computers in terms of currently available spin manipulation technology such as pulse-based electron-nuclear double resonance (pulsed or pulse ENDOR) devoted to QC/QIP. Firstly, a general introduction and introductory remarks to pulsed ENDOR spectroscopy as electron-nuclear spin manipulation technology is given. Super dense coding (SDC) experiments by the use of pulsed ENDOR are also introduced to understand differentiating QC ENDOR from QC NMR based on modern nuclear spin technology. Direct observation of the spinor inherent in an electron spin, detected for the first time, will be shown in connection with the entanglement of an electron-nuclear hybrid system. Novel microwave spin manipulation technology enabling us to deal with genuine electron-electron spin-qubit systems in the molecular frame will be introduced, illustrating, from the synthetic strategy of matter spin-qubits, a key-role of the molecular design of g-tensor/hyperfine-(A-)tensor molecular engineering for QC/QIP. Finally, important technological achievements of recently-emerging CD ELDOR (Coherent-Dual ELectron-electron DOuble Resonance) spin technology enabling us to manipulate electron spin-qubits are described.

Keywords: NMR; ESR; ENDOR; ELDOR; QC; QIP

1. Introduction

1.1. *General Introduction*

The last decade has witnessed that implementation of quantum computers, which can give all the advantages of quantum computing (QC) and quantum information processing (QIP), has been the focus of the contemporary issues in quantum science and related fields.[1,2] Among various physical systems for qubits, photon qubits have recently been utilized for quantum information communications in our ordinary life. Quantum cryptography has been used to protect Swiss Federal Election in autumn of 2007 against hacking into the database or accidental data corruption, exemplifying that QC/QIP is really emerging from the practical side. QC/QIP technology promises to solve problems that are intractable on currently available digital classical computers.

Quantum algorithms can reduce the CPU time for some important problems by many orders of magnitude. An important advantage of QC is the rapid parallel execution of logic operations carried out by quantum entangled or superposition states. For example, with the same input and output, the quantum processing of given information data represents exponential speed-up for factorization by Shor algorithm[3] and quadratic speed-up for search problems using Grover algorithm.[4] Also, by the implementation of the quantum information algorithms such as quantum teleportation[5] and quantum super dense coding,[6] some intrinsic advantages can be achieved compared with the classical information processing. From the theoretical side, quantum information processing and quantum computation have been established considerably well during the last decade.[2] A road map to the goal of building practical quantum computers (QCs) shows problems to be solved such as the establishment and possible utilization of the entangled states, the preparation of scalable qubits, the creation and storage of quantum data bases and the implementation of novel QC algorithms.

1.2. *An Electron Spin as an Inherent Matter Spin-Qubit*

In view of the implementation of QC/QIP, an electron spin itself as a matter qubit naturally given in the molecular frame has only lately appeared in the research field of QC/QIP. An electron-spin with the spin quantum number 1/2 is an inherent spin-qubit, which spinor property belongs to. The spinor is a physical quantity of quantum phase that plays an essential role in QC/QIP, but never experimentally nor explicitly observed so far in contrast to a nuclear spin-1/2. Direct observation of the spinor of an electron spin, detected for the first time, has been carried out in connection with the entanglement of an electron-nuclear hybrid system. Some highlighted parts of the observation will be shown in this chapter. Recently, electron-spin qubit systems in "organic" molecular frames have emerged as a hybrid spin-qubit system with a nuclear spin-1/2 qubit,[7–13] where "hybrid" denotes the physical qubits composed of electron and nuclear spin-qubits. Physical realization of genuine electron-spin qubit based QC/QIP in the molecular frame is the focus of contemporary issues for implementing QC/QIP.

Among many technologically promising candidates for physical matter qubits,[14–17] it is worth noting the reasons why an electron-spin qubit has been the latest comer in the field of QC/QIP. There have been two major drawbacks of electron spin-qubits in molecular frames, when the unpaired electron spins are utilized as matter qubits. The first one is a few but crucial technical difficulties intrinsic to decoherence time of the electron-spin qubit, compared with nuclear spin-qubits, and the second is difficulty in the preparation of the assemblies of the electron spin-qubit in terms of materials science: The decoherence time is three orders of magnitude shorter than that of nuclear spin-qubits. This is the case for both in-ensemble electron spin and single-molecule based QC. The issue of decoherence is one of the apparently intractable obstacles for the physical realization of practical electron spin-qubit based QCs. Taking advantages of long decoherence intrinsic to nuclear spin-qubits and their resonant interactions of radiofrequency pulses, as invoked by current spin manipulation technology, NMR based QC/QIP has successfully illustrated the most significant achievement in terms of implementation of quantum logic gates.[2] Recently, quantum entanglements based on electron-nuclear spin systems as matter qubits in ensemble have emerged from the experimental side,[7–13] referring to quantum information processing.[9,10,12] The entanglement between two 1/2-spins is essential in quantum information science. The occurrence of the entangled spin states in a crystalline solid becomes an important issue in solid-state quantum computing. We have applied pulse-based electron-nuclear and electron-electron multiple magnetic resonance (EMR/ENDOR/ELDOR) technique to molecular spin-based qubits in order to implement the ensemble quantum computing in solid,[7-13] where EMR is initialism for Electron Magnetic Resonance, and ENDOR and ELDOR are acronym for Electron Nuclear DOuble Resonance and ELectron electron DOuble Resonance, respectively. This chapter describes some parts of our work on ENDOR based QC/QIP and important QC-ENDOR related spin techniques carried out by other groups recently, referring to a short course to pulsed ENDOR in the beginning of the chapter.

The second drawback described above, i.e., difficulty in the preparation of the assemblies of matter electron spin-qubits in molecular frames is closely relevant to an issue of the scalability of qubits for practical true QCs. Indeed, the issue is a materials challenge for chemists and in this context chemistry for QC/QIP is a new field in terms of the synthetic strategy. This is an important issue for implementing realistic scalable QCs, but beyond the scope of this chapter.

This chapter will also deal with the latest achievements for genuine, not hybrid, molecular electron spin-qubit based QC/QIP in terms of technological innovations, i.e., coherent-dual (CD) frequency manipulation techniques in pulse microwave spin technology. By invoking this CD microwave spin technology, the relative phase of dual frequencies, i.e., the quantum phases of electron spin-qubits can be manipulated: This is the most significant step, ever made from the technological side, to the physical realization of molecular electron spin-qubit based QC/QIP for the first time. Matter spin-qubits can be accessed and treated by spin magnetic resonance in the presence of the static magnetic field as enabling technology. Here, spin magnetic resonance is generally defined as magnetic resonance phenomena arising from in-ensemble magnetic moments in which each moment is composed of spin angular momentum and orbital angular momentum in molecular entities.

1.3. *Advantages of Photon Qubits Compared with Matter Spin-Qubits in Terms of Enabling Technology*

Physical systems so far realized as qubits have their own right in terms of the implementation of QC/QIP. It is worth noting what are the advantages of photon qubits in marked contrast to matter spin-qubits in the present stage of implementing QC/QIP. Generally speaking, the advantages of photon qubits and photon systems for QIP are as follows: (1) Easy representation of qubits in terms of the relative phases, polarization and photon number (vacuum-single photon). (2) Very weak couplings with environments. (3) Available single-photon measurement laser technology. Totally, due to these advantages, photons provide "practical" physical realization and distinct testing grounds for theories for QC/QIP, contrasting with matter qubits. Matter spin-qubits, particularly electron spin-qubits in molecular frames, are not allowed to enjoy all of these advantages. Additionally photons allow us to utilize fruitful results of extensive researches and developments made in the current optical communication industry. It is worth developing the feasibility of photon-qubit based quantum information processing systems by the use of enabling technology of fiber and integrated optics. From the practical side, experimental tests of quantum cryptography have been carried out so far. On the other hand, the main drawback of photon qubits is less scalability of the qubits, thus not feasible to construct practical true QCs by arranging the qubit array. In some aspects, matter qubits promise to solve other problems that are intractable on currently available digital classical computers. The issue of the qubit scalability for practical true QCs can be accessed by molecular spin-qubits in terms of synthetic strategy. This is a true reason why electron spin-qubits in particular molecular frames have been a materials challenge for chemists and materials scientists for the past few years. In the sections following "2. A Short Course of Pulsed ENDOR for Non-Specialists," we will discuss why molecular spin-qubits promise the scalability.

2. A Short Course of Pulsed ENDOR for Non-Specialists

2.1. *Gyromagnetic Ratios of an Electron Spin and Nuclear Spin: Bohr Magneton for Electrons vs. Nuclear Magneton*

QC ENDOR technology features, as described above, hybrid nature of electron and nuclear spins as qubits. "Hybrid" terms electron spins as bus nature of qubits and nuclear spins as client ones. The gyromagnetic ratio of an electron is about 1800 times larger than that of a nuclear spin. Simply, it is believed that thus "Resonance based quantum computation involving electron spins in addition to nuclear spins, ENDOR" may overcome difficulties in "Quantum computation using only nuclear spins, NMR", particularly poor spin polarization due to the small gyromagnetic ratio of nuclear spins in an ordinary condition: It is noted that sophisticated polarization transfer such as dynamic nuclear polarization is implemented in the initialization process of QC NMR experiments. In view of implementing QC spin technology, this context should be understood with adding "In spite of apparent advantages from the intrinsic nature of electrons, electron spin-qubits may encounter shaded difficulties, referring to current microwave (MW) spin technology." The MW spin technology is still "shadowy" as spin manipulation technology compared with the counterpart of NMR in many important experimental aspects or events, because of 1800 times larger irradiation energy for resonance than that in QC NMR.

Reminding the ratio of the Bohr magneton vs. the nuclear magneton: Magnetic moments of nuclei and molecules are proportional to the angular momenta of these microscopic entities, conveniently described as a product of dimensionless g factor and a dimensioned factor called the magneton.

$$\boldsymbol{\mu} = \alpha g \beta \boldsymbol{J}, \qquad (1)$$

where g in the same units as vector $\boldsymbol{\mu}$: g denotes the magnitude of the electron Zeeman factor for the physical entities under study and $\boldsymbol{J}$ the general angular momentum vector in units of $h/2\pi$. $\alpha = +1$ or -1. For free electrons, $\alpha = \alpha_e = -1$ and the operator $\boldsymbol{J}$ is equivalent to the operator $\boldsymbol{S}$ for the intrinsic spin angular momentum, thus β becomes

$$\beta_e = e\frac{h}{2\pi}2m_e = 9.2740154(31) \times 10^{-24}\,\mathrm{JT}^{-1} \qquad (2)$$

which is called the Bohr magneton: The latest value is $9.27400915 \times 10^{-24}\,\mathrm{JT}^{-1}$. The difference mainly arises from the accuracy in determining h. The Zeeman splitting constant for the free electron is

$$g_e = 2.0002319304386(20). \qquad (3)$$

The symbol g is customarily utilized for the electrons interacting with other physical entities, in which $g = g_e$ for the free electron.

For nuclei, similarly the nuclear magneton is defined as

$$\beta_n = e\frac{h}{2\pi}2m_n = 5.0507866(17) \times 10^{-27}\,\mathrm{JT}^{-1}, \qquad (4)$$

where m_n is the mass of the proton (^{1}H). Customarily, g_n denotes nuclear g factors. Thus, the ratio of β_e/β_n is 1836. Thus, induced magnetic resonance for an electron requires microwave frequency when induced nuclear magnetic resonance NMR in MHz radiofrequency regions occurs in the presence of the same static magnetic field. Both main advantages and disadvantages of electron spin-qubits originate in this ratio, when they are utilized for QC/QIP. In terms of physics and chemistry of electrons in molecular frames, there is one thing to be noted: They generally have delocalization nature in contrast with that of any nuclear spins, except for electrons confined into molecular caged in molecular frames such as C_{60} fullerenes. The magnetic interactions due to electrons are more than three orders of magnitude larger than those due to nuclear spins. Furthermore, there are a variety of the interactions, i.e., the relevant interaction terms in the spin Hamiltonian, compared with those in nuclear spin assemblies dealt with NMR spectroscopy. This is due to the long-range/delocalization nature of molecular electrons and the nature of such molecular electronic wavefunctions as unpaired electrons are accommodated in. Indeed, modern quantum chemistry and molecular science based on quantum chemical calculations at various levels of approximation give access to the interpretation of the experimentally determined magnetic tensors and physical parameters.

Table 1. Microwave frequencies, band assignments, conventional microwave frequencies used for ESR spectroscopy and resonance magnetic fields for $g = g_e$. *The definition for the frequency range is ambiguous. Only typical definitions are given.

Band Assignment	Frequency Range* /GHz	Conventional ESR frequency/GHz	Corresponding ESR magnetic field B_r/mT
L	0.390-1.552	1.5	53.5
S	1.55-3.90 or 2.6-4.0	3.0	107
C	3.90-6.20 or 4.0-6.0	6.0	214
X	6.20-10.9 or 8.2-12.4	9.5	339
K	10.9-36.0 or 18-26.5	23	821
Q	36.0-46.0 or 33-50	36	1285
V	46.0-56.0 or 50-75	50	1784
W	56.0-100 or 75-100	95	3390

In Table 1 are given typical microwave frequencies utilized for conventional ESR spectroscopy. Definition for the frequency range designating the corresponding band is ambiguous. Table 1 only gives two typical definitions. The transition (= resonance) magnetic field B_r for $g = g_e$ is calculated for the MW frequency inducing the corresponding magnetic resonance (see Fig. 1). MW frequency becomes higher, electron spin polarization becomes rapidly larger, whereas difficulties in terms of spin technology increase in many experimental aspects. Simply speaking, the g-value in ESR spectroscopy corresponds to chemical shift in NMR spectroscopy. The g-value gives information on the electronic/molecular structure of an open-shell physical entity under study.

Any practical QCs require to fulfill DiVincenzo's five criteria.[18] Thus, in implementing molecule-based QC ENDOR, the five criteria are the starting point. Particularly, the first criterion, i.e., well defined and scalable qubits,

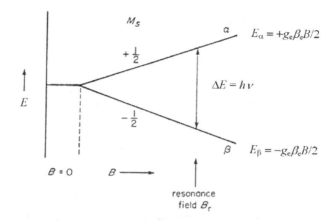

Fig. 1. Energy-level scheme for the simplest electron system as a function of swept static magnetic field B, depicting ESR absorption. B_r denotes the resonance field. The other symbols have usual meanings.

has to be amplified: Molecular hybrid spin-qubits are chemically identified and well characterized in view of molecular and electronic-spin structures. The magnetic tensors of the targeted qubits, i.e., the g or fine-structure tensor for bus electrons and/or hyperfine/quadrupolar tensors for client nuclei, should be determined at least experimentally prior to use as the defined qubits. Obviously, the anisotropic tensors are for solid-state QCs and isotropic principal values for QC/QIP in isotropic media like solution.

2.2. *What is Electron Spin Resonance/Electron Paramagnetic Resonance (ESR/EPR)? : A Basis for ESR/EPR Spectroscopy*

Among all forms of modern spectroscopy, ESR/EPR (abbreviated to ESR) has been termed a peculiar spectroscopy for a long time, whereas NMR and ESR methodologically share a common and similar origin. Ordinary spectroscopy sweeps its oscillating radiation energy for resonance by measuring the absorption/emission of continuous monochromatic radiation by a sample. ESR spectroscopy sweeps applied static magnetic fields instead of microwave energy sweep, except a few types of experiments. The peculiarity originates in the swept static magnetic field for resonance. This is due to a long-standing limitation to oscillating MW technology relevant to be β_e, underlying a notorious drawback to ESR spectroscopy. So far, MW tech-

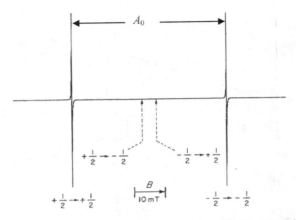

Fig. 2. A first-derivative field-swept ESR spectrum from the simplest chemical entity with one unpaired electron and nuclear spin-1/2 such as a hydrogen atom, assuming a positive g_n. The ESR allowed transitions are for $\Delta M_\mathrm{I} = 0$ and the forbidden ones for nonvanishing ΔM_I under the conventional experimental conditions, i.e., in the MW detection mode with $\boldsymbol{B}_1$ perpendicular to $\boldsymbol{B}$. Thus, the two ESR allowed transitions occur. The signal in the low field (left) corresponds to the transition denoted by the double-arrow k in Fig. 3 ($A = A_0$) and the one in the high field to m in Fig. 3, respectively.

nology does not allow us to sweep its alternating frequency in a wide range, instead applied static magnetic field has to be swept to identify an event of magnetic resonance, i.e., to find resonance which the swept static magnetic field, i.e., resonance field B_r matches the MW energy $h\nu$, as depicted by the single- and double-arrows in Fig. 1. On magnetic resonance, quantum mechanical selection rules for group-theoretically allowed transitions require the static magnetic field $\boldsymbol{B}_0$ (often written as $\boldsymbol{B}$) perpendicular to the oscillating MW magnetic field $\boldsymbol{B}_1$. Often, group-theoretically forbidden transitions appear even with the perpendicular excitation of microwave field $\boldsymbol{B}_1$. Usually, experimental conditions with high MW frequency hamper chance of the occurrence of the forbidden transitions. In QC ENDOR experiments, hyperfine forbidden transitions have a special significance. As interpreted in the later section, the preparation of an initial electro-nuclear state prior to any quantum operations in the ENDOR experiments, termed initialization, can be underlain by the use of the hyperfine forbidden transitions for particular hybrid qubits in molecular frames.[19]

The magnetic-field-derivative detection scheme is required to acquire both good signal-to-noise ratio and high spectral resolution in an ordinary continuous- wave (CW) ESR spectroscopy: Field modulation of a certain carrier frequency is applied parallel to the static swept magnetic field for

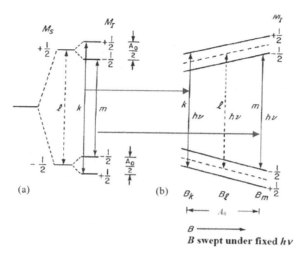

Fig. 3. (a) Energy diagram for electron-nuclear spin states of the simplest chemical entity given in Fig. 2 in the presence of the static magnetic field and (b) the variation of the states as a function of the swept magnetic field B under the fixed MW energy $h\nu$. An isotropic hyperfine coupling A_0 between an unpaired electron and the nuclear spin-1/2 is assumed.

these purposes. The field-derivative detection of high carrier-frequencies such as 100 kHz is usually carried out by invoking phase-sensitive detection (PSD) modes, in which only signals oscillating at the PSD carrier frequency with its phase are picked up. It is worth noting that the spectral resolution is subject to the higher limit to the carrier frequency due to the side-band production inherent in the field modulation. Also, due to the use of the field modulation, CW ESR spectroscopy suffers from relatively low time resolution of the signal detection. Short-lived paramagnetic entities with lifetime of lower than sub milliseconds are difficult to be accessed by the PSD of ordinary carrier frequencies, except for the occurrence of spin-polarized states due to some chemical or physical reasons. CW ESR spectroscopy requires a certain amount of electron spin concentration in the steady state. Figure 2 exemplifies a first-derivative field-swept ESR spectrum from the simplest chemical entity with one unpaired electron and one nuclear spin-1/2 such as a hydrogen atom, assuming a positive g_n. The ESR allowed transitions are for $\Delta M_I = 0$, while the hyperfine forbidden ones are for non-vanishing ΔM_I under the conventional experimental conditions, i.e., in the MW detection mode with $\boldsymbol{B}_1$ perpendicular to $\boldsymbol{B}_0$. The difference in these two transition schemes is group-theoretically interpretable. Thus, the two ESR

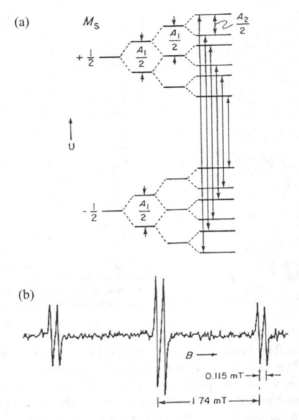

Fig. 4. (a) Energy diagram for an unpaired electron system with a pair of equivalent nuclei having a nuclear spin-1/2 and A_1 and another nucleus having a nuclear spin-1/2 and $A_2 < A_1$. (b) An X-band hyperfine ESR spectrum from CH_2OH radical in methanol observed at 299 K; $A_1 = 1.74$ mT, $A_2 = 0.115$ mT. [Adopted from R. Livingston and H. Zedes, J. Chem. Phys., **44**, 1245 (1966).]

allowed transitions occur. The signal in the low field (left) corresponds to the transition denoted by the double-arrow k in Fig. 3 and the one in the high field (right) to m in Fig. 3, respectively. The separation between the two signals gives the absolute value for the hyperfine splitting A. The splitting corresponds to a hyperfine coupling constant A_0 for an isotropic case, while the hyperfine splitting for an anisotropic case is as a function of the orientation of the static magnetic field B with respect to the crystal axis system of a paramagnetic sample such as a molecular crystal. The dashed lines denote the resonance fields for possible hyperfine forbidden transitions.

Figure 3 depicts the electron-nuclear spin states of the simplest chemical entity given in Fig. 2 in the presence of the static magnetic field B (b). The solid double-arrows k and m denote the hyperfine allowed transitions. In Fig. 3, an isotropic hyperfine coupling A_0 between an unpaired electron and the nuclear spin-1/2 is depicted. Figure 4(b) illustrates a well-resolved X-band hyperfine ESR spectrum from CH_2OH radical in methanol observed at 299 K; $A_1 = 1.74\,\text{mT}$, $A_2 = 0.115\,\text{mT}$. The spectrum is composed of a pair of equivalent protons and another proton with a larger hyperfine coupling constant A_1.

2.3. *Spectral Density of Electron Magnetic Resonance Transitions: High Resolution and High Sensitivity by Quantum Transformation in ENDOR Spectroscopy*

Let us start treating a simple electron-nuclear hybrid system by introducing the corresponding spin Hamiltonian. The appropriate spin Hamiltonian for a simple physical system of the hyperfine interaction of one electron spin S interacting with one nuclear spin I in an external static magnetic field B_0 is given in a straightforward manner as follows:

$$H = H_{\text{EZ}} + H_{\text{NZ}} + H_{\text{HFS}}. \tag{5}$$

Thus, we have the form

$$H = \beta_e B_0 \cdot g \cdot S - g_n \beta_n B_0 + h S \cdot A \cdot I, \tag{6}$$

where g and A are the g tensor for the electron and the hyperfine one for the nucleus, respectively. The tensor elements are given in units of frequency. h is Planck's constant. We note that the electron and nuclear spins are no longer independent via the hyperfine term.

For simplicity, assuming only isotropic interactions with B_0 chosen along the z axis, we have

$$H = g\beta_e B_0 S_z - g_n \beta_n g_n B_0 I_z + ha S \cdot I, \tag{7}$$

where a is Fermi contact coupling in units of frequency. The constant a is given by

$$a = h^{-1}(2\mu_0/3)g\beta_e g_n \beta_n |\psi(0)|^2, \tag{8}$$

where ha measures the magnetic interaction energy between the electron and the nucleus. In a strong external magnetic filed (typically 0.34 T at X-band MW ESR spectroscopy) the electron Zeeman term is the dominant term for usual non-high spin cases, noting that for high-spin entities fine-structure terms describing electron-dipolar and electron spin-orbit couplings are required. In this high-field limit, ignoring second-order hyperfine terms, we have the energy levels for the present simplest case as follows:

$$E(M_S, M_I) = g\beta_e B_0 M_S - g_n\beta_n B_0 M_I + h a M_S M_I, \tag{9}$$

where M_S and M_I denote the magnetic spin quantum number for the electron and nucleus, respectively.

Defining the electron and nuclear Zeeman terms in units of frequency as

$$\nu_e = g\beta_e B_0/h \tag{10}$$

$$\nu_n = g_n\beta_e B_0/h. \tag{11}$$

We have the energy levels by

$$E(M_S, M_I)/h = \nu_e M_S - \nu_n M_I + a M_S M_I. \tag{12}$$

The splittings of the unperturbed electronic spin energy levels E/h by the electronic Zeeman, the nuclear Zeeman and the hyperfine interaction will be schematically shown later in Figure 6. The four energy levels are designated by the signs of the simple product spin functions $|M_S, M_I\rangle$, e.g., $|-+\rangle$ for the case of $a > 0$ and $\nu_n > a/2$.

The selection rules for electron spin transitions (ESR/EPR cases), i.e.,

$$\Delta M_S = \pm 1 \quad \text{and} \quad \Delta M_I = 0 \tag{13}$$

give the two ESR allowed transitions, yielding the corresponding resonance frequencies

$$\nu_{ESR} = \nu_e \pm a/2, \tag{14}$$

where the contributions from the nuclear Zeeman term are cancelled out. Applying the selection rules for nuclear spin transitions (NMR cases), i.e.,

$$\Delta M_S = 0 \quad \text{and} \quad \Delta M_I = \pm 1, \tag{15}$$

we have the two NMR transitions at the frequencies

$$\nu_{NMR} = |\nu_n \pm a/2|. \tag{16}$$

These NMR transitions in open-shell physical entities are detected in Electron-Nuclear DOuble Resonance experiments via the change in the intensity of a particular (monitored) ESR transition, when the MW (ESR) and RF (NMR) coherent irradiations are simultaneously applied: The experimental setups will be shown later. The change is enhanced by quantum transformation from RF quantum to MW one, due to the ratio of β_e/β_n, 1836.

In deriving Eq. (9), we neglected the off-diagonal term such as $ha(S_+I_- + S_-I_+)/2$, where S_+ and S_- are the raising and lowering operators for the electron, and I_+ and I_- have the same meanings for the nuclear spin, respectively. The selection rules given by Eq. (13) are exact only when the kets $|M_S, M_I\rangle$ are eigenfunctions of the spin Hamiltonian given by Eq. (7), i.e., for the case that the off-diagonal term above vanishes. The second or higher-order perturbation treatments can provide more accurate solutions for the cases with non-vanishing off-diagonal terms. General analytical solutions for the spin Hamiltonian such as Eq. (6) including both fine-structure (for $S > 1/2$) and quadtrupolar (for $I > 1/2$) terms can be acquired even in an arbitrary coordinate system. In Appendix A given at the end of this chapter, general and analytical solutions for ENDOR transition frequencies and probabilities are described in arbitrary reference frame, in which only the unit vector h for the direction of the static magnetic field B_0 is defined.

The perturbation approaches, in which the electron Zeeman term in Eq. (6) is assumed to be the non-perturbed term, are frequently used and useful to interpret any ESR/ENDOR spectra. On the other hand, for the complete spectral analyses, the exact diagonalization of the spin Hamiltonian is required. There have been developed two dominant approaches: One is the conventional eigenenergy method for the diagonalization, in which resonance fields are searched numerically to match the difference between the calculated eigenenergies under the fixed MW $h\nu$. The eigenenergy method frequently suffers from notorious non-convergence problems in the matching procedure. The other is the eigenfield method, in which any resonance field can be obtained by the direct diagonalization of an eigenfield $n^2 \times n^2$ matrix: $n \times n$ denotes the dimension of the energy matrix of the spin Hamiltonian. The eigenfield method treats a generalized matrix associated with complex eigenvalues and requires much CPU time, particularly enormous CPU time for anisotropic media. To reduce the CPU time, we have developed a hybrid eigenfield method, in which eigenfunctions required for the calculated transition probabilities are obtained via the corresponding energy matrices with exact eigenfields substituted. In the hybrid method, complex numerical procedures for the eigenvalues derived directly from the generalized matrix are skipped, reducing the CPU time. Nevertheless, the eigenfield method based softwares for spectral simulation are available on the website (e.g., EasySpin or Hybrid Eigenfield) and the method has been emerging as a conventional tool in ordinary laboratories. If paramagnetic molecular entities have more than one magnetic nucleus interacting with the electron, the terms relevant to I_{zi} are additive and nuclear-nuclear interactions are negligible in conventional ESR spectroscopy. Summing over all magnetic nuclei, we have

$$H = g\beta_e B_0 S_z - \sum_i g_{ni}\beta_n B_0 I_{zi} + \sum_i ha_i \boldsymbol{S}\cdot\boldsymbol{I}_i. \qquad (17)$$

In Fig. 4, the six ESR absorption lines appear due to a doubling by a proton of the three allowed transitions from the two equivalent protons, as depicted in Fig. 4(a). In general, the maximum possible number of distinct ESR lines for sets of m and n equivalent protons in a molecular frame is given by $(m+1)(n+1)$. If there are an arbitrary number of N of sets of such equivalent protons, the total number of ESR lines is given by $\prod_k(nk+1)$, where $\prod_k$ denotes a product of over all values for k ($k = 1, 2, \ldots, N$).

With the increasing number of equivalent or non-equivalent nuclei having non-zero spins over which an unpaired electron is delocalized, the corresponding hyperfine ESR spectrum becomes extremely complicated. Often, even computer-assisted spectral simulations, reproducing the well-resolved and side-band production free experimental spectrum, cannot afford to derive a set of accurate hyperfine couplings. The spectral density of transitions involved in a hyperfine ESR spectrum is given by

$$\Omega_{\text{ESR}} = \frac{\prod_k(2n_k I_k + 1)}{2\sum_k a_k n_k I_k}, \qquad (18)$$

where n_k denotes the number of equivalent nuclei with a nuclear spin quantum number I_k and a hyperfine coupling constant a_k in units of frequency and k runs over $k = 1$ to K.

The hyperfine structure of the ESR spectrum is composed of the total number of $\prod_k(2n_k I_k + 1)$ resonance transition lines. $2\sum_k a_k n_k I_k$ is given by the difference between the maximum and minimum resonance frequencies. On the other hand, in ENDOR spectroscopy all the equivalent nuclei give the same NMR, i.e., ENDOR frequency. In solution ENDOR, quadrupolar interactions are vanishing and thus each set of the nuclei gives only two NMR transition lines, totally the number of $2K$ lines showing up. Therefore, the spectral density of transitions involved in an ENDOR spectrum is described as

$$\Omega_{\text{ENDOR}} = \frac{2K}{\max a_k}. \qquad (19)$$

In ENDOR spectroscopy, NMR transitions in a paramagnetic entity, in which nuclei with nonzero nuclear spins are involved, are detected through changes in ESR transition intensities (= ESR monitor) when the NMR transitions are induced by sweeping radiofrequency in the CW mode. In ENDOR spectroscopy, the NMR transitions are monitored by quantum transformation to microwave quanta from radiofrequency ones. This quantum transformation entertains high sensitivity of ESR transitions relevant to β_e in detecting NMR transitions. This is the reason why ENDOR techniques enhance sensitivity in on-resonance NMR transitions by orders of magnitude, compared with NMR spectroscopy. Indeed, ENDOR selection rules are equivalent to those of NMR transitions in paramagnetic systems, any hyperfine ESR spectra may be spectroscopically decomposed to give two or three orders of magnitude better resolution, as described by Eq. (19). Magnetic parameters such as hyperfine coupling constants are directly derived from the observed ENDOR lines accurately, to the first order in terms of perturbation theory. It is noteworthy that in many cases hyperfine internal fields arising from hyperfine couplings play an important role to interpret ENDOR spectra. We emphasize that the spectral density of transitions in ESR spectroscopy often amounts to two or three orders of magnitude greater than that in ENDOR spectroscopy. In paramagnetic molecular frames, the difference between hyperfine splittings is often small. If there are a_i and $a_j \approx ka_i$ of protons in the molecular system, where k is an integer or the reciprocal of an integer, the corresponding ESR spectrum has fewer than the expected number of absorption lines with the deviation from a binominal intensity distribution. For small hyperfine splittings, the side-band production gives lineshape distortion, hampering the spectral resolution due to the field modulation of high carrier frequency. The linewidth difference between hyperfine lines gives a further hazard of spectral misassignment. For such cases ENDOR spectroscopy plays an extremely powerful role, exemplifying Figure 5. In Fig. 5, $\Omega_{ESR}/\Omega_{ENDOR}$ exceeds 30.

Figure 6 shows again a schematic energy diagram for the physical system with an unpaired electron and one nucleus with $I = 1/2$ in the presence of a given static magnetic field. The symbols have usual meanings and the nuclear spin sublevels are exaggerated for clarity. ENDOR experiments are carried out with an ESR absorption line monitored (= pumped by strong MW on-resonance frequency) and radiofrequency for ENDOR transitions swept in the CW mode or irradiated. Thus, the vertical axis in the experimental ENDOR spectrum corresponds to any changes of MW quanta occurring in the monitored ESR line during the sweep or irradiation of radiofrequency for inducing ENDOR transitions, i.e., observing any changes of MW quanta as a function of the ENDOR frequency, i.e., any MW changes on NMR resonance. Because of the double resonance, the effect on the NMR resonance is transformed to the MW quanta. Thus the effect is enhanced, giving high sensitivity due to the MW quanta. This is the reason why ENDOR transitions give high sensitivity compared with genuine NMR transitions induced by RF irradiations. In Fig. 6, the two ENDOR transitions between level 1-2 and level 3-4 are shown, where ω_N ($\omega_N = 2\pi\nu_n$) denotes the NMR angular frequency corresponding to the nucleus under study and A the hyperfine coupling constant. Depending on the relative magnitude of

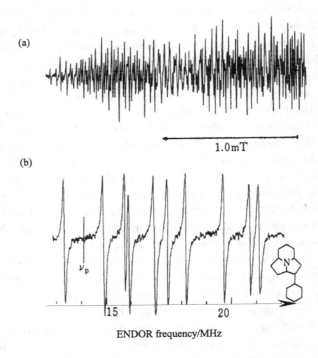

Fig. 5. Comparison between ESR and ENDOR spectral densities. (a) Spectral density of transitions in CW ESR spectroscopy and (b) that in CW ENDOR spectroscopy (b) for 2-phenyl[3,2,2]azine anion radical in dimethyl ether observed at 210 K. (a) Only the half of the ESR spectrum in the higher field, noting that the arrow indicates the direction of the field sweep. (b) Only the half of the ^{1}H ENDOR spectrum in the high frequency region. ν_p denotes the NMR frequency for free proton. The ENDOR spectrum was detected in the mode of ENDOR frequency modulation. This is why the first-derivative signals appeared in the ENDOR spectrum. Adopted from F. Gerson, J. Jackimowcz, K. Moebius, R. Bielhl, J. S. Hyde, and D. S. Leniart, J. Magn. Reson. **18**, 473 (1975).

$A/2$ and ω_N the center frequency between the two ENDOR lines gives, to the first order, either $A/2$ (case (a)) or ω_N (case (b)), as depicted in Fig. 7. It is noteworthy that in ENDOR spectroscopy in either liquid or solid state the spectral linewidths are of orders of magnitude of kHz.

2.4. Fourier-Transform ESR/ENDOR Spectroscopy: Pulse-Based ESR/ENDOR as Enabling Spin Technology

In the preceding section, the advantages of ENDOR technology have been described briefly, giving a basis for ENDOR spectroscopy mainly from the experimental side. We have noted that there are many important aspects from the theoretical side, closely associated with sophisticated CW ENDOR experiments and a diverse number of applications. Referred to quantum gate operations for QC/QIP, currently there is nothing to do with CW ENDOR in view of spin-qubit manipulation technology. We now come to totally different aspects of electron magnetic resonance technology, in other words the issues of spin manipulation in the time domain. The issues are to

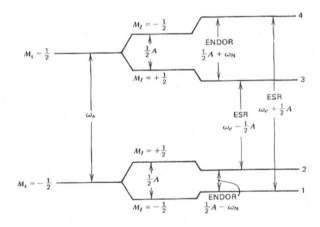

Fig. 6. Schematic energy diagram for the physical system with an unpaired electron and one nucleus with $I = 1/2$ in the presence of a given static magnetic field. ENDOR experiments are carried out with an ESR absorption line monitored (= pumped by strong MW on-resonance frequency) and radiofrequency for ENDOR transitions swept in the CW mode or irradiated. Thus, the vertical axis in the experimental ENDOR spectrum corresponds to any changes (MW quanta) occurring in the monitored ESR line during the sweep or irradiation of radiofrequency for inducing ENDOR transitions, i.e., observing any changes of MW quanta as a function of the ENDOR frequency.

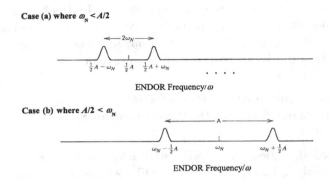

Fig. 7. ENDOR spectral patterns for the simplest system with an unpaired electron and one nuclear spin-1/2. The center frequency between the two ENDOR peaks depends on the relative magnitude of $A/2$ and angular frequency ω_N ($\omega_N = 2\nu_n$).

manipulate both electron and nuclear spins in molecular frames in terms of their time evolution and phases enabling us to discriminate any quantum spin states. The key technology is the introduction of Fourier-transform techniques to electron magnetic resonance, which make us utilize intense pulses of MW and RF (radiofrequency) irradiations to generate a coherent superposition of the relevant spin states in ensemble. In this context, the irradiation pulses link the magnetic resonance to matter spin-qubit based QC/QIP. Macroscopic magnetic moments, as the coherent superposition, precessing at frequencies ω_k, ($k = 1$ to K) and with amplitudes A_k emit radiation that corresponds to free induction decay (FID) signals. They are coherently detected and digitized for further procedure of Fourier-transform (FT) analyses. Pulse-based ENDOR spectroscopy utilizes both MW and RF pulses, in which MW radiation signals in FID or electron spin-echo scheme after on-resonance MW excitations are monitored when pulse-based NMR events occur.

As well known, since the first nuclear spin echoes were observed by Hahn in 1950, NMR developed rapidly especially upon the introduction of FT NMR in 1966 by Ernst and Anderson. For the past thirty years, NMR spectroscopy has benefited the most from the very introduction of Fourier transform techniques based on strong radiofrequency pulse generation and developments of both digital computers and various digital technologies. The introduction of RF multiple pulses into FT-NMR has achieved 2D NMR and n-dimensional spectroscopy.

The main reasons for this remarkable development in the time-domain

NMR were (1) the higher sensitivity that could not be acquired by the CW frequency-domain NMR and (2) the possible occurrence of many sophisticated experiments that CW techniques could not afford.

In 1958, only eight years after the nuclear spin echoes appeared, the electron spin echo (ESE) was reported by Blume on trapped electrons in liq. Ammonia. In contrast to time-domain NMR, the utilization of ESR in time-domain took a long time. There are several reasons for this: (1) The spectral linewidths in the order of 10-25% of the carrier frequency (g-anisotropy and/or large hyperfine or quadrupolar interactions) due to large energies involved in the electron spin interactions. (2) The electron-spin relaxation times in orders of magnitude shorter than those in NMR. These properties have put extreme demands on the experimental and technical conditions: In order to excite a 10% spectral region at X-band (9 GHz GHz), sub-nanosecond $\pi/2$ pulses are needed. This is the reason why the first pulsed ESR studies were confined to the physical entities with both long relaxation times and small linewidths by ESE techniques: Following Mims' pioneering work, as early as late 1980s, MW equipments for pulse-ESR experiments became available, thus the applicability of the pulsed ESR techniques having met a remarkable rise in many application fields, except for quantum information science until very recently.[7-13] The particular reasons for this were already discussed in the preceding section.

The theory for pulse-based electron magnetic resonance experiments is well established. The theoretical basis is given by the density matrix formalism. For non-specialists, however, classical descriptions are presented and here the description of ESE phenomena is based on an inhomogeneously broadened resonance line. An inhomogeneously broadened line is composed of homogeneous spin packets. In this model, the macroscopic ensemble under study is divided into sub-ensembles whose density matrices are $\rho_{\omega_\ni}$. Various sub-ensembles described by $\rho_{\omega_\ni}$ have different resonant frequencies, for example, in the presence of a magnetic field gradient. We also assume that the linewidth due to homogeneous processes is not greater than the one generated by the inhomogeneous broadening. In addition, we note that we have an ensemble of electron spin $S = 1/2$. In Figure 8, a spin packet model to understand an ESE phenomenon is given. Each spin packet denoted by i experiences different local magnetic field $\boldsymbol{B}_{\text{local}}$ which is added, as vector sum, to the applied static magnetic $\boldsymbol{B}_0$. As a result, the spin packet precesses at a Larmor frequency $\omega_i = \gamma B_i$ around the z axis, where $\gamma(h/2\pi) = g\beta_\text{e}$. The inhomogeneous broadening gives rise to a distribution of ω_i around the center frequency $\omega_0 = B_0$. In the rotating frame around the direction of B_0, defined as the z axis, with the frequency ω_0, the precession frequency of the spin packet is given by $\omega'_i = \gamma B_i - \omega_0$. All individual magnetic moments belonging to the spin packet denoted by i precess with the same frequency but with different phases. Figure 9 shows how the Hahn's two-pulse ($\pi/2$-τ-π-τ) sequence generates an ESE signal after the waiting time τ (twice). After the $\pi/2$ pulse which is applied along the $-x'$ axis and is strong enough to cover the spread in Larmor frequencies ω_i, the macroscopic magnetic moment composed of the assemblies of the spin packets starts spreading in the $x'y'$ plane due to their different Larmor frequencies. This process is termed dephasing, which will be discussed in more details in the later section. After the time 2τ spin packets are united along the $-y'$ axis with somewhat reduced magnitude.

In QC ENDOR experiments, magnetic moments of electron and nuclear

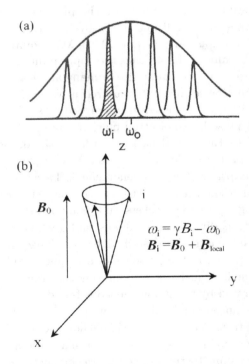

Fig. 8. Spin packet model. (a) An inhomogeneously broadened lineshape composed of spin packets with different Larmor frequencies, ω'_is and (b) the origin of the frequency due to local magnetic field B_{local} superimposing the static magnetic field B_0. The shaded spin packet in (a) precesses at a Larmor frequency $\omega_i = \gamma B_i$ around the z axis.

spin-qubits have to be manipulated in any desired orientations in the Bloch sphere, as depicted in Fig. 10 for illustrative purposes. When the $\pi/2$ pulse along the y axis operates on the $|0\rangle$ state, a superposition of the $|0\rangle$ and $|1\rangle$ states with the equivalent weight is generated, as schematically depicted in Fig. 11(a). Figure 11(b) depicts how the direct product composed of the two superpositions is represented in the macroscopic magnetization qubit scheme.

In order to illustrate the entanglement between an electron spin-qubit and nuclear spin-qubits, quantum phases belonging to the spin states are utilized. QC types of experiments, in which the quantum phases are controlled and manipulated, have never been carried out and neither necessary nor useful for ordinary pulsed electron magnetic resonance spectroscopy. In this context, the implementation of pulsed ENDOR QC/QIP may provide new aspects for electron magnetic resonance spectroscopy.

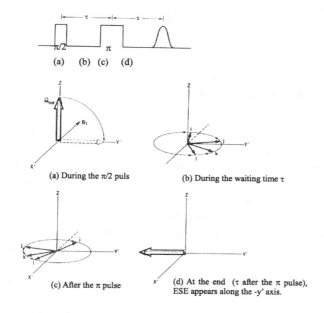

Fig. 9. Pulse sequence for Hahn's two- pulse electron spin echo (ESE) (top). The MW pulse B_1 is applied along the $-x'$ axis. In the rotating frame of the MW frequency, B_1 is the stationary field., defining the $-x'$ axis. The shaded arrow denotes the macroscopic magnetic moment. (a)-(d) correspond to the time in the pulse sequence (top). In (b) and (c), note that the time evolution of each magnetic moment continues in the same direction as original, as depicted in (b) and (c). Note that the π pulse applied along the same x axis after the time τ is not explicitly "depicted" for clarity.

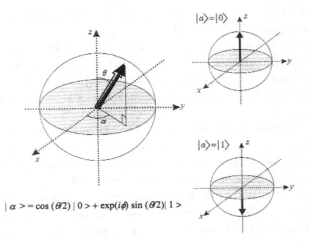

$|\alpha\rangle = \cos(\theta/2)|0\rangle + \exp(i\phi)\sin(\theta/2)|1\rangle$

Fig. 10. Magnetic moments represented as qubits in the Bloch sphere. The thick arrow denotes the moment. The moment in the xy-plane of the Bloch sphere is a superposition of the $|0\rangle$ and $|1\rangle$ state with the equivalent weight.

2.5. A Basis of Spin Manipulation Technology for QC/QIP in Pulse Electron Magnetic Resonance

Figure 12 shows how the single MW or RF pulses on resonance, acting on the magnetization aligning originally along the z axis, generate superpositions of the states or quantum gates. All the pulse operations are applied in the rotating frame of the oscillating irradiation field. In Fig. 12(a), $|\Psi\rangle$ is represented by two variables, θ and ϕ. For $\theta = \pi/2$ the cases of $\phi = 0$ and $\pi/2$ generate distinguishable superpositions of the states in terms of the phase. These situations can be achievable in an on-resonance $\pi/2$ pulse irradiation along the y or x axis. Importantly, for a spin-1/2 qubit, the $2pi$ rotation of the magnetization around the x axis does not recover the original state, but changes the sign of the phase and only the $4pi$ rotation does, denoting the spinor property for spin-1/2. In Fig. 12(b), the bold arrow along a particular axis denotes the axis along which the irradiation pulse is applied. All the situations depicted in Fig. 12 are achieved in terms of the rotating frame of the oscillating irradiation of MW or RF with the static magnetic field along the z axis. Practically, it is important to create the stable, narrow and strong pulses of desired shapes in spin manipulation technology. MW high-frequency technology still suffers from technical difficulties in power, multiple-frequency production and relative phase control

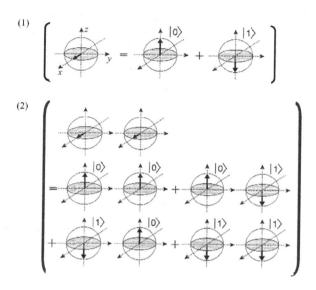

Fig. 11. Schematic drawing for macroscopic magnetizations of spins in pulse-based spin manipulation technology. (1) A superposition of the $|0\rangle$ and $|1\rangle$ states as generated by the $\pi/2$ pulse applied along the y axis in pulse magnetic resonance. The thick arrow denotes a macroscopic magnetization of spins. (2) A schematic representation of the direct product of the two superpositions in terms of the macroscopic magnetizations. In both (1) and (2), all the coefficients associated to the states are omitted for clarity.

between the multiple frequencies.

Figure 13 exemplifies quantum logic gates, the transformation of the states after the operation and the corresponding magnetic resonance pulses. The Hadamard gate in Fig. 13(b) is achieved by the use of two pulses, i.e., the first $\pi/2$ pulse along the y axis followed by the second π pulse along the x axis. The Hadamard transformation is performed by a π-rotation around the particular axis $\pi/4$-rotated from the z axis in the zx plane. The operation around this particular rotation axis corresponds conventionally to the combination of the two pulses given in Fig. 13(b) in pulse electron magnetic resonance. In Fig. 13, the gate (c) is a controlled not gate composed of two qubits, in which an electron- nuclear hybrid system with one electron spin-qubit and one nuclear spin-1/2 qubit are assumed. The corresponding energy level diagram and state designation are given in (c). The transition between the level 3 and 4 corresponds to the ENDOR one which converts the populations of the level 3 and 4 by a π RF pulse applied along the x

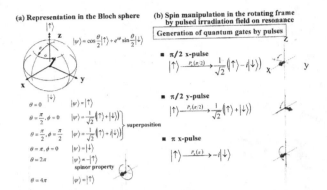

Fig. 12. Macroscopic magnetization represented in the Bloch sphere and the role of on-resonance pulses generating various superpositions of the states $|\uparrow\rangle$ and $|\downarrow\rangle$ and quantum gates. All the operations by the pulses are applied in the rotating frame of the oscillating irradiation field. The magnetization is denoted by the thick arrow in (a) and the arrow originally aligning along the z axis in (b). It is noteworthy that for a spin-1/2 qubit the 2π rotation of the magnetization around the x axis does not recover the original state, but the 4π rotation does, denoting the spinor property for spin-1/2.

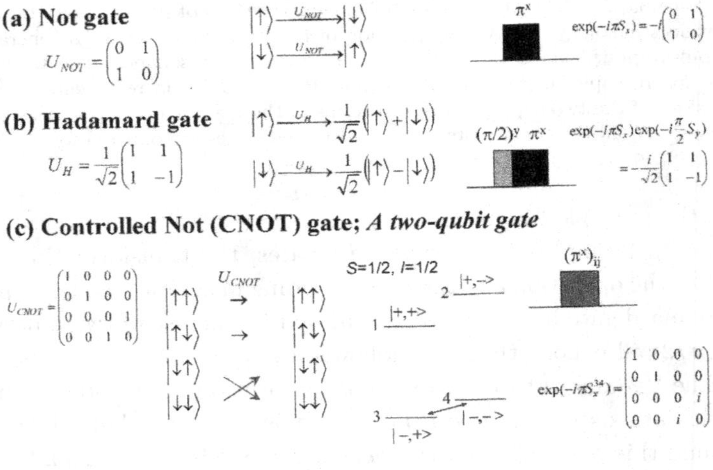

Fig. 13. Quantum logic gates, the transformation of the states after the operation and the corresponding electron magnetic resonance pulses. The NOT (a) and CNOT (c) gate are achieved by the operation of the corresponding single pulse. The Hadamard gate is achieved by the first $pi/2$ pulse applied along the y axis followed by the second π pulse along the x axis. The CNOT (c) gate is a two-qubit one, for which the corresponding energy level diagram with the state designation is given. In (c), an electron-nuclear hybrid system with one electron spin-qubit and one nuclear spin-1/2 is assumed. The state $|+-\rangle$ designates $|M_\mathrm{S} = -1/2, M_\mathrm{I} = +1/2\rangle$ and so on

axis. Again, all the quantum operations in Fig. 13 are achieved in the on-resonance rotating frames of the oscillating irradiation fields for electron magnetic resonance.

2.6. *An Electron-Spin Echo of Spin Packets Generated by Three-Pulse Sequence in Electron Magnetic Resonance*

An original Hahn echo in electron magnetic resonance is generated by the operation of two on-resonance pulses of the $\pi/2$-τ-π-τ sequence. There is an important variant of an electron spin echo of spin packets generated by a three-pulse sequence, as depicted in Fig. 14. The echo is termed a stimulated one. It should be noted that the characteristic time of the stimulated echo decay as a function of T is much longer than T_M since the phase information is stored in z'-axis, where T_M denotes a phase memory time of the spin system under study and T defined in Fig. 14. To understand the behavior of the stimulated ESE, we follow the description given by Keijzers *et al.* (see the caption for Fig. 15) and the primes for the x and y axes designate ones in the rotating frame of the MW irradiation frequency in Figs. 15 and 16.

In Fig. 15(c), it is shown that the x'-components of the spin packets give rise to an electron spin Hahn-echo at time t after the second $\pi/2$ pulse. The echo is called an eight-ball echo. The projections of the magnetization components, the spin packets i, j, k, l, and m onto the $x'z$-plane after the second $\pi/2$ pulse are depicted in Fig. 16(d). After a long waiting period T only the z'-components of the spin packets are left, as shown in Fig. 16(e). As described earlier, here the characteristic time of the stimulated echo decay as a function of T has to be much longer than T_M since the phase information is stored in z-axis, where T_M denotes a phase memory time of the spin system under study. The third $\pi/2$ along the x' axis brings these z-components back onto the $x'y'$-plane. After time τ they give rise to

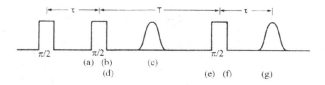

Fig. 14. Three-$\pi/2$-pulse sequence for generating a stimulated echo and timing chart. Behavior of the spin packets at the time denoted by (a)-(g) is depicted in Fig. 15

the stimulated echo. In Fig. 16(g), it is shown that the $x'y'$-magnetization components, i.e., the spin packets i, j, k, l, and m at this time are lying on a circle defined by $M_x^2 + (M_y + 1/2)^2 = 1/4$.

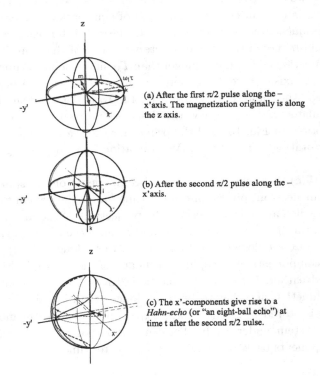

(a) After the first $\pi/2$ pulse along the $-x'$axis. The magnetization originally is along the z axis.

(b) After the second $\pi/2$ pulse along the $-x'$axis.

(c) The x'-components give rise to a Hahn-echo (or "an eight-ball echo") at time t after the second $\pi/2$ pulse.

Fig. 15. Behavior of the spin packets denoted by i, j, k, l and m during the operation of three-$\pi/2$-pulse sequence for generating a stimulated echo. The magnetization is originally aligned parallel to the z' axis with the static magnetic field along the z axis. The timing denoted by (a)-(g) is depicted in the timing chart given in Fig. 14. Note that the x'-components of the spin packets give rise to an electron spin Hahn-echo at time t after the second $\pi/2$ pulse. The original figures are adopted from "Pulsed EPR: A new field of applications," edited by C. P. Keijzers, E. J. Reijerse and J. Schmidt, Koninklijke Nederlandse Akademie van Wetenschappen, North Holland, Amsterdam (1989) and the figures are reconstructed and modified. The primes for the x and y axes designate ones in the rotating frame of the MW irradiation frequency.

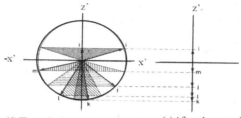

(d) The projection of the magnetization components i, j, k, l, and m onto the x'z'-plane after the second π/2 pulse.

(e) After a longer waiting time T than the phase memory time T_M only the z-components are left.

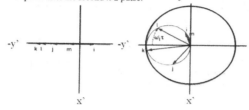

(f) The third π/2 pulse brings these components back into the x'y'-plane.

(g) After time τ they give rise to the stimulated echo. It is shown that the x'y'-magnetization components at this time are lying on a circle defined by $Mz^2 + (My+1/2)^2 = 1/4$.

Fig. 16. The projections of the magnetization components, i.e., the spin packets denoted by i, j, k, l, and m onto the $x'z$- or $x'y'$-planes at the timings of (d), (e), (f) and (g) depicted in Fig. 14. The original figures are adopted from "Pulsed EPR: A new field of applications," edited by C. P. Keijzers, E. J. Reijerse and J. Schmidt, Koninklijke Nederlandse Akademie van Wetenschappen, North Holland, Amsterdam (1989), and the figures are reconstructed and modified.

2.7. A Basis for Pulse-Based ENDOR Spin Technology: Two Types of Electron-Spin-Echo Detected ENDOR Spectroscopy

Now, we come to pulse-based ENDOR spin technology. There are typically two types of Electron-Spin-Echo (ESE) detected ENDOR techniques, i.e., Davis-type ENDOR and Mims-type ENDOR. Figure 17 illustrates their pulse sequences and timing charts, in which MW and RF pulse irradiations are for ESR and NMR transitions, respectively. ENDOR signals are detected (= monitored) by an ESE scheme. Here, for the understanding of the ESE-detected pulse ENDOR spectroscopy we follow the description given by Keijzers et al. (see the caption for Fig. 15).

In Davis-type ENDOR, as depicted in Fig. 17(a), P_{0S} and P_{2S} are MW π-pulses and P_{1S} is a MW π/2-pulse: P_{0S} with a strength $\omega_1 \ll A$ is applied

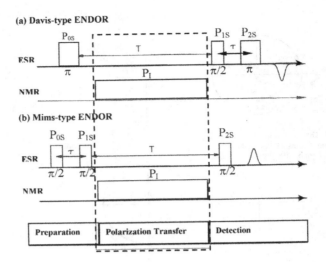

Fig. 17. Typical two types of Electron-Spin-Echo (ESE) detected ENDOR techniques: (a) Davis-type ENDOR and (b) Mims-type ENDOR. Their pulse sequences and timing charts are illustrated, in which MW and RF pulse irradiations are for ESR and NMR transitions, respectively. The period denoted by T is for the transfer of the polarization between particular electron-nuclear sublevels. It should be noted that any change of an ESE signal during the NMR transition driven by the RF pulse is monitored in the detection period. The original figures are adopted from "Pulsed EPR: A new field of applications," edited by C. P. Keijzers, E. J. Reijerse and J. Schmidt, Koninklijke Nederlandse Akademie van Wetenschappen, North Holland, Amsterdam (1989), and the figures are reconstructed and modified.

to the ESR transition 1-3 in Fig. 18 and P_I denotes the RF pulse applied during the waiting period T. A denotes a hyperfine splitting in the ESR spectrum. The first MW P_{0S} pulse interchanges the populations of the level 1 and 3, as is exemplified in Fig. 18. For simplicity, we neglect relaxation effects and the two-pulse ESE at a time T after the inversion pulse P_{0S} is inverted with respect to the one for an initial equilibrium state, as depicted in Fig. 17(a). Thus on NMR resonance, 1-2 or 3-4, where the RF pulse PI is applied during the waiting period T, the population change of the level 1 resp. 3 is detected as an increase in the "ESE amplitude." The maximum ENDOR effect is obtained when the nuclear sublevel population of the NMR transition 1-2 or 3-4 are inverted, i.e., for the nuclear flip angle $\theta = \omega_r t_p = \pi$ where ω_r is the effective nuclear Rabi-frequency (nutation frequency) and t_p is the RF pulse length. Figure 18 implies that if the ESE signal is observed via a pulse sequence which excites both component of the ESR spectrum, i.e., for the non-selective excitation the ENDOR effect will not appear. Also, when the preparation MW pulse P_{0S} is non-selective, i.e., $\omega_1 \gg A$, both components are inverted, resulting in no appreciable population difference of the hyperfine levels. An ESR excitation has to be hyperfine nuclear-level selective in Davis-type ENDOR. For QC/QIP ENDOR experiments, the hyperfine nuclear-level selective excitation by a MW pulse is essentially important for generating entanglements composed of electron-nuclear spin-qubits, as discussed in the later section.

Figure 17(b) shows the pulse sequence and timing chart for Mims-type ENDOR spectroscopy, where the MW three-pulse ESE scheme is utilized for the preparation and detection periods. Mims-type ENDOR spectroscopy is particularly useful for studying molecular information on nuclei with small hyperfine interactions and small nuclear Zeeman splittings. In the frequency domain, the two-pulses in the preparation period produce a periodic pattern $M_{zi} = M_{0,i} \cos(\Delta\omega_i \tau)$. Generally, the pattern M_{zi} after the second pulse P_{1S} with the length of time $t_{P_{1S}}$ is described by the equation $M_{zi}(t_{P_{0S}} + \tau + \tau_{P_{1S}}) = -M_{0,i} \sin\theta_{P_{0S}} \sin\theta_{P_{1S}} \cos(\Delta\omega_i \tau)$ in the rotating frame, where $\omega_1 t_{P_{0S}} = \theta_{P_{0S}}$ and $\omega_1 t_{P_{1S}} = \theta_{P_{1S}}$ with $\gamma B_1 = \omega_1$. The M_z-component of

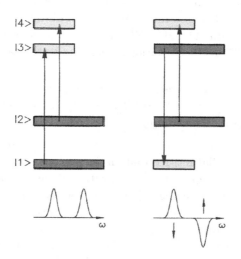

Fig. 18. Population inversion driven by the preparation π-pulse of MW for the transition between the level 1 and 3 in Davis-type ENDOR spectroscopy. The inversion enhances an ESE amplitude during the NMR transition between the level 1 and 2 or the level 3 and 4, as depicted on the right. The ESR spectrum is composed of two lines with a hyperfine splitting A.

a spin packet i depends on how its precession frequency, as defined by the frequency offset $\Delta\omega_i$ in the rotating frame fits in the waiting time τ. For example, some spin packets at resonance are oriented along the negative z axis whereas other spin packets with a frequency offset of $\Delta\omega_i = \pi/\tau$ along the positive z axis.

In the standard stimulated echo, the whole pattern refocuses at the time τ after the third pulse P$_{2S}$. When the RF pulse P$_I$, as applied between the second P$_{1S}$ and third P$_{2S}$ pulses, is resonant with a transition matching to a hyperfine interaction A, the polarization transfer shifts the whole M_z pattern up for some components and down for others in frequency by an amount of A. Thus, the pattern of M_z-components becomes less clear (blurred) and as a result the ESE intensity is reduced: In other word, this resonance shift causes a general smoothing of the pattern's feature and a reduction in the stimulated ESE intensity. Only if $A = n/\tau$ ($n = 0, 1, 2, \ldots$) pattern is retained, the effect disappears and the stimulated echo amplitude is unaffected. In other words, the echo amplitude will be modulated by a factor $\cos(2\pi A\tau)$. The only drawback of the Mims-type ENDOR technique is the occurrence of the blind spots for $A = n/\tau$. In order to avoid this drawback, ENDOR spectra can be detected at different τ values. This procedure provides us with a two-dimensional (2D) experiment, in which we observe ENDOR spectra as a function of τ and obtain a Fourier transformation of the τ domain. The 2D-ENDOR spectrum yields cross peaks that correlate different ENDOR transitions belonging to the same nucleus.

2.8. Generation of A Pseudo Pure State for Electron-Nuclear Spin-Qubit Systems by Pulse-Based ENDOR Spin Technology

Pulse-based ENDOR spectroscopy, carried out in the rotating frame of applied coherent (oscillating) irradiation fields, consists of main three operation periods, i.e., preparation, polarization transfer and detection, as illustrated at the bottom in Fig. 17. During the second period takes place mixing/evolution relevant to the spin states involved on resonance. The three periods correspond to those in QC/QIP processes in time in pulse-based electron magnetic resonance experiments. The first, second and third period correspond to initialization, manipulation/computing and readout (= detection), respectively, in pulse-EMR based QC/QIP experiments. The initialization prepares an either pure or pseudo-pure state necessary for executing any quantum computation. The manipulation/computing is for selective/non-selective excitation among the allowed or forbidden transi-

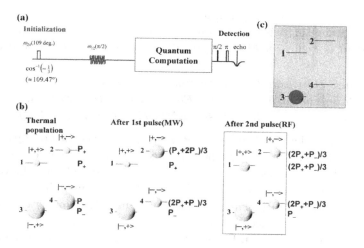

Fig. 19. Initialization of an electron-nuclear spin-qubit system by pulse ENDOR spin technology. (a) An ENDOR-based pulse sequence for preparing the initialization of an electron-nuclear system with one electron and one nuclear spin-1/2. (b) The change in the population among the state on each pulse operation of resonance. (c) Here, the level 3 as a pseudo-pure state of the electron-nuclear hybrid system is depicted.

tions between the spin states, or time evolution of the states involved for particular purposes. In the second period, the phases of the states can be manipulated in phase-controlled interferogram experiments for QC/QIP. The readout formally includes variants of electron-spin based detection such as Hahn ESE, three-pulse stimulated echo, refocused echo and FID.

Figure 19 shows an ENDOR-based pulse sequence for preparing the initialization of an electron-nuclear system with one electron and one nuclear spin-1/2 (a) and the population change among the states involved on each pulse operation at resonance (b). The first MW pulse for selective excitation is of a particular role that generates the population inversion between the level 3 and 4, assuming the equal Boltzmann distribution between the nuclear spin states in the same electron-spin M_S sub-level. The second RF $\pi/2$-pulse equates the population between the nuclear sub-levels, 1 and 2 of the electron-spin $M_S = +1/2$ level. The two pulses of the MW and RF irradiations on resonance redistribute $2P^+$ and PP^- equally among the levels, 1,2 and 4, making only the level 3 populated more by $2(P^- - P^+)/3$. Here, a pseudo-pure state of the electron-nuclear hybrid system with the four spin states is generated, as the level 3 generated after the second RF $\pi/2$ pulse in Fig. 19(b). It is shown in Fig. 19(c) that the level 3 is obtained as a pseudo-pure state of the electron-nuclear hybrid system. In Fig. 19(a), any quantum computing with operations is followed in the second period, and the readout is exemplified in a Hahn ESE scheme as the detection period.

2.9. *Generation and Identification of Quantum Entanglement between An Electron and One Nuclear Spin-Qubit by Pulse-Based ENDOR Spin Technology*

Quantum entanglement plays the most important role in QC/QIP. The generation of quantum entanglement between an electron and one nuclear spin-1/2 qubit in a molecular entity has for the first time been achieved by Mehring's group,[7] and the establishment of the entanglement has been identified by invoking TPPI (Time-Proportional-Phase-Increment) technique, which enables us to detect quantum phases belonging to particular spin states in the hybrid system. Following the line shown by Mehring's group, we illustrate how to generate the entangled state in an electron-nuclear spin-qubit system by the use of the pulse ENDOR technique, as given in Fig. 20.

In Fig. 20(a), after the initialization of an electron-nuclear spin-qubit system, e.g., the level 3 as an initialized state (pseudo-pure state), a sequence of RF2 and MW on-resonance pulses required for generating entangled states are schematically given and resulting state-transformation by the pulses. The role of each pulse is indicated. For entangling the electron-nuclear spin states, we inevitably utilize MW pulses forcing electron spin sublevels involved in the entanglement process. This process technically gives rise to some difficulties in QC/QIP experiments. The levels 2 and 3 are entangled in the present pulse scheme. Figure 20(b) depicts the four spin states involved and their population changes by the pulses. The level 1 apparently is not involved during the processes. Any relaxations are ignored here. In Fig. 20(c), the pulse sequence for establishing the entanglement between an electron and a nuclear spin-1/2 is schematically given in time. The first period is for the preparation of the pseudo-pure state, as discussed above, and the RF2 $\pi/2$-pulse and π-MW one generate a pair of the entangled states. This second period corresponds to manipulation of the spin-qubits involved in any quantum operation. The third period depicted as vague figures in Fig. 20(c) is for the readout of the manipulation, which is discussed in detail in the later section. The third part is a highlighted one to illustrate QC/QIP experiments in terms of the phase detection of the entangled states. In Fig. 20(d), are depicted quantum logic gates for generating the entanglement between one spin-qubit and another spin-qubit.

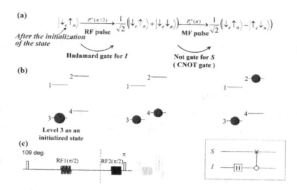

Fig. 20. Generation of entanglement between an electron and a nuclear spin-1/2 qubit by pulse ENDOR technique. (a) After the initialization of an electron-nuclear spin-qubit system, e.g., the level 3 as an initialized state (pseudo-pure state), a sequence of RF2 and MW on-resonance pulses required for generating entangled states and resulting state-transformation by the pulses. The role of each pulse is indicated as gate operations. The levels 2 and 3 are entangled in the present pulse scheme. (b) The four spin states involved and their population changes by the pulses. The level 1 apparently is not involved during the processes. Any relaxations are not considered. (c) The pulse sequence for establishing the entanglement between an electron and a nuclear spin-1/2 is given. The first period is for the preparation of the pseudo-pure state and the RF2 $\pi/2$-pulse and p-MW one generate a pair of the entangled states. The final period as depicted vaguely denotes the readout. (d) Quantum logic gates for generating the entanglement between one spin-qubit and another spin-qubit. Here, S and I stand for an electron and nuclear spin-qubit, respectively.

Here, S and I stand for an electron and nuclear spin-qubit.

In order to identify the establishment of the entanglement between an electron and one nuclear spin-1/2 qubit, TPPI technique has been introduced by Mehring et al. to pulse ENDOR technology for QC/QIP experiments, for the first time. The method introduced really convinces us of the occurrence of the entanglement between the matter spin-qubits. Basically, TPPI is general technique to enhance spectroscopic information by increasing the number of spectral dimensions. This can be achieved by introducing more than one numbers of phases belonging to pulses such as MW and/or RF irradiations relevant to magnetic transitions. The phase increment can be described in terms of angular frequency by time increments, in which the frequency is composed of difference between two relevant frequencies. By invoking this technique, spectroscopic information is multi-dimensional. For example, TPPI has been applied in order to allow on-resonance excitation and the spreading of n-quantum-transition spectra by $n\Delta\omega$. The phase of the pulse is shifted by $\delta\phi = \Delta\omega\delta t_1$, i.e., $\delta\phi = 2\pi\Delta\nu\delta t_1$, each time the evolution time is incremented by δt_1, resolving multiple-quantum coherence in pulse-based ENDOR spectroscopy in solution by Hoefer et al., for the first time 20.

Figure 21 shows experimental approaches to the evaluation of the quantum entanglement between the electro-nuclear hybrid system. One is a variant of the technique for rotating a MW phase in the case that the real MW phase rotation is unable, in which MW pulse channels can be utilized as a substitute "quasi-rotation" technology for the MW part. This variant has been used in our first previous experiments for identifying the occurrence of the entanglement between an electron and a nuclear spin-1/2 qubit.[9,10] For the entangled states (Bell states) generated by an electron and one nuclear spin-1/2 qubit, each Bell state is characterized by its own phase originally composed of both the quantum numbers, M_S and M_I. Thus, the introduction of the corresponding two phases, ϕ_{MW} and ϕ_{RF} for the MW π-pulse and the RF $\pi/2$-pulse, respectively, enables us to discriminate the Bell states. The desired phase shift can be controlled by time increment described above. Any possible quantum phase-interference between an electron and the nuclear spin-qubit in time can be experimentally acquired via interferograms transformed into the frequency domain. Examples for interferograms will be given with particular QC/QIP experiments in the later section. An ESE intensity is described by an equation, as given in Fig. 21, depending on the difference between the two phases in the case of the occurrence of the entanglement. Otherwise, the ESE signal is constant in the time increment.

2.10. *Inter-Conversion of Entangled States by Pulse ENDOR Technique*

Once the entangled state between an electron and a nuclear spin-1/2 qubit is generated by a particular pulse protocol, the pulse ENDOR technique enables us to interchange the entangled states by manipulating the nuclear spin sublevels by RF pulses. The procedures for the inter-conversion between the Bell states are schematically depicted in Figure 22, noting the difference in the phase belonging to each Bell state. Inter-conversion between the entangled states (Bell states) can be achieved by one of the unitary transformations denoted by X, Y and Z, each of which corresponds to the curved thick arrow, as depicted in Fig. 22(a). The Z operations designate 2π rotation for a half-integral spin-qubit. Thus, the inter-conversion between the electron-nuclear hybrid spin-qubit system can be implemented

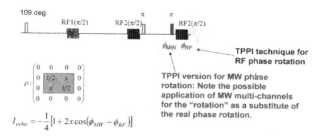

Fig. 21. Enabling pulse-ENDOR spin technology for identifying the establishment of the entanglement between an electron and nuclear spin-1/2 qubit. A pulse sequence enabling us to manipulate quantum phases of the matter spin-qubits is depicted schematically. There are two approaches to the phase rotations of the MW and/or RF pulse. One is a variant of the real MW phase rotation termed "MW-phase quasi-rotation," as a substitute rotation technology in the case that the real rotation of the MW pulse phase is technically unable. In the quasi-rotation, different MW channels for the MW different phases are utilized, achieving quasi-rotation of the MW pulse. I_{echo} is derived in Appendix 2 given at the end of this chapter.

by pulse ENDOR spin technology. In Fig. 22(b), the procedures depicted in Fig. 22(a) are represented by a quantum logic gate, in which a unitary operation U_i is applied after generating any entangled state. In the later section, real hybrid qubit systems in molecular frames will be treated from the experimental side. Our pulse ENDOR spectrometer devoted to QC/QIP experiments operating at Q-band has been designed for the MW phase so as to be rotated at any arbitrary orientations in three dimensions.

2.11. *TPPI Detection of the Entanglement between Electron-Nuclear Hybrid Spin-Qubits by Pulse ENDOR Technique*

Figure 23 summarizes TPPI detection of the entanglement between an electron and a nuclear spin-1/2 qubits by pulse ENDOR technique. The pulse sequence for the TPPI procedure is depicted at the top along with the role of the three periods. The TPPI procedure is carried out in the second period of quantum operation/manipulation of the qubits, where the phase of the MW π-pulse and that of the RF2 $\pi/2$-pulse are controlled in the time order in pulse ENDOR technology. In order to manipulate genuine electron spin-qubits in contrast to the hybrid qubit system, we have to introduce the corresponding phase control technique, say, for two electron qubits with non-equivalent g-factors in molecular frames. Current MW spin technology requires two MW sources with their relative phases locked electronically.

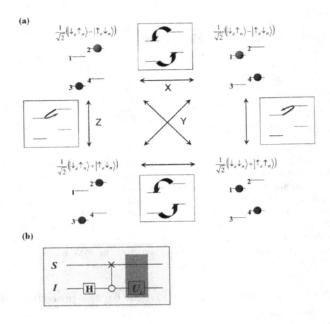

Fig. 22. Inter-conversion between the entangled states (Bell states) in the electron-nuclear hybrid spin system by unitary transformations denoted by X, Y and Z. (a) The unitary transformation denoted by X, Y or Z corresponds to the curved arrow. The Z operations designate 2π rotation for a half-integral spin-qubit. The inter-conversion between the Bell states can be achieved by pulse ENDOR spin technology. (b) The procedures depicted in (a) are represented by a quantum logic gate, in which a unitary operation U_i is applied after generating the entangled state.

Such MW phase locking technology has never been implemented so far, nor necessary for ordinary pulse-based EMR/ESR spectroscopy.

The TPPI frequency denoted by ν_{TPPI} corresponds to the phase information on which electron-nuclear sublevel is entangled with other sublevel, as exemplified in the table in Fig. 23. It should be noted that the two pairs among the four Bell states give the same ν_{TPPI} as seen in Table in Fig. 23. The transition frequencies, ω_{12} and ω_{34}, between the level 1 and 2 and the level 3 and 4, respectively, correspond to ENDOR ones.

3. Implementation of Molecular-Spin Based QC/QIP by the Use of Pulse ENDOR Spin Technology

The section 3 is an independent one that deals with molecular spin-qubit based QC/QIP by the use of pulse ENDOR spin technology. A basis for the pulse ENDOR technology has been given in the preceding sections. In this section, we are involved in a bit extended application of the ENDOR spin technology. Further extended applications by the use of multi-qubit systems in molecular frames such as quantum teleportation experiments, data storage processing or realistic applications of quantum algorithms have been underway, noting that prior to any such QC/QIP experiments syntheses associated with elaborate molecular designs are required. We emphasize that the field of QC/QIP becomes interdisciplinary over chemistry and modern materials science as well, and a new area in chemistry is emerging. We start answering a fundamental query "Why is molecular spin-qubit based QC/QIP by using pulse spin technology emerging?"

3.1. *Why Is Molecular Spin-Qubit Based QC/QIP by Using Pulse ENDOR Spin Technology Emerging?*

A very well-known physical system for the realization of QC/QIP is based on modern pulsed NMR spectroscopy.[14,15] In this scheme, the quantum information is stored in nuclear spins, as qubits, of particular molecules. Liquid-state NMR spectroscopy has been widely used for implementations of even considerably complicated quantum non-local algorithms and the experimental outcomes apparently represent the capability of NMR as the

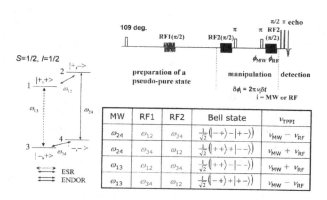

Fig. 23. TPPI detection of the entanglement between an electron and one nuclear spin-1/2 qubits by pulse ENDOR technique. The pulse sequence for the TPPI procedure is depicted at the top along with the role of the three periods. The transitions are denoted by ω_{ij}. The longer solid and dotted arrows correspond to the ESR transitions. It should be noted that the two pairs among the four Bell states give the same ν_{TPPI} as seen in the table. The relationships between the Bell state and the corresponding TPPI frequency are derived in Appendix B given in the end of this chapter.

physical system for QC/QIP.[21–24] Nevertheless, liquid-state NMR at ambient temperature suffers from its intrinsic low spin polarization, making the initial state in a highly mixed state, whereas the accessibility to a pure initial state is one of the major requirements that any physical qubit system fulfills to be a valid candidate for demonstrating a quantum computer[18] as discussed above in part. In order to overcome this drawback, in a conventional approach, pseudo-pure states have been introduced[25,26] and widely used for NMR-based QC/QIP experiments. Electron-based molecular spin-qubits in ensemble exploit a similar initialization condition above a certain critical temperature depending on the microwave resonance energy for selective excitation of the qubits.

3.1.1. *Pseudo-pure states and quantum entanglements*

A pseudo-pure state is composed of two terms, the one belonging to the highly mixed state of the unitary part and the pure state term with a coefficient, which is related to experimental conditions. Referred to the current technology of the NMR spectroscopy at ambient temperature the corresponding value is as small as the magnitude of order of typically 10^{-5}. Since all the observables in NMR are traceless, the mixed state term is hardly detectable through the NMR procedure. Weak signal intensities, however, through the other term still give some information through quantum processing and then the whole pseudo-pure state looks as if an initial pure state has been used for quantum information processing.

Quantum entanglement is known as a prerequisite for any quantum non-local algorithm.[27,28] Thermal mixed states used for NMR QC/QIP have been proved to be separable,[29,30] therefore not formidable for the quantum non-local processing. Pseudo-pure states that apparently are capable of resolving the problem of the admixture of the initial state are not useful any more referred to the separability of the state, since the pseudo-pure state can also be represented as a convex combination of thermal states. The entanglement is a convex function. Then, the pseudo-pure states are at least as separable as the thermal states. In this context, liquid-state NMR with low spin polarization cannot afford entanglement-based advantages of QC/QIP. In order to solve this drawback, nuclear spin polarization should be increased to the extent expected by the theoretical criteria for the existence of the relevant entanglement. The enhancement of the nuclear spin polarization has been one of the recent focuses in the field of quantum information science.[31,32]

Despite the drawbacks of QC-NMR as discussed, it is no doubt that NMR spectroscopy can be such a superb physical system as represents very important advantages for the realization of QCs in some crucial aspects. QC-NMR has brought reality for QC/QIP and driven the rapid progress in this field. The existence of nuclear spin-qubits with long decoherence time is proper physical realization of matter spin-qubits since the spin manipulation can easily be performed by introducing a series of many sophisticated RF pulses with relevant resonance frequencies. In this context, to retain the advantage of the long decoherence time is crucial for looking for novel matter spin-qubit systems in which easy-to-access spin manipulations are performed. In our latest study, pulse-based Electron Nuclear DOuble Resonance (ENDOR),[33] as well as pulsed ELectron electron Double Resonance (ELDOR) for electron spin-qubits, has been examined as a novel candidate to approach QCs by invoking ensemble molecular entities with both electron and nuclear spins in the solid state. Since the physical system under study involves nuclear spin-qubits, pulsed ENDOR spin-manipulation technology retains the main advantage inherent in NMR-based QC systems. In addition, it seems easier to circumvent the drawbacks of the NMR systems for the realization of QCs, because additional molecular electron spin-qubits originating in open-shell molecular entities are also incorporated in the matter spin-qubit systems.

Nevertheless, an ENDOR based QC event as electron-nuclear multiple magnetic resonance spectroscopy is much heavier and formidable experimental task compared with QC-NMR, as some crucial parts discussed above, but QC-ENDOR rewards back essential aspects and results by adding up the advantages of NMR and ESR spin technology. It is worthwhile to notice that an elaborate total design of the QC-ENDOR experimental setup should be associated with molecular design for open-shell entities, as frequently emphasized. This is because the number of client nuclear spin-qubits, the kinds of the nuclear spin-qubits available for RF irradiation, i.e., their gyromagnetic ratios, the magnitudes of their hyperfine coupling/fine-structure tensors and tuning relaxation times of molecular open-shell entities are crucially important to study foe the QC-ENDOR setup. These aspects underlie the current stage of pulsed MW technology for QC/QIP.

3.1.2. *Molecular spin-qubit ENDOR based quantum computers*

For any physical system as a candidate for the realization of a quantum computer, there are some fundamental criteria, known as DiVincenzo's five criteria that should be met.[18] The molecule-based ENDOR system is also expected to meet these criteria in order to be a realistic physical system for QC/QIP; see Table 2 for a list of DiVincenzo's five criteria and the corresponding properties of the ENDOR system. Molecular spin-qubit QC ENDOR is developing as novel electron spin manipulation technology. Particularly, the high frequency versions are preferable, but need further technological developments for QC/QIP. As matter spin-qubits, synthetic efforts based on new molecular designing are also required, as discussed in the later section. The table cannot be complete and only outlines them.

In the ENDOR-based QC/QIP, molecular electron spins in addition to nuclear spins have been introduced as quantum bits (qubits) that play as a bus spin while the nuclear spins as client qubits. In thermal equilibrium, the populations in the ground states of the molecular electron spins are more than 10^3 times larger than the corresponding excited states with different M_S-manifolds in the presence of conventional static magnetic field or the ones with zero-field splittings, compared with QC-NMR. Therefore, by using ground-state ENDOR systems, the required experimental conditions for preparing the pure initial state for QC/QIP seems substantially easier to achieve by the current electron magnetic resonance technology. Alternatively, for the exact and complete preparation of pure initial states, the possible manipulation of single-molecule based systems is formidable, for which electron magnetic resonance or Larmor precession detection has been considered and even the experimental equipments by the use of electric detection schemes seem to be accessible in the near future.

Any physical/molecular system for QC/QIP should be chemically stable during computational processing. We have prepared robust organic open-shell entities against long and high-power irradiations of both radiofrequency and microwave at ambient temperature, as some of those exemplified in our work. In addition, because of the hydrocarbon-based organic entities the corresponding decoherence time of the qubits is expected to be long enough compared with the computational or gate operational times. As a result of the existence of the nuclear spin-qubits in molecular open-shell entities, there is a wide possibility to work with samples having long decoherence times due to the nuclear spin-qubits. Proper samples with the long decoherence time and their synthetic procedure should be considered, in advance. In this context, open-shell metal cationic complexes with available nuclear client qubits have intrinsic advantages in terms of synthetic strategy whereas relatively strong spin-orbit couplings hamper their decoherence time in the solid state.[34] Long enough decoherence times for

Table 2. QC-ENDOR systems regarding the satisfactions of the DiVincenzo's five criteria.

	DiVincenzo's Criteria[18]	QC-ENDOR
Qubits	Identifiable; well characterized and scalable qubits are required.	Molecule-based electrons & nuclear spins in molecular open-shell entities, in which hyperfine couplings play an essentially important role for selective excitations of both the electron spin- and nuclear spin-qubits. Molecular designs, syntheses and identifications of spin properties are required. The scalability of client nuclear spin-qubits in electron spin bus QCs is limited from the synthetic viewpoint. Proto-types of 1D periodic electron spin-qubit systems are designed and synthesized.[34]
Initialization	Possibility to be initialized to simple and fiduciary states.	Pseudo-pure states can be used in this context, whereas in order to avoid them the high spin polarizations of electron spins can be coherently transferred to nuclear spins by applying relevant pulse sequences followed by proper waiting times.
Decoherence time	Long relevant decoherence times, much longer times than gate operations are necessary.	Long decoherence times of nuclear spins and electron spin in organic radical qubits in the solid state have been available for the demonstration of quantum operations between the bi- or tri-partite qubits. Proper molecular entities with long decoherence times for multi-qubit operations are not out of reach, for which stable isotope-labeled open-shell molecules have been designed and synthesized.[34] QC-ENDOR experiments in solution also are not out of reach.
Quantum operation	A universal set of quantum gates is required.	Quantum gates between a single electron and a single nuclear spin have been demonstrated experimentally. Multi-qubit operations in terms of ENDOR spin Hamiltonians are underway. Particularly, a protocol for tri-partite QC operations has been implemented.[40]
Measurement	The ability of measurements on quantum qubits to obtain the result of the computation is required.	The current measurement scheme is ensemble-based, in which an individual client nuclear spin-qubit is readout via the electron bus spin-qubit. A field gradient approach for the readout is proposed. On the other hand, single electron spin detections may be available by the use of STM-based electron magnetic resonance detection.

samples involving two qubits have been measured during the course of the work. For particular molecular entities the feasibility of the ENDOR based QC/QIP has been examined from both experimental and theoretical sides, as in part reported in the following sections.

In QC-ENDOR, manipulation and computation processing on the qubits as well as the readout processing can be realized by introducing both MW and RF pulses in an approach different from genuine NMR-based QC experiments, i.e., ENDOR (NMR) resonant pulses on the client nuclear spin-qubits and/or by the MW frequency resonant pulses on the electron bus spin-qubits, as simply exemplified in Fig. 24 where the state notation is modified for the later section.

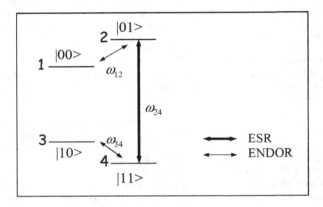

Fig. 24. Energy levels and corresponding ESR and ENDOR resonance transitions in the presence of a static magnetic field. Note that the nuclear sublevels correspond to the case for the system with a positive hyperfine coupling. The splittings of the nuclear sublevels are exaggerated for clarity. The definition for the level is as follows; $|00\rangle(=|++\rangle)$ and $|10\rangle(=|-+\rangle)$ denote $|M_S = +1/2, M_I = +1/2\rangle$ and $|M_S = -1/2, M_I = +1/2\rangle$, respectively. Similarly, $|01\rangle(=|+-\rangle)$ and $|11\rangle(=|--\rangle)$ denote $|M_S = +1/2, M_I = -1/2\rangle$ and $|M_S = -1/2, M_I = -1/2\rangle$, respectively. Also, see the text for the use of the notation of the sublevels. Both the notations for the electron-nuclear sublevels are used and for later section the Arabic numerals are used for convenience.

It is known that the realization of particular quantum gates being known as universal gates can be enough for the implementation of any other quantum gates.[2] One-qubit gates in addition to a non-trivial two-qubit gate, e.g. a Controlled-Not (CNOT) gate, give a universal set of quantum gates. For the implementation of the quantum gates in terms of resonance concepts, it is possible to perform this task by introducing the relevant pulses. From the experimental side, for QC-ENDOR systems particular quantum gates have been demonstrated by using the two-qubit system composed of one electron spin-qubit and one nuclear spin-qubit, as it is described in the following section. Multi-qubit gates for molecular systems involving a larger number of qubits should be considered for relevant physical systems in terms of the particular form of the spin Hamiltonian for the corresponding sample.[7-13] Tri-partite spin-qubit experiments have for the first time been demonstrated for a separable set of bi-partite entangled states in our work, as described in the following section. There are several typical quantum operations based on multi-qubit gates, depending upon spin Hamiltonians for realistic QC-ENDOR experiments: A QC-ENDOR pulse protocol for quantum teleportation differs from the counterpart of QC-NMR.[40]

In the ENDOR based QC, the readout or the measurement has been implemented by introducing RF and MW pulses on nuclear spin- and electron spin-qubits, respectively. Nevertheless, as is in a similar approach to QC-NMR, the measurement scheme is ensemble-based and rather different from exact measurements that are required for QC. In terms of QC-ENDOR, to practically solve this issue is still an open problem. This is due to the fact that molecular electron spin-qubits play a special role in QC-ENDOR, in which the role is such as bus or the selective excitation/readout processes are carried out by electron-spin echo detection, instead of FID detection in QC-NMR.

It has been proved that the entanglement for pure states is necessary requirement for quantum exponential speed-up over the classical counterpart. For mixed states, the statement has not completely been proved yet, but highly believed correct. Thus, one very important issue that should be examined for any physical system for QC/QIP is the entanglement status. Realization of the entanglement between an electron spin- and a nuclear spin-qubit has for the first time been reported in an ENDOR experiment by using the pseudo-pure states.[7,8] So far, we have been mainly engaged in two experimental tasks for the realization of QCs by using molecule-based ENDOR. One has been an attempt for the preparation of the experimental requirements for demonstrating the true entanglement between a molecular electron spin- and a nuclear spin-qubit by the use of a simple organic radical in the single crystal and avoiding the use of pseudo-pure states. High spin polarizations on both the electron and nuclear spins are essentially required to achieve the true entanglement between the two spin-qubits in the molecular frame. The required entanglement for the ENDOR system composed of only two electron and nuclear spins can be established below 0.8 K in the presence of the static magnetic field for the microwave transition frequency of 95 GHz, as obtained by the negativity criterion.[35,36] Whereas, if the pulses can be applied for the transfer of the high spin polarization, the

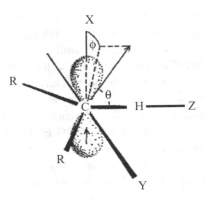

Fig. 25. The principal axes of the magnetic tensors in malonyl radical. The principal axes of the g-tensor and the hyperfine tensor of the α-proton are collinear. The principal Z axis is along the C-H bond and the X axis parallel to the symmetry axis of the pπ-orbital at the carbon site.

required temperature under the same magnetic field is nearly 5.1 K, which is well in reach with the current technology with a W-band (95 GHz; see Table 1) ENDOR spectrometer operating at liquid Helium temperature.

Besides preparing the instrumentation requirements above, the efforts can be made on some other aspects of the research. One thing is materials challenge to design and synthesize stable or scalable open-shell molecular entities suitable for QC/QIP ENDOR experiments. Novel molecular open-shell systems with stable isotope labels suitable for our purposes have been designed and synthesized.[40] Also, the critical temperature can be tuned by invoking stable high-spin ($S > 1/2$) molecular entities. Some of the crucial aspects for scalable molecular electron spin-qubits are given elsewhere, emphasizing a road-map to the synthetic strategy for them.[34] We have exploited a supramolecular chemistry approach to embed, in a one-dimensional manner, metal cationic open-shell entities with non-equivalent g-tensors in the solid state. This has been an attempt to synthesize the 1D periodic electron spin-qubit array as a scalable matter spin-qubit, for which Lloyd's theoretical proposal is described in Sec. 5. We note that such g-tensor engineering enables us to achieve selective microwave excitations for the preparation of initial states, manipulation and readout procedures.

Also, the efforts have been made on the investigation of the credibility of the pulsed ENDOR based QC/QIP to develop necessary quantum gates and entangling unitary operations. It is clear that for dealing with these issues by fulfilling the experimental conditions for entanglement and pure

HOOC−CH₂−COOH → HOOC−CH−COOH

Scheme 1. Generation of malonyl radical from malonyl acid in the single crystal by X- or γ-ray irradiation.

Table 3. The spin Hamiltonian parameters of malonyl radical. The corresponding principal axes are depicted in Fig. 25.

S	Principal g-values			$^{\alpha}A$/MHz		
	xx	yy	zz	xx	yy	zz
1/2	2.0026	2.0035	2.0033	−61	−91	−29

states there would be no need to have an additional experimental processing to establish the pseudo-pure state.

In order to check both the credibility of the ENDOR physical system for QC/QIP and the feasibility of the molecule-based QC-ENDOR with the current technology, the implementation of super dense coding (SDC)[6] has been revisited in our experiments. The pulsed ENDOR technique has been applied to a molecular electron- and nuclear-spin system, i.e., malonyl radical in the single crystal of malonic acid[37] to implement the SDC in the electron-nuclear spin-qubit system. Additionally, we have shown that non-selective microwave excitation in QC-ENDOR experiments gives a set of separable states composed of the entangled states. The attempt has been QC experiments on tri-partite spin-qubits. This can be identified in terms of the phases of the entangled spin states involved. In this experiment, we have chosen a multi-nuclear client organic stable radical such as diphenyl nitroxide-h_{10} magnetically diluted in diamagnetic host lattices.

3.1.3. *Preparation of a molecular entity for QC-ENDOR: The simplest case*

Malonyl radicals incorporated in the single crystal of malonic acid were generated by X-ray irradiation at ambient temperature, as shown in Scheme 1. Spin Hamiltonian parameters of malonyl radical under study, as summarized in Table 3, have been reported by McConnell and coworkers.[37] The principal axes, X, Y and Z of the g-tensor and the hyperfine tensor of the α-proton are collinear, as depicted in Fig. 25. A typical echo-detected field-swept Q-band ESR spectrum of malonyl radical observed at 50 K is

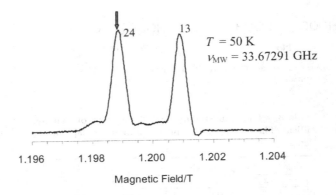

Fig. 26. A typical ESE-detected field-swept Q-band ESR spectrum of malonyl radical in the single crystal of malonic acid observed at 50 K. The numbers at the peaks denote the ESR allowed transition assignments (see the energy diagram in Fig. 24).

shown in Fig. 26. Pulsed ENDOR measurements for malonyl radical were performed with a Davies-type pulse sequence[38] and an observed Davies-ENDOR spectrum is shown in Fig. 27.

Existence of one electron spin on the carbon atom and one α-proton with a large hyperfine coupling gives two spins as the required two qubits for super dense coding (SDC). The large hyperfine interaction is primarily required to enable us to make a selective microwave excitation within the framework of the current spin technology for QC/QIP experiments. Energy levels and the corresponding resonance frequencies of the RF and MW frequencies are shown in Fig. 24, noting that the order of the nuclear sublevels should be reverse for malonyl radical with the negative hyperfine coupling (see Table 3). Detection of the entanglement by using the pseudo-pure state of this system has been already reported[7] and in a different approach we demonstrate the implementation of SDC by the use of the pseudo-entangled states and pulse ENDOR spin technology.

3.1.4. *Implementation of super dense coding by pulsed QC-ENDOR and a direct detection of the spinor of the spin-1/2 proton*

Super dense coding (SDC) introduced by Bennett and Wiesnercite[6] is a non-local quantum algorithm in which two classical bits of information are transformed from Alice to Bob by sending only a single qubit. The scheme is based on the fact that the entangled initial states have been shared between the two involved parties, i.e., Alice and Bob. The efficiency of the scheme is two times compared with the classical counterpart, since maximum one bit of information can be transferred through a single use of an information channel.

SDC can be simply explained as follows, referring to Fig. 28: Two qubits which are originally entangled each other have been shared between the two involved parties, say Alice and Bob. Alice encodes the qubit by applying a unitary transformation out of the four choices of {I, X, Y, Z} (refer Fig. 22 and Table 4 for the definitions for I, X, Y, and Z) and then sends the encoded qubit to Bob, who has been also initially given a qubit entangled to the Alices's one. After receiving the encoded qubit from Alice, Bob applies a measurement in the Bell basis (see Table 4) on both of the qubits. The result of the measurement makes it clear for Bob what the Alice's choice has been in the encoding part. Therefore, he extracts the information on the Alices's choice, which means that a two-bit message has been transferred by sending only a single qubit.

Concepts for the ENDOR based experimental setup for two-qubit SDC are depicted in Fig. 29, in which S and I denote the electron-spin part and nuclear-spin one. Figure 29 shows one of quantum circuits implementing SDC. The quantum circuit for SDC consists of Hadamard, denoted by H in Fig. 29, and controlled NOT (CNOT) gates. U_i stands for one of the unitary transformation for encoding to be carried by Alice. In Fig. 29,

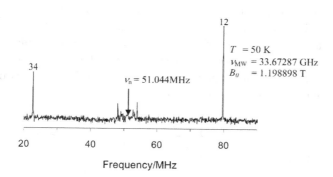

Fig. 27. A Davis-type pulsed Q-band ENDOR spectrum from malonyl radical with the magnetic field set as indicated by the arrow in Fig. 26. A selective MW pulse for the preparation and non-selective MW pulses for the detection were applied. The numbers indicate the ENDOR transitions (see Fig. 24) due to the α-proton of malonyl radical with a negative hyperfine coupling. The only strong peaks appearing in the ENDOR spectrum are indicative of the crystal alignment properly made in the crystallographic axis system.

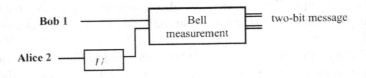

Fig. 28. Scheme for super dense coding (SDC). U denotes a unitary transformation.

Table 4. The unitary operations and corresponding pulse sequences for encoding in the SDC experiments.

U	Initial state	Necessary operation	Required pulses	Encoded State
I	$\frac{|++\rangle+|--\rangle}{\sqrt{2}}$			$\frac{|++\rangle+|--\rangle}{\sqrt{2}}$
X	$\frac{|++\rangle+|--\rangle}{\sqrt{2}}$	$\exp(-i\pi I_x)$	$P_x^{34}(\pi)P_x^{12}(\pi)$	$\frac{|+-\rangle+|-+\rangle}{\sqrt{2}}$
Y	$\frac{|++\rangle+|--\rangle}{\sqrt{2}}$	$\exp(-i\pi I_y)$	$P_x^{34}(2\pi)P_x^{34}(\pi)P_x^{12}(\pi)$	$\frac{|+-\rangle-|-+\rangle}{\sqrt{2}}$
Z	$\frac{|++\rangle+|--\rangle}{\sqrt{2}}$	$\exp(-i\pi I_z)$	$P_x^{34}(2\pi)$	$\frac{|++\rangle-|--\rangle}{\sqrt{2}}$

an alternative unitary operation carried by the Bob side is depicted in the dotted box. From the experimental side, the choice of encoding is important, exemplified in the later section. Also, from the technical viewpoint, both encoding procedures may not be possible in pulse ENDOR spin technology. The first Hadamard and CNOT gates generate an entangled state between the electron and nuclear spin-qubits. Following the unitary transformation, the CNOT and Hadamard gates back-transform the entangled state, being responsible for extracting the encoding result. Except for a phase factor which doses not affect any signal, selective $\pi/2$ and π pulses are available for the Hadamard gate for a single qubit and the CNOT gate for two qubits in pulsed magnetic resonance spectroscopy, respectively.

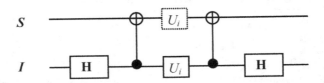

Fig. 29. Quantum circuit implementing super dense coding (SDC). U_i denotes a unitary transformation. U_i can be carried out also by the S side instead of the I side.

Super dense coding has been implemented using some quantum physical systems including NMR.[14] This contribution is the first report on the implementation of SDC by ENDOR based spin technology. However, the main idea is only to test the ENDOR system for QIP rather than total implementation of SDC, giving a testing ground for QC/QIP to molecular-spin based ENDOR. We have noted that there must be an argument about whether the physical system under study is truly realizable in manipulating the entangled state. In our experiments, the initial states have been prepared as pseudo-pure ones in a similar approach as reported by Mehring's group.[7,8]

The pulse sequence used for implementation of SDC with our ENDOR experiments is represented in Fig. 30. As generally discussed in the preceding sections, there are three main parts of the sequences, i.e., A; the preparation of the pseudo-pure states, B; the manipulation of spin-qubits for quantum operations and finally C; the detection according to the customs in magnetic resonance spectroscopy. Note that the part B consists of the three sub-parts, B1, B2 and B3, corresponding to entangling, quantum operation and phase rotation to measure phase interference in the resulting entangled states. In the ENDOR experiment described in Fig. 30, the readout is simply carried out by an electron-spin Hahn echo detection scheme.

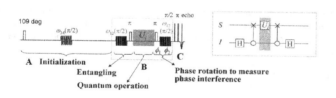

Fig. 30. A pulse sequence for implementation of super dense coding (SDC) by pulse ENDOR spin technology. The sequence is composed of the three periods: A; the preparation period for the initialization, B; the manipulation period for any quantum operations including entangling procedure, quantum phase rotations by the use of TPPI technique, or additional unitary operations denoted by U_i, U_i can be performed on the I side. The part B consists of three sub-parts, B1, B2 and B3 in the order of time. C; the detection period for the readout of the results obtained by QC/QIP operations.

In contrast to the preparation period in ordinary magnetic resonance spectroscopy, the left part labeled by "A" is for preparing the pseudo-pure state required as initialization for quantum operations or computing. Two pulses on electron and nuclear spins with additional waiting times in order to make the off diagonal term of the density matrix vanishing are required to acquire the pseudo-pure state. The first two pulses in the central part of the sequence are for entangling and the last two pulses are for the detection of the entanglement, as reported by Mehring's group.[7,8] Two phases of ϕ_1 and ϕ_2 for the pulses in the detection part are required for discriminating the entangled states from the simple superposition states. In the central part between the entangling and detecting the entanglement, which is labeled by the second sub-part B2 ($= U_i$), one of the qubits, the electron spin-qubit (or the client nuclear spin) in our experiments, is encoded by randomly applying one of the four pulses of $\{I, X, Y, Z\}$. The necessary pulses for encoding are described in Table 4.

Finally, there are pulses for the detection by an electron-spin Hahn echo signal, as carried out in the detection period denoted by C in Fig. 30. For the measurement part, the situation has been modified for some detection considerations. In this study, we have used the electron spin echo detection. The echo intensities have been detected for different angular dependencies of the pulses in the encoding part. As a result, there are four sets of angular dependencies for the RF pulses which have been used for encoding, as given in Table 5. The angular dependences of the intensities of the detected ESE signals in Table 5, with the definition for the angles, are derived in Appendix given at the end of this chapter.

Table 5. Angular dependence of the intensities of the ESE signals in the SDC experiments. The encoding pulse is defined by the angles θ and ϕ. The angular dependences in Table 5, with the definition for the angles, are derived in Appendix B given at the end of this chapter.

U	Angular dependent RF pulses for encoding	Detected angular dependent echo intensity
I		$-\frac{1}{4}[1 + \cos(\phi_{\text{MW}} + phi_{\text{RF}})]$
X	$P_x^{34}(\pi)P_x^{12}(\pi)$	$\frac{1}{16}[-3\cos\theta - \cos\phi - 4\cos\frac{\theta}{2}\cos\frac{\phi}{2}\cos(\phi_{\text{MW}} + \phi_{\text{RF}})]$
Y	$P_x^{34}(2\pi)P_x^{34}(\theta)P_x^{12}(\phi)$	$\frac{1}{16}[-3\cos\theta - \cos\phi + 4\cos\frac{\theta}{2}\cos\frac{\phi}{2}\cos(\phi_{\text{MW}} + \phi_{\text{RF}})]$
Z	$P_x^{34}(\theta)$	$\frac{1}{16}[-1 - 3\cos\theta - 4\cos\frac{\theta}{2}\cos(\phi_{\text{MW}} + \phi_{\text{RF}})]$

In QC-ENDOR, unitary operations are realized by some particular pulse operations with their controlled phases and polarizations. Phase manipulations are crucial in the molecular-spin based ENDOR experiments for QC/QIP. Table 5 shows the angular dependence of the electro-spin-echo (ESE) signal intensities in the detection part, depending also on experimental conditions for the measurements. The ESE signal intensities as functions of the angles including the phase rotations are derived, with the definitions for the angles, in Appendix given at the end of this chapter. In the SDC experiments, the detected ESE signal intensities incorporate the terms characteristic of 4π period, as shown in Table 5. This 4π period originates from spinor nature intrinsic to spin-1/2, but never shows up under ordinary experimental conditions. The very entanglement condition allows the spinor nature to experimentally appear. We have detected this salient behavior of the intensities, exemplifying the case for $U_i =$ X as given in Fig. 31, in which the phase manipulation was carried out for the client nuclear spin-1/2. Thus, the spinor observed in Fig. 31 is due to the client α-proton. In contrast to the 2π period of the population, the observed 4π period originates from the spinor property[20] as the intrinsic nature of spins under study with the experimental condition of the selective microwave excitation. The sign difference appearing between Fig. 31 and Table 5 is due to the difference between the experimental setup. The case for non-selective microwave excitations is of interest and under study.

In Fig. 31, the spinor nature appears in one of the Bell states $(|-+\rangle+|+-\rangle)/\sqrt{2}$ at the origin of time. At 20 μs corresponding to the angle $\theta = 2\pi$, the original state is converted to $(-|-+\rangle+|+-\rangle)/\sqrt{2}$, not transformed back to the original state due to the spinor nature of the nuclear spin-1/2. The observed ESE intensity is well reproduced by the theoretical function of the angle θ, as shown in the dotted curve in Fig. 31. The salient features of the ESE intensity appearing in this SDC experiment gives a direct evidence of the spinor nature of the proton spin. This is the first direct evidence that the spinor nature shows up in QC/QIP experiments. Noticeably, the spinor of an electron has never directly been illustrated so far, whereas the spinor is unequivocally established from the theoretical side. Thus, any direct detection of the spinor is challenging from the experimental side. A QC/QIP-based experimental version for the spinor of an electron will be described in the following section. In addition, it is noteworthy that the spinor nature of spin-1/2 is intrinsically quantum mechanical and of double-group property to be treated in group theory. For spectroscopic purposes, the spinor nature of protons has been utilized in pulse ENDOR spectroscopy by Hoefer and Mehring for the first time, implementing spinor-ENDOR spectroscopy to achieve an enormous amount of sensitivity enhancement in detecting proton ENDOR transitions.[20]

3.2. *TPPI Detection and Inter-Conversion of the Bell States by Pulse ENDOR*

A pulse sequence applied for the QC-ENDOR experiments has already been given in Fig. 31. The first microwave and second RF pulses are applied to

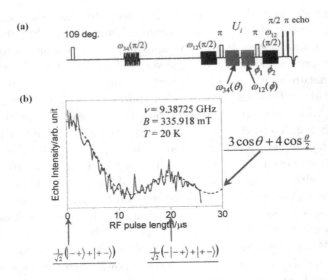

Fig. 31. Spinor of the spin-1/2 proton of malonyl radical in the single crystal detected in the SDC experiments by pulse ENDOR spin technology. (a) A pulse sequence protocol for the phase manipulation of the nuclear spin-qubit, i.e., manipulating the nuclear spin-1/2 qubit by invoking variable angles, θ and ϕ(b) 4π Periodic dependence of the electron spin echo readout as a function of the RF pulse length. Microwave frequency: $\nu = 9.38725$ GHz. Static magnetic field set for ENDOR: $B = 335.918$ mT. Temperature: $T = 20$ K. The phase angles, ϕ_1 and ϕ_2 stand for the MW pulse and the RF pulse, respectively, during the TPPI procedure. The phase angles are composed of two virtual frequencies introduced in the TPPI procedure. The theoretical curve above is derived in Appendix B given at the end of this chapter.

make the pseudo-pure state described above. Some waiting times between the first and second pulses and between the second and third pulses are required to diminish any coherence generated by the supplied pulses. After making the pseudo-pure state, we apply further microwave and RF pulses in order to manipulate electron and nuclear spin-qubits. This procedure belongs to quantum operations. When applying a microwave π-pulse following an RF $\pi/2$ pulse, we can generate a quantum entangled state which is one of the Bell states and the most important event in the QC experiments. Changing a combination of the MW and RF pulses with different resonance conditions, different Bell states can be constructed (see the table in Fig. 23 and Table 4). In order to detect the entangled states and to make the establishment of the entanglement clear from "conventional" or ordinary superposition states, we employ time proportional phase increment (TPPI) technique mainly for the nuclear spin during our QC/QIP experiments. As it is intractable to perform the TPPI technique for electron spin-qubits because of currently available MW technological difficulties, the phase rotation by the use of four different pulse channels (x, y, $-x$ and $-y$) for MW irradiation has been applied in our experiment. Implementation of the TPPI version of MW technology devoted to QC/QIP, mimicking NMR paradigm, is underway, and some parts of the latest innovation are described later in this contribution.

The TPPI technique in pulse ENDOR spectroscopy is known for the separation of multiple quantum coherences.[20] On the other hand, we have used the scheme in a way that it can give global information on the state of electron and nuclear spins in terms of the phases of spin-qubits in the Bloch sphere. This is for resolving the restriction of ENDOR spectroscopy that any measurement can only be done via ESE signals. Especially, for the case that information on the state of entanglement is required, the TPPI technique is an essential part for distinguishing the entangled states from the simple "conventional" superposition states. The TPPI procedure is one of the highlighted parts in QC/QIP experiments. Dynamical parts of ESE signals as a function of TPPI procedure gives the time structure of entanglement, which is useful for QC/QIP experiments for multi-partite spin-qubits. An essential part of the trick used in the TPPI technique applied for novel ELDOR spectroscopy is described in the later section (see Fig. 38), and the technique is applicable to increase the number of dimensions by using virtual frequencies in any pulse-based spectroscopic measurements. In this context, the TPPI technique is a general and formidable tool.

Figure 32 shows that the TPPI-detected phase spectra from malonyl radical in the single crystal of malonic acid allow us to distinguish between the entangled Bell states composed of one electron spin- and one nuclear spin-qubit. The phase spectra illustrated in Fig. 32 have been achieved in a Q-band pulse ENDOR spectrometer in which both the MW and RF pulses can be applied in any desired orientations in the Bloch sphere. The phases ϕ_1 and ϕ_2 in Figs. 30 and 31 correspond to ν_{MW} and ν_{RF}, respectively, for the phase rotation as given in Figs. 21 and 23. Combination signals between ν_{MW} and ν_{RF}, which are due to the entangled states between an electron spin and a proton nuclear spin, were observed, indicating that the corresponding entangled Bell states are discriminated in terms of the phase. The appearance of ν_{MW} or ν{RF} is due to imperfectness of the TPPI procedure. The line widths of the phase spectra reflect decay times appearing in the time-domain interferograms.

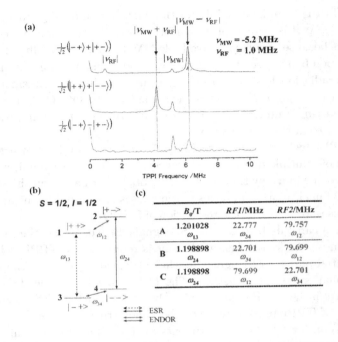

Fig. 32. Phase spectra of malonyl radical observed by Q-band pulsed ENDOR technique with the TPPI detection. Combination signals between ν_{MW} and ν_{RF}, which are due to the entangled states between an electron spin and one proton nuclear spin, were observed. $|-+\rangle, |+-\rangle, |++\rangle$ and $|--\rangle$ denote the four entangled Bell states composed of one electron spin-qubit and one nuclear spin-qubit. See the caption of Fig. 1 for the notations of $|-+\rangle, |+-\rangle, |++\rangle$ and $|--\rangle$. ϕ_1 and ϕ_2 in Figs 30 and 31 correspond to ν_{MW} and ν_{RF} for the phase rotation, respectively. Also, see Fig. 23.

3.3. *The First Direct Detection of Spinor of an Electron Spin-Qubit by Phase Manipulation*

Another measurement was carried out in order to examine the quantum spin entangled states. If the pulses generate the entangled states, they are expected to show 4π periodicity originated from the spinor property, as predicted in Table 5. In Fig. 31, the TPPI-based phase manipulation was made on the client nuclear qubit spin-1/2 by rotating ϕ_2 with an X-band pulse ENDOR spectrometer. Recently, we have designed and set up a novel Q-band pulsed electron magnetic resonance (EMR) based QC/QIP apparatus. Figure 33 shows a protocol for the MW/RF pulse sequence for the spinor behavior of an electron spin-qubit in the electron-nuclear spin entangled state and a readout time-domain profile as a function of the MW pulse length applied to the entangled state, observed by invoking the phase rotation of ϕ_1, i.e., microwave TPPI operated on the electron spin-qubit in the malonyl radical system. Figure 33(a) shows the pulse sequence that generates the time evolution of the electron-nuclear spin entangled state as a function of θ (the MW pulse length; also see Appendix at the end of this chapter), illustrating the electron-spin spinor in a straightforward manner. U_i denotes the quantum operation of the MW excitation pulse length as a function of time t. Figure 33(b) clearly shows that the echo-signal intensity oscillates with 4π periodicity. The 4π periodicity doses not appear in usual magnetic resonance experiments because of the inherence of the phase property of the electron spin. In order to observe the periodicity, it is necessary to make an interferometric experiment for at least a two-spin system. The findings of the periodicity support the generation of the en-

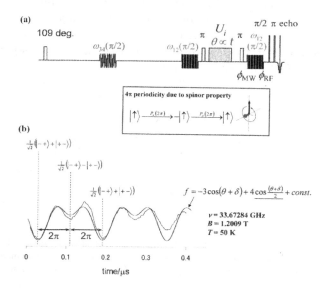

Fig. 33. 4π-Spinor behavior due to electron spin-1/2 appeared in the electron-spin echo intensity when applying a selective microwave pulse. (a) The pulse sequence which generates the time evolution of the electron-nuclear spin entangled state illustrating the electron-spin spinor in a straightforward manner. U_i denotes the quantum operation of the MW excitation pulse length as a function of time t. (b) The Hahn ESE intensity was readout after the operations as a function of the MW pulse duration time t. The curves in thin and thick black denote the experimental and theoretical one, respectively. The former shows a relaxation effect due to decoherence during the experiment. The theoretical curve in (b) is derived in Appendix B given at the end of this chapter.

tangled states in an alternative way. When applying a microwave 2π pulse in this experiment, the spin manipulation corresponds to inter-conversion between the entangled states, e.g.,

$$\frac{1}{\sqrt{2}}(|++\rangle - |--\rangle) \xleftrightarrow{2\pi} \frac{1}{\sqrt{2}}(|++\rangle + |--\rangle).$$

It is shown that the spin manipulation is successfully performed by the pulsed ENDOR technique. In our QC/QIP experiments, it is also possible to operate the nuclear spin using the RF pulse. When applying the RF pulse for the generated entangled state instead of the microwave pulse, the spinor behaviour due to the nuclear spin-1/2 appeared directly, as illustrated in Fig. 31. It is indicated that application of the microwave and RF pulses enable us to manipulate the electron-nuclear spin-qubit system, and the pulsed ENDOR technique is available for performing the fundamental quantum computation in the molecular spin system. We emphasize that the direct detection of the spinor from an electron spin-qubit has for the first time been achieved only by building the entangled state involving the electron spin-qubit. Schematic picture of the inter-conversion between the entangled states in the electron-nuclear two-spin system is given with corresponding quantum operations in Fig. 34.

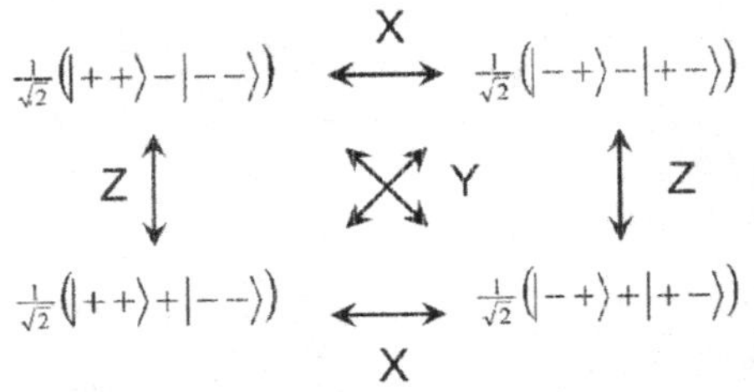

$$\begin{array}{ccc}
\frac{1}{\sqrt{2}}(|++\rangle - |--\rangle) & \xleftrightarrow{X} & \frac{1}{\sqrt{2}}(|-+\rangle - |+-\rangle) \\
Z \updownarrow & \underset{Y}{\times} & \updownarrow Z \\
\frac{1}{\sqrt{2}}(|++\rangle + |--\rangle) & \xleftrightarrow{X} & \frac{1}{\sqrt{2}}(|-+\rangle + |+-\rangle)
\end{array}$$

Fig. 34. Inter-conversion between the four Bell (entangled) states composed of the two spin-1/2 qubit system. X, Y and Z denote the quantum transformations converting the one Bell state to another. See also Table 4 for the definition of the operations, X, Y and Z.

3.4. *Tri-partite Electron-Spin Nuclear-Qubits Experiments; Identification of Separable States Decomposed into the Bi-partite Entanglements*

The first electron-involved tri-partite QC experiment, in which one electron bus spin and two client nuclear spins are involved, has been carried

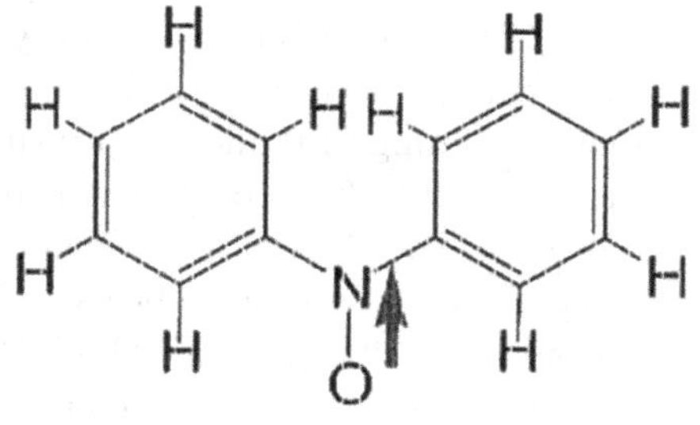

DPNO-h_{10}

Fig. 35. Stable non-deuterated diphenyl nitroxide (DPNO-h_10) with multi-nuclear spin-qubits. DPNO-h_{10} is diluted in diamagnetic benzophenone-d_{10} crystals at any desired concentration ratios.

out under the condition of non-selective microwave excitation. In this experiment, we also demonstrate the appearance of the entanglement between an electron spin and two nuclear spins, nitrogen and proton nuclei in the pseudo-pure state of stable non-deuterated diphenyl nitroxide (DPNO-h_{10}) diluted in the lattice of diamagnetic benzophenone-h_{10} crystals, as is given in Fig. 35. DPNO-h_{10}, which is a typical organic radical, has multi-nuclear client spin-qubits of protons and extremely stable in the benzophenone lattice under the exposure of strong MW and RF irradiations. The α-protons of the phenyl rings are capable of nuclear client qubits in the solid state, simply because anisotropic nature of the proton hyperfine coupling appears. All the magnetic tensors of DPNO-h_{10} have experimentally been determined by CW-ENDOR/ESR spectroscopy prior to use for the molecular-spin qubit and isotope-labeled qubits.[39,40] In order to precisely determine the magnetic tensors including the g-tensor, appropriate isotope labeling has been achieved. The perdeuterated benzophenone-h_{10} host lattice is chosen to reduce the line-width of the ESR transitions, the spectral density due to the protons of neighboring host molecules in pulse ENDOR spectra and to suppress dephasing effects arising from the protons of neighboring and surrounding host benzophenone molecules.

The pulse ENDOR spectroscopy for DPNO-h_{10} magnetically diluted in the benzophenone-h_{10} lattice can be carried out at any temperature in the range of ambient to liquid helium temperatures. Figure 36 shows (b) an ESE detected Q-band ESR spectrum of DPNO-h_{10} diluted in the perdeuterated benzophenone crystal and (c) a Davies-type ENDOR spectrum observed with the static magnetic field along a crystallographic axis of the benzophenone crystal. In Fig. 36(c), the ESR transition in the lowest field was excited, as the corresponding ESR allowed transition is given in the electron-nuclear sublevels depicted in Fig. 36(a). The three detected peaks in Fig. 36(b) are due to the hyperfine allowed transitions of the ^{14}N nucleus, where proton hyperfine splittings are unresolved because of the overlapping of many hyperfine transitions of small couplings. The MW selective hyperfine excitation with DPNO-h_{10} is not feasible in many orientations of the static magnetic field with respect to the crystal. It is noted that in the ESE detected scheme MW selective excitations in hyperfine levels are necessary for most cases of further quantum operations in current QC/QIP experiments. In order to achieve effective MW selective excitations, well-resolved hyperfine ESR lines with appropriate magnitudes of the splittings and narrow line-widths are required, having made us design and synthesize a series of isotope-labeled qubit molecules. Detailed descriptions of this issue are beyond this chapter. In Fig. 36(c), only the proton ENDOR transitions from the DPNO-h_{10} guest molecules are seen around the free proton NMR frequency, ν_p determined by the static magnetic field. Any proton ENDOR signals from the neighboring or distant host benzophenone-d_{10} molecules are vanishing due to the perdeuteration of the host molecule.

The phase spectra relevant to the entangled states in the electron-nuclear spin-qubit system of DPNO-h_{10} have been obtained by using the TPPI technique. As described above, the microwave excitation in the ESE-detected ESR spectrum are not selective in this particular QC/QIP experiment, because DPNO-h_{10} we used does not give highly resolved hyperfine splittings due to many protons in any orientations of the static magnetic field. Thus, the ^{14}N nucleus spin-qubit is apparently involved in the ESR transition, but not in the entanglement process and only one client proton entangled with an electron bus qubit. Figure 37 shows the separable tri-partite states that can be decomposed into a pair of the bi-partite entangled states. In the tri-partite QC/QIP experiments, the well-resolved proton ENDOR peaks were selected for RF excitation.

The entangled states established here involve those in which only the two particles, i.e., an electron spin and one proton nucleus are entangled. Seemingly, the Q-band pulsed ENDOR-based phase interferogram measurements involve tri-partite particles, but the situation in terms of entanglement is bi-partite. The nitrogen nuclear states are borrowed and play only a role of the electron sublevels; the initialization of the pseudo-pure state was estab-

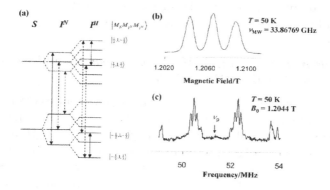

Fig. 36. An energy level diagram for tri-partite QC experiment, an ESE detected Q-band ESR and Davies-type ENDOR spectra of DPNO-h_{10} magnetically diluted in the diamagnetic benzophenone-d_{10} lattice. In (a), the energy levels relevent to the ESR and ENDOR transitions involved are represented with only two nuclei, i.e., one nitrogen and one proton nucleus, considerd. (b) An ESE-detected Q-band ESR spectrum observed with the static magnetic field along a crystallographic axis of the benzophenone crystal. (c) A Davies-type ENDOR spectrum observed with the same orientation of the static magnetic field. In (c), the ESR transition in the lowest field shown in (b) was excited, as the corresponding ESR allowed transition is given in the electron-nuclear sublevels depicted in (a).

lished in the preparation period of our QC experiment. In our three-qubit experiment, unresolved hyperfine splittings due to the protons were excited by the ^{14}N-hyperfine selective microwave irradiation, as denoted by the broken double-arrows in Fig. 37(a). In this context, the microwave excitation was not completely selective in terms of proton nuclear spin states, nevertheless the experiments are still appealing because the double bi-partite pseudo-entanglement between an electron and one nucleus-1/2 qubit was for the first time carried out by means of pulsed ENDOR applied to the real three-qubit physical systems. True tri-partite entanglement is realizable by the use of partially deuterated DPNO properly oriented in the benzophenone lattice.[40] Such electron-bus qubit systems give us a chance to achieve such quantum teleportation experiments as electron spin-qubits are involved in molecular-spin levels.[40] The teleportation experiments by using molecular-spins with nuclear client qubits are distinct from the counterpart of NMR in some physical aspects. Detailed QC/QIP experiments on molecular-spin quantum teleportation will be given elsewhere with the full analysis of electronic and molecular structures of DPNO-h_{10} from both experimental and theoretical sides.

4. Development of Coherent-Dual ELDOR Technique for Molecular Electron Spin-Qubits Based Quantum Computers

4.1. *Electron Spin Manipulation Technology beyond Pulse ENDOR: CD-ELDOR Technique*

In this section, we refer to the spin manipulation of an electron-electron exchange-coupled system for molecular-spin QCs. This is an important issue to extend prototypical working model systems for QC/QIP to realistic scalable molecular-spin based QCs. In pulsed EMR measurements for the multi-electron spin system, it is necessary to control at least two ESR transitions. In this context, ELectron-electron DOuble Resonance (ELDOR) gives a minimal but crucial technological basis for electron spin-qubits based QC/QIP. The ELDOR technique in the current stage potentially enables us to carry out the spin manipulation for QC/QIP in the multi-electron spin system such as an organic biradical, in which two electron spins are weakly coupled. Conventional ELDOR technique, however, does not allow

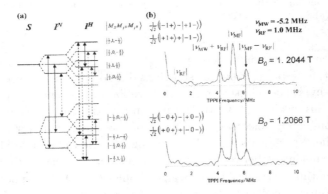

Fig. 37. Phase spectra obtained from the tri-partite "apparent entanglement" in DPNO-h_{10} magnetically diluted in the diamagnetic benzophenone-d_{10} lattice. (a) Energy diagram of the tri-partite spin-qubits under study in DPNO-h_{10}. The MW excitations and ENDOR transitions involved in the tri-partite experiments are denoted by the arrows. The longest arrow denoting the ESR allowed transition corresponds to the ESE peak appearing in the lowest field ($B_0 = 1.2044$ T). The middle one corresponds to the peak at $B_0 = 1.2066$ T. (b) Phase spectrum of DPNO-h_{10} observed by Q-band pulsed ENDOR technique with the TPPI detection. Phase frequency values, ν_1 and ν_2 are shown in the corner of this figure. The appearance of both these combination frequencies, i.e., $\nu_1 + \nu_2$ and $\nu_1 - \nu_2$, is experimental evidence for the occurrence of the superposition of the two entangled states.

us to distinguish between two electron spin-qubits in terms of the phase in the Bloch sphere because individual microwave sources are independent during its double resonance experiment, particularly in terms of microwave phase. It is noteworthy that pulse-based electron multiple resonance pulse technology in terms of NMR standards has not been achieved yet, much less electron spin technology for its phase manipulation attempted.

In order to control the phase of electron spin-qubits, we have developed novel pulsed Q-band ELDOR technique equipped with a coherent dual-frequency unit (termed coherent-dual ELDOR). Using the coherent-dual ELDOR (CD ELDOR) technique, it is possible to perform the TPPI measurements for electron spins. High frequency microwave such as Q-band is preferable for the establishment of quantum entanglements between two electron spin-qubits based on the pure states below 1.5 K. In contrast to pulsed ENDOR based QCs, to build true QCs by invoking molecular electron spin-qubits, several essential but intractable problems must be solved. Our efforts so far have enabled us to solve only a few of them in the course of this work.

A basic pulse sequence for the phase control of a microwave pulse by the TPPI technique is shown in Fig. 38. When the phase of the second π-pulse is modulated, a Hahn ESE signal intensity oscillates as a function of 2ϕ. The corresponding equation for the signal intensity is derived in Appendix at the end of this chapter. When the $\pi/2$ pulse of the frequency ν_1 and the π-pulse of the frequency ν_2 are setup, the quantity ϕ can be expressed by $2\pi(\nu_2-\nu_1)t$ with t introduced as the time separation between the two pulses, as depicted in Fig. 38. Now we have two variables to manipulate the phase of the ESE signal: One is $\Delta\nu = (\nu_2 - \nu_1)$, and the other time t. The latter allows us to execute a TPPI technique of the electron spin-qubit version. Referred to the former, the two desired frequencies can be provided by two "different MW sources," but the dual frequency generation is controlled with an external locking unit: This contrasts with the current status of pulse NMR technology, in which arbitrary-frequency network generation is available. In this context, the two MW frequencies are coherent. Each MW frequency can excite a particular ESR transition of a different MW frequency with a controlled phase angle. This is the reason why this novel electron spin manipulation technique is termed Coherent-Dual ELDOR. Thus, this CD ELDOR technique enables us to manipulate the phases of electron spin-qubits on resonance in the Bloch sphere.

We show representative TPPI-detected time-domain profiles for Q-band ELDOR spectra based on the CD ELDOR technique in Fig. 39. An ESE-

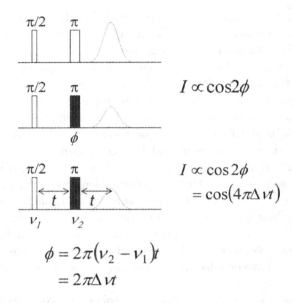

Fig. 38. A basic pulse sequence for the phase control of microwave pulses by the TPPI technique. The ESE signal as a function of 2ϕ is derived in Appendix B given at the end of this chapter.

detected signal of coal sample monitored at $\nu_1 = 33.594081\,\text{GHz}$ clearly oscillates as a function of a given frequency of $\Delta\nu = \nu_2 - \nu_1 = 1\,\text{MHz}$ in Fig. 39(c) and 10 MHz in Fig. 39(d), respectively, where ν_1 and ν_2 stand for the two different MW frequencies supplied by the two MW generators. The off-resonance ESE signal component is coherently detected in terms of the dual frequency technique.

In the TPPI experiment, the Hahn-echo signal intensity, as shown in Fig. 39, oscillates with a frequency of $4\pi\Delta\nu t$. Figure 40 shows a fourier-transformed spectrum from the time-domain profile of Fig. 39(d), clearly illustrating that the oscillation frequency of the echo signal corresponds to the difference between ν_1 and ν_2.

$$I \propto \cos 2\phi = \cos 4\Delta\nu t = \cos|4\pi(\nu_2 - \nu_1)t|.$$

In Figs. 38-40, we demonstrate the phase control of microwave pulses by the TPPI technique for the electron spin based on CD ELDOR, only discussing the necessity of CD ELDOR as novel ELDOR technique, compared with the conventional ELDOR spectroscopy. Current applications of the CD ELDOR technique to QC experiments for the molecular spin system with multi-electron spins give potentialities of scalable matter spin-qubits in molecular frames. It is noteworthy that the CD ELDOR technique is capable of providing novel ELDOR-detected NMR spectroscopy, in which $\Delta \nu$ resolution is as high as kHz. Figure 41 exemplifies a prototypical experiment for such enabling spectroscopy.

In the following section, we emphasize how to design specific molecular spin qubits for QC/QIP in connection with currently available MW spin technology. It is important to precisely control the quantum coherence of the quantum multi-spin system at a desired level of advanced MW technology. Applying the CD ELDOR spin technology to organic biradicals, in which two "non-identical" electron qubits are weakly exchange-coupled in solution or oriented media, has been the focus of current issues in the field of QC/QIP.

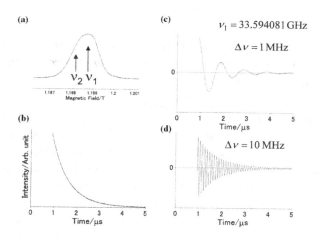

Fig. 39. Representative TPPI-detected time-domain profiles for Q-band ELDOR spectra based on the CD ELDOR technique. A field-swept Hahn ESE signal of a coal sample observed at Q-band and its ESE decay are shown in (a) and (b), respectively. An ESE-detected signal of the coal sample monitored at $\nu_1 = 33.594081$ GHz clearly oscillates as a function of a given frequency of $\Delta \nu = \nu_2 - \nu_1 = 1$ MHz in (c) and 10 MHz in (d), respectively, where ν_1 and ν_2 stand for the two different MW frequencies supplied by the two MW generators. A frequency-domain spectrum fourier-transformed from the time-domain profile given in (d) is shown in Fig. 40. The off-resonance ESE signal component is coherently detected in terms of the coherent-dual frequency technique. All the measurements were carried out at ambient temperature.

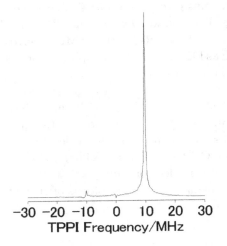

Fig. 40. A Fourier-transformed spectrum from the coal sample in a representative TPPI measurement using coherent-dual ELDOR technique. The corresponding time domain spectrum is given in Fig. 39(d).

4.2. *Materials Challenge for Biradical Qubits with Small Spin-Orbit Interactions: g-Engineering Approach to Molecular Design*

Referred to the current electron magnetic spin resonance technology, the MW selective excitation procedure for electron spin-qubits is crucial, contrasting with the paradigm of QC-NMR, whereas pulse ESR variants of the QC/QIP NMR experiments are underway.[19,41] From the viewpoint of molecular design for electron spin-qubits in the molecular frame, this requires g-tensor engineering and/or hyperfine (A) tensor engineering. For executing any quantum operations, weak exchange interactions or spin dipolar ones between the electron spin-qubits are required in a controllable manner. Besides decoherence and time resolution problems, currently available MW pulses covering two ESR transitions are subject to crucial limitations in terms of the magnitude of the interactions. Robust exchange interactions between electron spin-qubits in molecular frames are beyond the current pulse-based MW spin technology, as frequently described: The MW pulse covers only small parts of the ESR spectra from such strongly exchange-coupled systems. This requires elaborate molecular designing for desired molecular-spin QC biradicals. Such biradicals play a significant role in demonstrating the first proof-in-principle experiment for quantum entanglement between genuine two electron spin-qubits in molecular frames, controlling the electron spin-qubit phases in the Bloch sphere.

Figure 42 schematically exemplifies one of the g-tensor engineered biradical qubits, and some of realistic ones have been designed and synthesized in our group. The X-ray crystallographic analyses of the synthesized biradicals, one of them given in Fig. 42, have illustrated that the dihedral angle between the two TEMPO moieties with unpaired electrons at the NO sites ranges from more than 45 to 90 degrees, in consistent with those obtained in the molecular design, showing that the g-tensor engineering for the molecular design can be successfully achieved. The biradical qubit 1 in Fig. 42 has the perpendicular conformation at the central bridge of the duryl moiety, giving the 63 degrees bent conformation between the two g-tensors at the TEMPO moieties. The perpendicular conformation is capable of the complete truncation of the π-conjugation which give rise to the vanishing electron exchange interaction. In such biradical qubits, spin-dipolar interactions are utilized for quantum operations in the QC/QIP procedure, in addition to the possible selective MW excitation of the individual spin-qubit in the molecular frame.

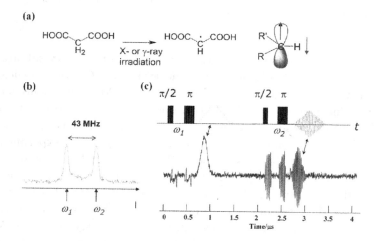

Fig. 41. Novel coherent-dual pulsed ELDOR-detected spectroscopy. (a) Malonyl radical generated from malonic acid in the crystal. (b) ESE detected ESR spectrum of malonyl radical with the magnetic field along a crystallographic axis of the crystal. (c) Only a prototypical experiment on malonyl radical in the single crystal of malonic acid is shown for the well-resolved hyperfine ESR lines in (b). Note that the pulse lengths are not optimized in the Hahn ESE type experiment in (c).

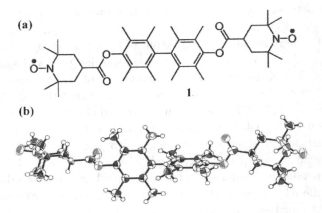

Fig. 42. Molecule-based biradical qubits designed in terms of the g-engineering. g-Tensor engineered duryl-bridged bis(TEMPO) and its molecular structure determined by X-ray crystallographic analysis are shown in (a) and (b), respectively. The dots at the NO sites in (a) denote unpaired electrons. The dihedral angle between the two TEMPO moieties is 63 degrees: TEMPO denotes the moiety with an unpaired electron localized at the NO site. The exchange interaction is truncated by the duryl bridge of the perpendicular conformation. The exchange interaction is extremely weakened, but spin dipolar interaction is non-vanishing. The biradical qubit has been designed in terms of homo-spin g-tensor engineering.

The biradical qubit is successfully incorporated into the crystal lattice of the corresponding diamagnetic molecule with the NO sites chemically reduced or with the ketone functions. The latter should be chemically more stable in ordinary conditions in the corresponding crystal lattices, to our knowledge. The corresponding bis(ketone) host molecules also have been synthesized. For the biradical to be suitable for QC/QIP experiments, methyl and methylene protons of the TEMPO moiety should be deuterated, giving well-resolved hyperfine ESR lines, as the deuteration depicted in Fig. 43. If homo-spin biradical qubits such as bis(TEMPO) have inversion symmetry, the g-tensor engineering approach does not work in electron magnetic resonance experiments. This inversion symmetry can be broken by introducing ^{15}N nucleus at one NO site, lifting the nuclear degeneracy due to Bosonic permutation symmetry, thus giving rise to the breakdown of the total inversion symmetry. The different hyperfine coupling due to ^{15}N nucleus (A-tensor engineering) and nuclear spin quantum number apparently gives non-equivalent resonance fields, enabling us to discriminate between the two TEMPO moieties even in solution. This feature has been

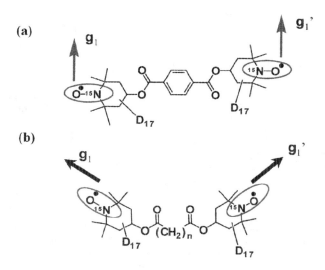

Fig. 43. Molecule-based biradical qubits designed in terms of the g-engineering and perdeuteration of the TEMPO meoieties. The ^{15}N labeling also contributes to the narrowing the ESR hyperfine lines from the ^{15}N nuclei due to the lack of the nuclear quadrupolar interaction. For the pseudo g-engineering, only the N nucleus at one TEMPO site has to be labeled.

exploited for implementing bi-partite molecular spin-qubit based QC/QIP experiments in solution or in the single crystals of diamagnetic hosts.[19] This approach is termed "pseudo g-tensor engineering."

Biradicals **2** and **3** depicted in Fig. 44 are successfully incorporated in the diamagnetic host lattice of the corresponding reduced form or the ketone form at the NO sites. Biradical **3** has inversion symmetry in the host lattice. Because of the linear conformation, radical **3** has extremely weak intramolecular exchange interaction in the crystalline state. We have synthesized the A-tensor engineered biradical **3** for CD ELDOR based QC experiments by the use of two electron spin-qubits with non-equivalent pseudo g-tensors. The genuine g-tensor engineering approach includes hetero-spin biradicals, in which one monoradical moiety differs from the other in terms of principal g-values. In the hetero-spin g-tensor engineering, hetero-spin moieties with an electron spin-1/2 have non-equivalent spin-orbit interaction, giving g-values due to the different electronic-spin structures and undergoing selective microwave excitation for the initialization for QC/QIP. The current MW spin technology imposes crucial restrictions on the use

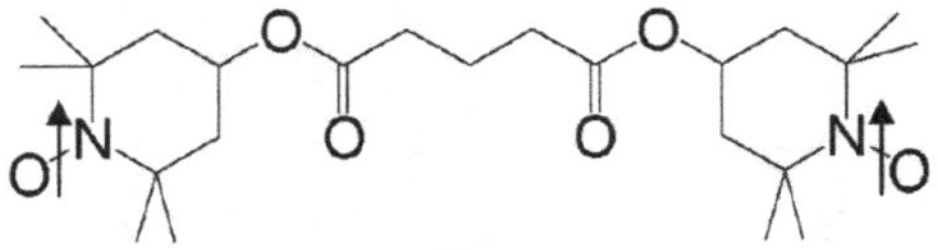

Fig. 44. Biradicals incorporated into the diamagnetic host lattice of the corresponding reduced or ketone form at the NO sites. The arrows denote unpaired electrons nearly isolated at the NO sites. Radical 3 has inversion symmetry in the host lattice. Thus, pseudo g-tensor engineering is required for biradical 3 so to be suitable for CD ELDOR based QC/QIP experiments.

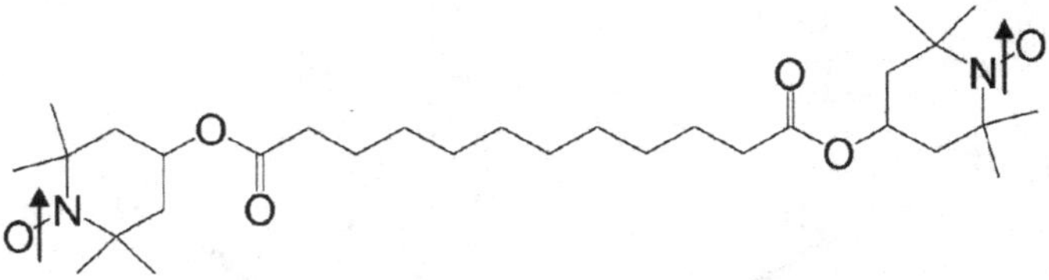

of molecular electron spin-qubit systems having large spin-orbit couplings. Only an ELDOR based approach to QC/QIP experiments by using electron spin-qubits in the weak exchange-coupling regime is feasible. In addition to the crucial limitation to time resolution in ELDOR experiments, this is the main reason why organic biradical qubits have been utilized for the first proof-in-principle and prototypical QC/QIP experiments based on electron spin-qubits in this work. Obviously, the time resolution is associated with the issue of decoherence inherent in molecular electron spins. The long coherence time is preferable in terms of the implementation of QC/QIP ELDOR spin technology, whereas repetition procedures for quantum operations are hampered from the experimental side.

5. Scalable Molecular-Spin Based Electron Spin-Qubits: Spin-Qubits Systems with 1D Periodic Spin Structure Proposed by Lloyd

In this section, first we introduce an approach to scalable molecular-spin based electron spin-qubit array featuring in the periodic spin structure of one dimension, which is an electron spin-qubit version of the famous Lloyd's proposal from the theoretical side.[42,43] Originally, the Lloyd's proposal is for a linear chain composed of nuclear spin-1/2 qubits such as $(ABC)_n$, in which

A, B and C denote nuclear spin-1/2 qubits with non-equivalent chemical shifts and only the nearest neighboring interaction (Ising-type interaction) such as isotropic indirect exchange coupling inside the chain is occurring among the nuclear spin-qubits, assuming that any interaction between the chains is negligible. Following the description by Lloyd, we extend this one-dimensional (1D) nuclear spin-qubit model to an electron spin-qubit version by replacing the nuclear spins to molecular electron spin-qubits. There are some polymeric molecular arrays of 1D nuclear spin-1/2 qubits suitable for the Lloyd's original model. In terms of molecular frames appropriate for the Lloyd's model, we believe the electron spin-qubit version is more promising for practically scalable QCs. Particularly, this point is true in terms of synthetic strategy. The synthetic strategy can be strengthened by supramolecular chemistry, if molecular designing is elaborately made.[34] In supramolecular approaches, complexes embracing open-shell metal cations with non-equivalent g-tens (g-engineering) can be molecular frames and the metal cation plays a role of an electron spin-qubit, whether electronic spin quantum numbers are 1/2 or greater than 1/2. In the latter, an effective spin-1/2 is utilized.

Hereafter, A, B and C denote electron spin-qubits with non-equivalent g-tensors in molecular frames. The chain is placed in the presence of a static magnetic field whose direction is designated by the z axis. The interaction inside the chain can be either electron exchange couplings or direct dipolar spin interactions in a 1D electron spin-qubit molecular array. The Hamiltonian describing the whole spin system is simply written as

$$\mathcal{H} = -\frac{h}{2\pi} \sum_k \left(-\omega_k S_k^z + 2 J_{k,k+1} S_k^z S_{k+1}^z \right),$$

where the summation is carried out over all the spin-qubits in the chain. The periodicity imposes

$$\omega_k = \omega_{k+3} \quad (k = 1, 2, \ldots),$$

and

$$-\omega_1 = \omega^A, \quad -\omega_2 = \omega^B, \quad -\omega_3 = \omega^C.$$

For the interaction term, also the periodicity holds as

$$J_{k,k+1} = J_{k+3,\,k+4} \quad (k = 1, 2, \ldots),$$

and

$$J_{1,2} = J_{4,5} = \cdots = J^{AB},$$
$$J_{2,3} = J_{5,6} = \cdots = J^{BC},$$
$$J_{3,4} = J_{6,7} = \cdots = J^{CA}.$$

The Hamiltonian has only diagonal terms. Thus, the eigenstates of the Hamiltonian are described in the form of the spin states such as

$$|010\,100\,110\ldots\rangle,$$

where some of the spin-qubits in any eigenstate are pointing "up," termed the state $|0\rangle$ and the rest of them pointing "down," termed the state $|1\rangle$. Let assume the following situation for the spin structure of the whole system: In a given state of the whole system, a spin, say B, is oriented "up," i.e., in the state $|0\rangle$, in another state of the whole system this spin B oriented "down," i.e., in the state $|1\rangle$, and all other spins keep their orientations unchanged. The energy difference ΔE between the two states of the whole system can be expressed as

$$\Delta E = \frac{h}{2\pi}(\omega^B \pm J^{AB} \pm J^{BC}),$$

where the upper (+) sign in front of J^{AB} denotes the situation with the state $|0\rangle$ of the neighboring spin A. The lower sign (−) in front of J^{AB} designates the one with the state $|1\rangle$ of the neighboring spin A. The signs for J^{BC} have the similar meaning for the neighboring spin C. Thus, with respect to the spin B, we have the four eigenfrequencies (= resonance frequencies) of the Hamiltonian as follows:

$$\omega_{00}^B = \omega^A + J^{AB} + J^{BC},$$
$$\omega_{01}^B = \omega^A + J^{AB} - J^{BC},$$
$$\omega_{10}^B = \omega^A - J^{AB} + J^{BC},$$
$$\omega_{11}^B = \omega^A - J^{AB} - J^{BC}.$$

It is noted that the eigenfrequency above corresponds to the energy required for the electron spin allowed transitions in which one spin flips, i.e., one spin is inverted by the resonant oscillating MW irradiation, exemplifying that ω_{00}^B corresponds the energy required for the inversion of the spin B from $|0_A 0_B 0_C \ldots\rangle$ to $|0_A 1_B 0_C \ldots\rangle$. ω_{ik}^A and ω_{ik}^C ($i, k = 0$ or 1) have a similar structure of the resonant frequency. The allowed transitions expected for the $(ABC)_n$ array are schematically depicted in Fig. 45, where the edged effect appearing at the end of the chain is avoided. If the spin A and C

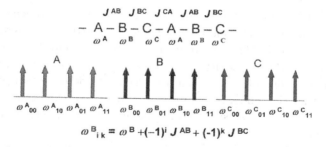

Fig. 45. Resonance frequencies for the (ABC)$_n$ spin-qubit array of the 1D periodic spin structure, proposed by Lloyd. The edge effect is avoided for clarity. Only the nearest neighboring interactions between the spin qubits and Ising-type interaction are assumed. A, B and C denote electron spin-qubits with non-equivalent g-tensors in molecular frames.

are at the left and right edge, the eigenfrequency (= resonance frequency) required for the inversion of the spin A is as follows, depending on the state ($|0\rangle$ or $|1\rangle$) of the neighboring spin B;

$$\omega_0^A = \omega^A + J^{AB},$$
$$\omega_1^A = \omega^A - J^{AB}.$$

For the spin C at the right edge of the 1D spin chain, the resonance frequency to invert the spin C is given as

$$\omega^{C_0} = \omega^C + J^{AB},$$
$$\omega^{C_1} = \omega^C - J^{BC}.$$

The inversion of a particular electron spin-qubit corresponds to the operation of a NOT gate which can be performed by the relevant resonant MW π-pulse. By using the (ABC)$_n$ array of the Lloyd's model, we can carry out the quantum operations composed of the corresponding swap logic circuit executing the exchange of the neighboring qubits, or the Controlled SWAP circuit (= Fredkin gate) in which the exchange takes place only if another neighboring spin is in the state $|1\rangle$. By the use of the SWAP gates, data can be transferred through the 1D spin structure. Both the edge spin-qubits have potentialities for the output and input of data. We have noted that building up pulse protocols for the SWAP gates by using the (ABC)$_n$ array of electron spin-qubits is formidable task from both the theoretical and experimental sides.

The synthetic strategy of the electron spin-qubit version of the Lloyd's model for scalable spin-qubit based QCs has never been documented so far, nor realistic materials suitable for the model reported. The Lloyd's model from the theoretical side is a promising physical system of matter spin-qubits in view of the scalability. Due to the low dimensionality of the spin system, the periodicity has its own advantage in the detection sensitivity. It is no doubt that the model is a materials challenge for chemists and materials scientists in many aspects. Besides the limitations of current MW spin manipulation technology, decoherence problems relevant to electron spin-qubits in molecular frames are crucial because there are many underlying physical origins to shorten coherence of the molecular spin systems. Internal freedom of motions is one of the issues to be elaborately considered in molecular designing of the electron qubit counterpart of the Lloyd's model. Now that the g-tensor engineering approach is essential and applicable to the electron spin-qubit version, elaborate molecular design should be achieved in view of avoiding decoherence origins as much as possible. In this context, double- or triple-stranded helicates with robust molecular structures embedding paramagnetic metal cations with non-equivalent g-tensors and pseudo octahedral symmetry at the cation site can be competent candidates for the proposed 1D periodic QC/QIP model. For an organic version of the 1D periodic spin-qubit array, DNA backbones incorporating g-engineered stable π-radicals such as triangulene-based radicals or triplet excited-state quinoxalines into the pitches give potentialities to solve the issue of the scalability. It is noteworthy that the DNA backbone is not robust enough to achieve strictly rigid 1D spin-structures which underlie long coherence time.

6. Conclusions

Electron bus spin-qubits with nuclear client spin-qubits in molecular frames are emerging as candidates for QC/QIP, which have recently appeared among a variety of physical systems in quantum information science. In terms of hidden qubits in molecular frames, molecular chirality associated with circularly polarized optical nature is available. Molecular electron-spin based qubits are only in the cradle and thus challenging materials targets in quantum information science and related fields. There are a rich variety of interactions, as utilized for quantum operations, between qubits in such molecular frames as hydrocarbon based open-shell systems, and their modes and magnitudes can be designed in terms of rapidly developing sophisticated quantum chemistry. Molecular materials that are capable of giving testing grounds for QC/QIP experiments or theoretical considerations can be synthesized in many cases. Because of space, we have only introduced the Lloyd's theoretical model and its variant of electron spin-qubit version for scalable QCs. Synthetic efforts for prototypical examples of such scalable QCs have been made by invoking sophisticated supramolecular chemistry, novel functionality molecular entities appearing and encouraging us to develop novel pulse-based MW spin manipulation technology. In particular, time resolution associated with manipulating the electron spin-qubits in ensemble is crucial, in addition to multi-frequency capabilities with coherency. In this context, spin manipulation technology itself for electron spin-qubits in any molecular frames are in the cradle and very immature. Electrical detection of magnetic resonance for single molecular spin-qubit systems is also an important issue, which is associated with two-dimensional array versions of the Lloyd's model. Any 2D versions also are a materials challenge in chemistry and materials science.

This chapter dealt with a short course to pulse ENDOR for QC/QIP experiments, omitting detailed or more general descriptions of eigenenergies or resonance eigenfields involved in electron magnetic resonance: There are a wealth of science behind them. For non-specialists, some detailed derivations of the basic equations appearing in this chapter are given in Appendices at the end of the chapter. Also, there have been omitted significant pulse technology dealing with decoherency in molecular frames, such as the bang-bang pulse method by Morton et al.[11] and the first important experimental results of the density-matrix based tomography for the electron-nuclear entangled states achieved for the C_{60} system incorporating ^{15}N nucleus ($S = 3/2, I = 1/2$).[13]

The QC/QIP experiments on molecular spin-qubits described in this chapter are based on electron-nuclear magnetic resonance methods for all the QC processes in ensemble solids. Some advantages of the use of the organic molecular spin-qubits are described referring to both NMR QC and the current pulse MW spin technology. The heart of our matter spin-qubit based QC/QIP experiments is phase manipulation technology for the entanglement between molecular electro-spin and client nuclear spin-qubits. The spin-qubit manipulation technology for quantum gate operations in this work is based on the time-proportional-phase-increment (TPPI) technique, enabling us to discriminate between the phases of the entangled Bell states. Pulsed ENDOR-based QC/QIP experiments for super dense coding (SDC) have for the first time been carried out by the use of stable malonyl radical as matter spin-qubits. The TPPI technique has illustrated the establishment of quantum entanglement between electron- and nuclear-spin states and mutual inter-convesion between the Bell states. The electron-spin 4π-periodicity in the rotation of the electron spin in the Bloch sphere has been explicitly observed in the QC experiments for the first time, illustrating the electron-spin spinor nature. A basis for tri-partite QC experiments involving electron and nuclear spin-qubits has been given for the first time.

This chaper dealt with a topic of another important relevance in molecular spin-qubit based QC/QIP experiments, describing that the pulsed coherent-dual ELDOR for electron-qubit based QC/QIP has for the first time been implemented by invoking a novel microwave coherent-dual phase-rotation technique. Thus, applications of the coherent-dual ELDOR to molecular electron spin-qubit systems are also discussed, emphasizing designing the molecular two electron-qubit systems appropriate for QC/QIP. g- and/or hyperfine A-tensor engineering approaches are proposed for preparing the two- and multi-electron-qubit systems, as a materials challenge.

The chapter implies that general requisites for scalable true electron spin-qubit systems such as 1D periodic robust spin structures, i.e., the electron spin-qubit variant of the Lloyd's model, in addition to DiVincenzo's five criteria, are necessary in terms of the synthetic strategy. In order to implement real QCs and QIP systems, the scalability of qubits is essential. In this context, the synthetic feasibility for the scalability of particularly chosen ligands is essentally important, when the supramolecular chemistry approach is adopted as the synthetic strategy. According to the requsites, double- or triple-stranded helicates embedding open-shell metal cations are suggested instead of organic molecular spin-qubits. The reasons why weakly exchange-coupled electron spin-qubits such as organic multi-radicals are suitable for present-stage QC/QIP experiments are emphasized referred to currently available MW spin technology. Nevertheless, this is an only temporal approach in order to circumvent the experimental limitations.

Acknowledgments

This work has partially been supported by "Implementation of Molecular-Spin Quantum Computers/Quantum Information Processing Systems," A Project of Core Research for Evolutional Science and Technology (CREST), Japan Science and Technology Agency (JST) and by Grand-in-Aids by the Ministry of Education, Culture, Sports, Science and Technology, Japan.

Appendix A. General Solutions for ENDOR Transition Frequencies and probabilities of Arbitrary Electron and Nuclear Spin Systems: A Perturbation Approach to ESR/ENDOR Spectroscopy

Appendix A.1. *General Solutions for ENDOR Transition Frequencies*

In Appendix A, general solutions for only ENDOR transition frequencies and probabilities with ESR resonance fields in electron magnetic resonance spectroscopy are described. All the solutions are analytical in terms of the second-order perturbation theory and general in spin quantum numbers S, I and an arbitrary coordinate-axes system.

The spin Hamiltonian is given as

$$\mathcal{H} = \beta \tilde{\boldsymbol{S}} \cdot \boldsymbol{g} \cdot \boldsymbol{B}_0 + \tilde{\boldsymbol{S}} \cdot \boldsymbol{D} \cdot \boldsymbol{S} + \tilde{\boldsymbol{S}} \cdot \boldsymbol{A} \cdot \boldsymbol{I} - g_n \beta_n \tilde{\boldsymbol{I}} \cdot \boldsymbol{B}_0 + \tilde{\boldsymbol{I}} \cdot \boldsymbol{P} \cdot \boldsymbol{I}, \quad (A.1)$$

where the second term denotes fine-structure interaction and the fifth term quadrupolar interaction for an arbitrary S and I. The other terms that have already appeared have usual meanings. The quadrupolar interaction is assumed much smaller than the hyperfine + nuclear Zeeman interactions. Here, β denotes the Bohr magneton: The subscript e is omitted. In the perturbation treatment, the first term is an unperturbed term and the rests of the terms perturbed ones. The treatment is to a good approximation in the presence of high static magnetic field. We first derive the eigenenergies up to the second order for the eigenstate $|S, M_S, I, M_I\rangle = |m_S, m_I\rangle$ as follows:

(1) The zeroth-order energy, $E^{(0)}(m_S, m_I)$;

$$E^{(0)}(m_S, m_I) = g\beta B_0 m_S, \quad (A.2)$$

which arises from the first term, the electron Zeeman interaction, and the g-value is given as

$$g^2 = \tilde{\boldsymbol{h}} \cdot \tilde{\boldsymbol{g}} \cdot \boldsymbol{g} \cdot \boldsymbol{h} = \tilde{\boldsymbol{h}} \cdot \boldsymbol{g}^2 \cdot \boldsymbol{h}, \quad \boldsymbol{u} = \boldsymbol{g} \cdot \boldsymbol{h}/g, \quad (A.3)$$

where the unit vector $\boldsymbol{u}$ defines the quantization axis for $\boldsymbol{S}$ and the unit vector $\boldsymbol{h}$ is given by $\boldsymbol{B}_0 = B_0 \boldsymbol{h}$. B_0 is the static magnetic field. Then, the first-order energy is described as follows:

(2) The first-order order energy, $E^{(1)}(m_S, m_I)$;

$$E^{(1)}(m_S, m_I) = -\frac{1}{2}(\tilde{\boldsymbol{u}} \cdot \boldsymbol{D} \cdot \boldsymbol{u})[S(S+1) - 2m_S^2] - K(m_S)m_I(m_S)$$
$$-\frac{1}{2}(\tilde{\boldsymbol{k}} \cdot \boldsymbol{P} \cdot \boldsymbol{k})[I(I+1) - 3m_I^2], \qquad (A.4)$$

where $\boldsymbol{K}(m_S)$ is a tensor defined as follows;

$$\boldsymbol{K}(m_S) = (\boldsymbol{A} \cdot \boldsymbol{g}/h)m_S - g_n\beta_n B_0 \boldsymbol{E}, \qquad (A.5)$$
$$K^2(m_S) = \tilde{\boldsymbol{h}} \cdot \boldsymbol{K}^2(m_S) \cdot \boldsymbol{h}, \quad \boldsymbol{k}(m_S) = -\boldsymbol{K}(m_S) \cdot \boldsymbol{h}/K(m_S). \quad (A.6)$$

In the case of the isotropic g-tensor, the second term of Eq. (A.4) is written as

$$E_{hf}^{eff(1)}(m_S, m_I) = -K(m_S)m_I(m_S)$$
$$= -[\tilde{\boldsymbol{h}} \cdot (Am_S - g_n\beta_n B_0 \boldsymbol{E})]^{1/2} m_I(m_S). \qquad (A.7)$$

The second-order energy is written as follows:

(3) The second-order energy, $E^{(2)}(m_S, m_I)$;

$$E^{(2)}(m_S, m_I) = -\frac{1}{2g\beta B_0} m_S \Big\{ |D_1|^2 [4S(S+1) - 8m_S^2 - 1]$$
$$- \frac{|D_2|^2}{4}[2S(S+1-2m_S^2-1)] \Big\}$$
$$- \frac{1}{g\beta B_0} D_A [S(S+1) - 3m_S^2] m_I$$
$$+ \frac{1}{2g\beta B_0} \Big\{ |A_1|^2 m_S m_I^2 - A_2[S(S+1) - m_S^2] m_I$$
$$+ \frac{A_3}{2} m_S[I(I+1) - m_I^2] \Big\}$$
$$+ \frac{1}{2k(m_S)} m_I \Big\{ |P_1|^2 [4I(I+1) - 8m_I^2 - 1]$$
$$- \frac{|P_2|^2}{4}[2I(I+1) - 2m_I^2 - 1] \Big\}, \qquad (A.8)$$

where the first term of Eq. (A.8) arises from the fine-structure term, and the second one from the cross term between the fine-structure and hyperfine structure term appearing in the spin Hamiltonian (A.8), respectively. The third term arises from the hyperfine interaction. $|D_1|$, $|D_2|$, and D_A are expressed as follows:

$$\left. \begin{array}{l} |D_1|^2 = (\tilde{\boldsymbol{u}} \cdot \boldsymbol{D}^2 \cdot \boldsymbol{u}) - (\tilde{\boldsymbol{u}} \cdot \boldsymbol{D} \cdot \boldsymbol{u})^2 \\ |D_2|^2 = 2\mathrm{Tr}\boldsymbol{D}^2 + (\tilde{\boldsymbol{u}} \cdot \boldsymbol{D} \cdot \boldsymbol{u})^2 - (\tilde{\boldsymbol{u}} \cdot \boldsymbol{D}^2 \cdot \boldsymbol{u}) \end{array} \right\}, \qquad (A.9)$$

and

$$\left.\begin{array}{l} D_A = \frac{1}{2}(D_1 A_1^* + D_1^* A_1) = (\tilde{u} \cdot D \cdot \tilde{A} \cdot k) - (\tilde{u} \cdot D \cdot u)(\tilde{k} \cdot A \cdot u) \\ \tilde{u} \cdot D \cdot \tilde{A} \cdot k = \tilde{h} \cdot \tilde{g} \cdot D \cdot \tilde{A} \cdot K(m_S) \cdot h/[gK(m_S)] \end{array}\right\}.$$
(A.10)

A_i's ($i = 1, 2, 3$) are expressed in the following:

$$\left.\begin{array}{l} |A_1|^2 = (\tilde{k} \cdot A^2 \cdot k) - (\tilde{k} \cdot A \cdot u)^2 \\ A_2 = \det A\,(\tilde{u} \cdot A^{-1} \cdot k) \\ A_3 = \text{Tr}\,A^2 - (\tilde{u} \cdot A^2 \cdot u) - (\tilde{k} \cdot A^2 \cdot k) + (\tilde{k} \cdot A \cdot u)^2 \end{array}\right\}.$$
(A.11)

The terms in Eq. (A.11) are expressed by the h-representation as follows:

$$\left.\begin{array}{l} \tilde{k} \cdot A^2 \cdot k^2 = \tilde{h} \cdot K(m_S) \cdot A^2 \cdot K(m_S) \cdot h/K^2(m_S) \\ \tilde{k} \cdot A \cdot k = \tilde{h} \cdot K(m_S) \cdot A \cdot g \cdot h/[gK(m_S)] \\ \tilde{u} \cdot A^{-1} \cdot k = -[m_S - g_n \beta_n B_0 (\tilde{h} \cdot \tilde{g} \cdot A^{-1} \cdot h)/g]/K(m_S) \\ \tilde{u} \cdot A^2 \cdot u = \tilde{h} \cdot \tilde{g} \cdot A^2 \cdot g \cdot h/g^2 \end{array}\right\}.$$
(A.12)

The forth term in Eq. (A.8) arises from the quadrupolar interaction with

$$\left.\begin{array}{l} |P_1|^2 = (\tilde{k} \cdot P^2 \cdot k) - (\tilde{k} \cdot P \cdot k)^2 \\ |P_2|^2 = \text{Tr}\,P^2 + (\tilde{k} \cdot P \cdot k)^2 - 4(\tilde{k} \cdot P^2 \cdot k) \\ \tilde{k} \cdot P \cdot k = \tilde{h} \cdot \tilde{K}(m_S) \cdot P \cdot K(m_S) \cdot h/K^2(m_S) \\ \tilde{k} \cdot P^2 \cdot k = \tilde{h} \cdot \tilde{K}(m_S) \cdot \tilde{P} \cdot P \cdot K(m_S) \cdot h/K^2(m_S) \end{array}\right\}.$$
(A.13)

In Eq. (A.8), the other cross terms are omitted and they can be evaluated similarly in terms of the second-order perturbation treatment.

The energy up to the second-order is given as

$$E(m_S, m_I) = E^{(0)}(m_S, m_I) + E^{(1)}(m_S, m_I) + E^{(2)}(m_S, m_I). \quad (A.14)$$

Thus, a given transition frequency, $\omega_{m_S \to m_S'}(m_I \leftrightarrow m_I')$ for $|m_S, m_I\rangle \leftrightarrow |m_S', m_I'\rangle$ can be analytically obtained in an arbitrary and general coordinate axes system. From Eq. (A.14), any ESR transitions (resonance) fields can be obtained as analytical forms under the resonance condition in a straightforward manner.

An ENDOR allowed transition frequency $\omega_{m_S}(m_I - 1 \leftrightarrow m_I) = 2\pi\nu_{m_S}(m_I - 1 \leftrightarrow m_I)$ for $|m_S, m_I - 1\rangle \leftrightarrow |m_S, m_I\rangle$ is expressed under the resonance condition as follows:

$$h\nu_{m_S}(m_I - 1 \leftrightarrow m_I) = E(m_S, m_I - 1) - E(m_S, m_I)$$
$$= K(m_S) - \frac{3}{2}(\tilde{\boldsymbol{k}} \cdot \boldsymbol{P} \cdot \boldsymbol{k})(2m_I - 1)$$
$$+ \frac{1}{g\beta B_0} D_A[S(S+1) - 3m_S^2]$$
$$- \frac{1}{g\beta B_0} \Big\{ |A_1|^2 m_S(2m_I - 1) - A_2[S(S+1) - m_S^2]$$
$$- \frac{A_2}{2} m_S(2m_I - 1) \Big\}$$
$$+ \frac{1}{2K(m_S)} \left[8|P_1|^2 - \frac{|P_2|^2}{2} \right] (2m_I - 1). \qquad (A.15)$$

The first and second terms in Eq. (A.15) are the first-order contribution, and the other terms denote the second-order contribution. Equation (A.15) is the most general expression for the ENDOR transition frequency, where only assumption that the quadrupolar interaction is much smaller than the hyperfine + nuclear Zeeman interactions is made. We note that Eq. (A.15) is not valid for the case of $m_S = 0$ for $I \geq 1$. The right side of Eq. (A.15) should be positive.

The most typical ENDOR transition frequency for the case of $S = 1/2$ with $I = 1/2$ is given as follows:

$$h\nu_{m_S}(m_I - 1 \leftrightarrow m_I) = K(m_S) - \frac{\det \boldsymbol{A}}{4g\beta B_0 K(m_S)}$$
$$\times [m_S - g_n \beta_n B_0 (\tilde{\boldsymbol{h}} \cdot \tilde{\boldsymbol{v}} g \cdot \boldsymbol{A}^{-1} \cdot \boldsymbol{h})/g], \qquad (A.16)$$

where $m_S = \pm 1/2$, $m_I = 1/2$, and $K^2(m_S)$ is described by Eq. (A.6). According to Eq. (A.16), when the ESR transition $|m_S = 1/2\rangle \leftrightarrow |m_S = -1/2\rangle$ is monitored, two ENDOR allowed transition, i.e., $|m_I = -1/2\rangle \leftrightarrow |m_I = +1/2\rangle$ and $|m_I = +1/2\rangle \leftrightarrow |m_I = -1/2\rangle$ are expected to observe. The one belongs to $|m_S = 1/2\rangle$ and the other to $|m_S = -1/2\rangle$.

Neglecting the second-order contribution in Eq. (A.16), we have

$$h\nu_{m_S}(-1/2 \leftrightarrow +1/2) = K(m_S) = [\tilde{\boldsymbol{h}} \cdot \boldsymbol{K}^2(m_S) \cdot \boldsymbol{h}]^{1/2}. \qquad (A.17)$$

For the case fo small g-anisotropy,

$$h\nu_{m_S}(-1/2 \leftrightarrow +1/2) \simeq [\tilde{\boldsymbol{h}} \cdot (\boldsymbol{A} m_S - g_n \beta_n B_0 \boldsymbol{E})^2 \cdot \boldsymbol{h}]^{1/2}. \qquad (A.18)$$

Thus, introducing $\nu_n = g_n \beta_n B_0/h$ for the free nuclear spin of I, we have

$$\nu_{m_S}(-1/2 \leftrightarrow +1/2) \simeq [\tilde{\boldsymbol{h}} \cdot (m_S \tilde{\boldsymbol{A}}/h - \nu_n \boldsymbol{E}) \cdot (m_S \boldsymbol{A}/h - \nu_n \boldsymbol{E}) \cdot \boldsymbol{h}]^{1/2}, \qquad (A.19)$$

where $m_S = \pm 1/2$.

From Eq. (A.17), we have a useful relationship between the two ENDOR transition frequencies as follows:

$$(\nu_{-1/2})^2 - (\nu_{1/2})^2 = K^2(-1/2) - K^2(+1/2)$$
$$= \frac{1}{h}\tilde{\bm{h}} \cdot (\tilde{\bm{A}} \cdot \bm{g} + \bm{A} \cdot \bm{g}) \cdot \bm{h}\,(\nu_n/g). \qquad (A.20)$$

From Eq. (A.20) is linear with repsect to $\bm{A}$ and thus is useful to analyze angular dependence of proton ENDOR spectra to approximately determine A-tensor from organic open-shell entities with $S = 1/2$.

Equation (A.15) is general and valid for any arbitrary reference frame, because we only define the unit vector $\bm{h}$ for the static magnetic field B_0. Let us exemplify a simple case of $S = 1/2$ and a ^{14}N nucleus ($I = 1$) with B_0 applied parallel to the direction of one of the principal value of the g tensor, e.g., g_X, where we assume that the g tensor, the A and P tensors of the nucleus are coaxial for simplicity. Then, the ENDOR transition frequencies for the ^{14}N nucleus are given as follows:

$$\nu_{m_S}(-1 \leftrightarrow 0) = \left| A_X m_S - \nu_n + \frac{3}{2} P_X \right|, \qquad (A.21a)$$

$$\nu_{m_S}(0 \leftrightarrow 1) = \left| A_X m_S - \nu_n - \frac{3}{2} P_X \right|, \qquad (A.21b)$$

where $m_S = \pm 1/2$, and A_X and P_X are in the units of MHz. The four ENDOR frequencies are observed according to Eqs. (A.21a) and (A.21b) and both the principal values for A_X and P_X are experimentally derived to the first order. The second-order corrections are easily calculated by the equations derived above.

Appendix A.2. *General Solutions for ENDOR Transition Probabilities*

Here, we describe general and analytical expressions for the ENDOR transition probability in terms of the second-order perturbation theory. The ENDOR transition probability for $|m_{S,m_I} - 1\rangle \leftrightarrow |m_{S,m_I}\rangle$ is generally described as

$$I_{m_S}(m_I - 1 \leftrightarrow m_I) = \frac{2\pi}{\hbar^2} |\langle m_I - 1 | \mathcal{H}_{\text{rad,n}} | m_I \rangle|, \qquad (A.22)$$

where the interacting Hamiltonian with the oscillating RF irradiation field B_2 is given as

$$\mathcal{H}_{\text{rad,n}} = B_2 [\beta \tilde{\bm{S}} \cdot \bm{g} - g_n \beta_n \tilde{\bm{I}}] \cdot \bm{x}. \qquad (A.23)$$

In Eq. (A.23), B_2 denotes the amplitude of the oscillating field with $\boldsymbol{x} \perp \boldsymbol{h}$, i.e., $\boldsymbol{B}_2 \perp \boldsymbol{B}_0$. The zeroth-order transition probability, $I_{m_S}^{(0)}(m_I - 1 \leftrightarrow m_I)$ is given as follows:

$$\begin{aligned} I_{m_S}^{(0)}(m_I - 1 \leftrightarrow m_I) &= \frac{\pi}{2}\left(\frac{g_n \beta_n}{h}\right)^2 B_2^2 P^2(I, m_S) \\ &\quad \times \left\{1 - [\tilde{\boldsymbol{x}} \cdot \boldsymbol{K}(m_S) \cdot \boldsymbol{h}]^2 / K^2(m_S)\right\} \\ &= \frac{\pi}{2}\gamma_n^2 B_2^2 P^2(I, m_S) \\ &\quad \times \left\{1 - [\tilde{\boldsymbol{x}} \cdot \boldsymbol{K}(m_S) \cdot \boldsymbol{h}]^2 / K^2(m_S)\right\}, \end{aligned} \quad \text{(A.24)}$$

with

$$P^2(I, m_S) = I(I+1) - m_I(m_I - 1). \quad \text{(A.25)}$$

In terms of the second-order perturbation treatment adopted here, the first-order transition probability, $I_{m_S}^{(0)}(m_I - 1 \leftrightarrow m_I)$ is obtained by using the eigenstates to the first order as follows:

$$\begin{aligned} I_{m_S}^{(0)}(m_I - 1 \leftrightarrow m_I) = \frac{\pi}{2}\gamma_n^2 B_2^2 P^2(I, m_I)\nu_n^{-2} &\Big\{\tilde{\boldsymbol{x}} \cdot \boldsymbol{K}^2(m_S) \cdot \boldsymbol{x} \\ &- [\tilde{\boldsymbol{x}} \cdot \boldsymbol{K}(m_s) \cdot \boldsymbol{h}]^2 \\ &+ [\tilde{\boldsymbol{x}} \cdot \boldsymbol{K}(m_s) \cdot \boldsymbol{h} - \tilde{\boldsymbol{x}} \cdot \boldsymbol{g}^2 \cdot \boldsymbol{h}\,(\nu_n/g^2)]^2 \\ &- \frac{1}{K^2(m_S)}[\tilde{\boldsymbol{x}} \cdot \boldsymbol{K}^2(m_S) \cdot \boldsymbol{h} \\ &- (\tilde{\boldsymbol{x}} \cdot \boldsymbol{g}^2 \cdot \boldsymbol{h})[\tilde{\boldsymbol{h}} \cdot \boldsymbol{K}(m_S) \cdot \boldsymbol{h}](\nu_n/g^2)]^2\Big\}. \end{aligned} \quad \text{(A.26)}$$

In Eq. (A.26), the contribution from the fine-structure term is not included. In the case of a small anisotropic g tensor, $\tilde{\boldsymbol{x}} \cdot \boldsymbol{g}^2 \cdot \boldsymbol{h}/g^2 \simeq 0$. Then, we have

$$\begin{aligned} I_{m_S}^{(1)}(m_I - 1 \leftrightarrow m_I) &\simeq \frac{\pi}{2}\gamma_n^2 B_2^2 P^2(I, m_I)\nu_n^{-2}\Big\{\tilde{\boldsymbol{x}} \cdot \boldsymbol{K}^2(m_S) \cdot \boldsymbol{x} \\ &\quad - [\tilde{\boldsymbol{x}} \cdot \boldsymbol{K}^2(m_S) \cdot \boldsymbol{h}]^2/K^2(m_S)\Big\} \\ &= \frac{\pi}{2}\gamma_n^2 B_2^2 P^2(I, m_I)\nu_n^{-2}\Big\{\tilde{\boldsymbol{x}} \cdot \boldsymbol{K}^2(m_S) \cdot \boldsymbol{x} \\ &\quad - [\tilde{\boldsymbol{k}} \cdot \boldsymbol{K}(m_S) \cdot \boldsymbol{x}]^2\Big\}. \end{aligned} \quad \text{(A.27)}$$

We note that the hyperfine enhancement factor defined as $E_{\text{hyper}} = (I_{m_S}^{(1)} - I_{m_S}^{(0)})^{1/2}$ greatly affects the ENDOR transition probability if a nucleus with non-vanishing I and small γ_n has a large A. However, generally speaking, the transition intensities of the ENDOR signals appearing in the low frequency region are weak because the probabilities depend on m_S. If a nucleus with a non-zero I and small γ_n has a small A, the corresponding hyperfine enhancement factor is less, thus the ENDOR transition probability becomes less. Equation (A.26) shows that the ENDOR transition probability depends on both the anisotropy of the A tensor and m_S. Thus, strong ENDOR signals are often observed with $\boldsymbol{B}_0$ along the principal axes of A tensors.

Appendix B. QC-ENDOR Based on Density Matrix Formalism

Appendix B.1. *Spin Dynamics Based on Density Matrix Formalism*

In Appendix B, we describe bases for the understanding of QC-ENDOR in terms of density matrix formalism, and we derive the key equations appearing in this chapter. Some of them are important for the essential understanding of the spinor behavior, which shows up in the QC/QIP experiments. The time development of the spin system is described by the equation of motion which is called the Liouville-von Neumann equation:

$$\frac{\partial}{\partial t}\rho(t) = -i[H(t), \rho(t)], \qquad (B.1)$$

where $\rho(t)(=|\psi(t)\rangle\langle\psi(t)|)$ stands for a time-dependent density matrix. Although the Hamiltonian $H(t)$ in Eq. (B.1) generally includes any irreversible processes such as relaxation, we ignore them in this appendix. The Liouville-von Neumann equation without the relaxation terms is derived from time dependent Schrödinger equation:

$$i\frac{\partial}{\partial t}|\psi(t)\rangle = H(t)|\psi(t)\rangle. \qquad (B.2)$$

The formal solution of Eq. (B.1) is expressed by

$$\rho(t) = U(t)\rho(0)U^{-1}(t), \qquad (B.3)$$

where

$$U(t) = P\exp\left[-i\int_0^t tH(t')dt'\right]. \qquad (B.4)$$

$U(t)$ denotes a propagator of the density matrix and P is the Dyson time-ordering operator. If the Hamiltonian is time-independent, the propagator is given by

$$U(t) = \exp[-iHt]. \qquad (B.5)$$

When we consider the thermal equilibrium state as the initial state at time $t = 0$, the density matrix for the initial state is

$$\rho(t=0) = \frac{\exp(-H/k_\mathrm{B}T)}{\mathrm{Tr}\left[\exp(-H/k_\mathrm{B}T)\right]}, \tag{B.6}$$

where k_B stands for Boltzmann constant.

In the pulsed magnetic resonance QC experiments, the time evolution of the spin system is considered in the presence of a static magnetic field. The pulsed experiments consist of irradiation by excitation pulses to the spin system and time-evolution of the spin system afterwards. The pulsed experiments are considered as a sequence of the propagators.

$$\rho(0) \xrightarrow{U^1(t_1)} \xrightarrow{U^2(t_2)} \cdots \xrightarrow{U^n(t_n)} \rho(t = t_1 + t_2 + \cdots + t_n) \tag{B.7}$$

The density matrix $\rho(t)$ at time t is calculated by

$$\rho(t) = U_n(t_n)\ldots U_2(t_2)U_1(t_1)\rho(0)U_1^{-1}(t_1)U_2^{-1}(t_2)\ldots U_n^{-1}(t_n). \tag{B.8}$$

Then we have the expectation value of any operator O at time t in the form of

$$\langle O \rangle = \mathrm{Tr}[O\rho(t)]. \tag{B.9}$$

Appendix B.2. *Spin Angular Momentum Operators and Rotation Operators*

Here we summarize spin angular momentum operators. We denote the operators of three components of the spin angular momentum, S_x, S_y and S_z. The operators of the spin angular momentum have the following cyclic commutation relationships:

$$[S_x, S_y] = iS_z, \tag{B.10a}$$
$$[S_y, S_z] = iS_x, \tag{B.10b}$$
$$[S_z, S_x] = iS_y. \tag{B.10c}$$

The typical spin angular momentum operators, S^2, S_z, $S_+(= S_x + iS_y)$, and $S_- = (S_x - iS_y)$, are expressed by

$$S^2|S, M_\mathrm{S}\rangle = S(S+1)|S, M_\mathrm{S}\rangle, \tag{B.11a}$$
$$S_z|S, M_\mathrm{S}\rangle = M_\mathrm{S}|S, M_\mathrm{S}\rangle, \tag{B.11b}$$
$$S_\pm|S, M_\mathrm{S}\rangle = \sqrt{S(S+1) - M_\mathrm{S}(M_\mathrm{S} \pm 1)}|S, M_\mathrm{S} \pm 1\rangle, \tag{B.11c}$$

in the spin Zeeman basis $\{|S, M_S\rangle \equiv |M_S\rangle, M_S = -S, -S+1, \ldots, S-1, S\}$. The nonzero matrix elements of S_x, S_y, S_z and $S^2 (= S_x^2 + S_y^2 + S_z^2)$ are

$$\langle S, M_S + 1|S_x|S, M_S\rangle = \frac{1}{2}\sqrt{S(S+1) - M_S(M_S+1)}, \quad \text{(B.12a)}$$

$$\langle S, M_S - 1|S_x|S, M_S\rangle = \frac{1}{2}\sqrt{S(S+1) - M_S(M_S-1)}, \quad \text{(B.12b)}$$

$$\langle S, M_S + 1|S_y|S, M_S\rangle = -\frac{i}{2}\sqrt{S(S+1) - M_S(M_S+1)}, \quad \text{(B.12c)}$$

$$\langle S, M_S - 1|S_y|S, M_S\rangle = \frac{i}{2}\sqrt{S(S+1) - M_S(M_S-1)}, \quad \text{(B.12d)}$$

$$\langle S, M_S|S_z|S, M_S\rangle = M_S, \quad \text{(B.12e)}$$

$$\langle S, M_S|S^2|S, M_S\rangle = S(S+1). \quad \text{(B.12f)}$$

We define S_e as a $(2S+1) \times (2S+1)$ identity matrix. For a spin-1/2 system, matrix representation of the angular momentum operators, S_x, s_y, S_z and S_e are written as

$$S_x = \begin{pmatrix} 0 & 1/2 \\ 1/2 & 0 \end{pmatrix}, \quad \text{(B.13a)}$$

$$S_y = \begin{pmatrix} 0 & -i/2 \\ i/2 & 0 \end{pmatrix}, \quad \text{(B.13b)}$$

$$S_z = \begin{pmatrix} 1/2 & 0 \\ 0 & -1/2 \end{pmatrix}, \quad \text{(B.13c)}$$

$$S_e = \begin{pmatrix} 1 & 0 \\ 0 & 1 \end{pmatrix}. \quad \text{(B.13d)}$$

In the case of two-coupled spins (S_1 and S_2), the matrix representations of the angular momentum operators of the coupled spins are obtained by tensor products of single-spin angular momentum operators given by Eqs.(B.11a)-(B.11c). In the case of $S_1 = S_2 = 1/2$, the spin matrices in the tensor product basis $|S_1, S_2, M_{S_1}, M_{S_2}\rangle = |S_1, M_{S_1}\rangle \otimes |S_2, M_{S_2}\rangle \equiv$

$|M_{S_1}, M_{S_1}\rangle$ are represented by

$$S_{1x} = S_{1x} \otimes S_{2e} = \begin{pmatrix} 0 & 0 & 1/2 & 0 \\ 0 & 0 & 0 & 1/2 \\ 1/2 & 0 & 0 & 0 \\ 0 & 1/2 & 0 & 0 \end{pmatrix}, \quad \text{(B.14a)}$$

$$S_{1y} = S_{1y} \otimes S_{2e} = \begin{pmatrix} 0 & 0 & -i/2 & 0 \\ 0 & 0 & 0 & -i/2 \\ i/2 & 0 & 0 & 0 \\ 0 & i/2 & 0 & 0 \end{pmatrix}, \quad \text{(B.14b)}$$

$$S_{1z} = S_{1x} \otimes S_{2e} = \begin{pmatrix} 1/2 & 0 & 0 & 0 \\ 0 & 1/2 & 0 & 0 \\ 0 & 0 & -1/2 & 0 \\ 0 & 0 & 0 & -1/2 \end{pmatrix}, \quad \text{(B.14c)}$$

$$S_{2x} = S_{1e} \otimes S_{2x} = \begin{pmatrix} 0 & 1/2 & 0 & 0 \\ 1/2 & 0 & 0 & 0 \\ 0 & 0 & 0 & 1/2 \\ 0 & 0 & 1/2 & 0 \end{pmatrix}, \quad \text{(B.14d)}$$

$$S_{2y} = S_{1e} \otimes S_{2y} = \begin{pmatrix} 0 & -i/2 & 0 & 0 \\ i/2 & 0 & 0 & 0 \\ 0 & 0 & 0 & -i/2 \\ 0 & 0 & i/2 & 0 \end{pmatrix}, \quad \text{(B.14e)}$$

$$S_{2z} = S_{1e} \otimes S_{2z} = \begin{pmatrix} 1/2 & 0 & 0 & 0 \\ 0 & -1/2 & 0 & 0 \\ 0 & 0 & 1/2 & 0 \\ 0 & 0 & 0 & -1/2 \end{pmatrix}. \quad \text{(B.14f)}$$

In the multi-spin system, descriptions based on single transition operators are efficient if we can perform a selective excitation. The single transition operators, $S_x^{r\text{-}s}$, $S_y^{r\text{-}s}$, and $S_z^{r\text{-}s}$ indicate fictitious spin-1/2 operators for the single transition between $|r\rangle$ and $|s\rangle$ spin sublevels using. The matrix representations of the single transition operators are given below.

$$S_x^{1\text{-}3} = \begin{pmatrix} 0 & 0 & 1/2 & 0 \\ 0 & 0 & 0 & 0 \\ 1/2 & 0 & 0 & 0 \\ 0 & 0 & 0 & 0 \end{pmatrix}, \quad \text{(B.15a)}$$

$$S_y^{1\text{-}3} = \begin{pmatrix} 0 & 0 & -i/2 & 0 \\ 0 & 0 & 0 & 0 \\ i/2 & 0 & 0 & 0 \\ 0 & 0 & 0 & 0 \end{pmatrix}, \quad \text{(B.15b)}$$

$$S_z^{1\text{-}3} = \begin{pmatrix} 1/2 & 0 & 0 & 0 \\ 0 & 0 & 0 & 0 \\ 0 & 0 & -1/2 & 0 \\ 0 & 0 & 0 & 0 \end{pmatrix}, \quad \text{(B.15c)}$$

$$S_x^{2\text{-}4} = \begin{pmatrix} 0 & 0 & 0 & 0 \\ 0 & 0 & 0 & 1/2 \\ 0 & 0 & 0 & 0 \\ 0 & 1/2 & 0 & 0 \end{pmatrix}, \quad \text{(B.16a)}$$

$$S_y^{2\text{-}4} = \begin{pmatrix} 0 & 0 & 0 & 0 \\ 0 & 0 & 0 & -i/2 \\ 0 & 0 & 0 & 0 \\ 0 & i/2 & 0 & 0 \end{pmatrix}, \quad \text{(B.16b)}$$

$$S_z^{2\text{-}4} = \begin{pmatrix} 0 & 0 & 0 & 0 \\ 0 & 1/2 & 0 & 0 \\ 0 & 0 & 0 & 0 \\ 0 & 0 & 0 & -1/2 \end{pmatrix}, \quad \text{(B.16c)}$$

$$S_x^{1\text{-}2} = \begin{pmatrix} 0 & 1/2 & 0 & 0 \\ 1/2 & 0 & 0 & 0 \\ 0 & 0 & 0 & 0 \\ 0 & 0 & 0 & 0 \end{pmatrix}, \tag{B.17a}$$

$$S_y^{2\text{-}4} = \begin{pmatrix} 0 & -i/2 & 0 & 0 \\ i/2 & 0 & 0 & 0 \\ 0 & 0 & 0 & 0 \\ 0 & 0 & 0 & 0 \end{pmatrix}, \tag{B.17b}$$

$$S_z^{2\text{-}4} = \begin{pmatrix} 1/2 & 0 & 0 & 0 \\ 0 & -1/2 & 0 & 0 \\ 0 & 0 & 0 & 0 \\ 0 & 0 & 0 & 0 \end{pmatrix}, \tag{B.17c}$$

$$S_x^{3\text{-}4} = \begin{pmatrix} 0 & 0 & 0 & 0 \\ 0 & 0 & 0 & 0 \\ 0 & 0 & 0 & 1/2 \\ 0 & 0 & 1/2 & 0 \end{pmatrix}, \tag{B.18a}$$

$$S_y^{3\text{-}4} = \begin{pmatrix} 0 & 0 & 0 & 0 \\ 0 & 0 & 0 & 0 \\ 0 & 0 & 0 & -i/2 \\ 0 & 0 & i/2 & 0 \end{pmatrix}, \tag{B.18b}$$

$$S_z^{3\text{-}4} = \begin{pmatrix} 0 & 0 & 0 & 0 \\ 0 & 0 & 0 & 0 \\ 0 & 0 & 1/2 & 0 \\ 0 & 0 & 0 & -1/2 \end{pmatrix}. \tag{B.18c}$$

In the above descriptions, 1, 2, 3, and 4 stand for the spin sublevels of $|1/2, 1/2\rangle (\equiv |+, +\rangle)$, $|1/2, -1/2\rangle (\equiv |+, -\rangle)$, $|-1/2, 1/2\rangle (\equiv |-, +\rangle)$, and $|-1/2, -1/2\rangle (\equiv |-, -\rangle)$, respectively.

The exponential operators of $e^{-i\alpha S_x}$, $e^{-i\alpha S_y}$, and $e^{-i\alpha S_z}$, generate rotations by angle α about the x, y and z axes, respectively. The exponential operators are called rotation operators.

$$R_x(\alpha) = \exp(-i\alpha S_x), \tag{B.19a}$$
$$R_y(\alpha) = \exp(-i\alpha S_y), \tag{B.19b}$$
$$R_z(\alpha) = \exp(-i\alpha S_z). \tag{B.19c}$$

The matrix representations of the rotation operators are obtained by expanding the exponential operators in a power series. For $R_x(\alpha)$ of the spin-1/2,

$$R_x(\alpha) = \exp(-i\alpha S_x)$$
$$= 1 + (-i\alpha)S_x + \frac{(-i\alpha)^2}{2!}S_x^2 + \frac{(-i\alpha)^3}{3!}S_x^3 + \cdots$$
$$= 1 - \frac{1}{2!}\left(\frac{\alpha}{2}\right)^2 (2S_x)^2 + \frac{1}{4!}\left(\frac{\alpha}{2}\right)^4 (2S_x)^4 - \frac{1}{6!}\left(\frac{\alpha}{2}\right)^6 (2S_x)^6 + \cdots$$
$$- i\left\{\frac{\alpha}{2}(2S_x) - \frac{1}{3!}\left(\frac{\alpha}{2}\right)^3 (2S_x)^3 + \frac{1}{5!}\left(\frac{\alpha}{2}\right)^5 (2S_x)^5 + \cdots\right\}$$
$$= S_{\text{e}}\left\{1 - \frac{\alpha^2}{2!} + \frac{\alpha^4}{4!} - \frac{\alpha^6}{6!}\right\} - 2iS_x\left\{\frac{\alpha}{2} - \frac{1}{3!}\left(\frac{\alpha}{2}\right)^3 + \frac{\alpha^5}{5!}\left(\frac{\alpha}{2}\right)^2 + \cdots\right\}$$
$$= S_{\text{e}}\cos\frac{\alpha}{2} - 2iS_x\sin\frac{\alpha}{2}.$$

The matrix representations of the rotation operators of the spin-1/2 are given as follows:

$$R_x(\alpha) = S_{\text{e}}\cos\frac{\alpha}{2} - 2iS_x\sin\frac{\alpha}{2}$$
$$= \begin{pmatrix} \cos(\alpha/2) & -i\sin(\alpha/2) \\ -i\sin(\alpha/2) & \cos(\alpha/2) \end{pmatrix}, \quad \text{(B.20a)}$$
$$R_y(\alpha) = S_{\text{e}}\cos\frac{\alpha}{2} - 2iS_y\sin\frac{\alpha}{2}$$
$$= \begin{pmatrix} \cos(\alpha/2) & -\sin(\alpha/2) \\ \sin(\alpha/2) & \cos(\alpha/2) \end{pmatrix}, \quad \text{(B.20b)}$$
$$R_z(\alpha) = S_{\text{e}}\cos\frac{\alpha}{2} - 2iS_z\sin\frac{\alpha}{2}$$
$$= \begin{pmatrix} \cos(\alpha/2) - i\sin(\alpha/2) & 0 \\ 0 & \cos(\alpha/2) + i\sin(\alpha/2) \end{pmatrix}$$
$$= \begin{pmatrix} \exp(-i\alpha/2) & 0 \\ 0 & \exp(i\alpha/2) \end{pmatrix}. \quad \text{(B.20c)}$$

We summarize some useful expressions about the rotation operators applied to the spin operators.

$$R_x(\alpha)S_x R_x^{-1}(\alpha) = S_x, \tag{B.21a}$$
$$R_y(\alpha)S_x R_y^{-1}(\alpha) = S_x \cos\alpha - S_z \sin\alpha, \tag{B.21b}$$
$$R_z(\alpha)S_x R_z^{-1}(\alpha) = S_x \cos\alpha + S_y \sin\alpha, \tag{B.21c}$$
$$R_x(\alpha)S_y R_x^{-1}(\alpha) = S_y \cos\alpha + S_z \sin\alpha, \tag{B.21d}$$
$$R_y(\alpha)S_y R_y^{-1}(\alpha) = S_y, \tag{B.21e}$$
$$R_z(\alpha)S_y R_z^{-1}(\alpha) = S_y \cos\alpha - S_x \sin\alpha, \tag{B.21f}$$
$$R_x(\alpha)S_z R_x^{-1}(\alpha) = S_z \cos\alpha - S_y \sin\alpha, \tag{B.21g}$$
$$R_y(\alpha)S_z R_y^{-1}(\alpha) = S_z \cos\alpha + S_x \sin\alpha, \tag{B.21h}$$
$$R_z(\alpha)S_z R_z^{-1}(\alpha) = S_z. \tag{B.21i}$$

In the two-coupled spins ($S_1 = 1/2$ and $S_2 = 1/2$), the rotation operators for individual spins are expressed by

$$\begin{aligned}R_{xS_1} &= \exp(-i\alpha S_{1x}) \\ &= S_{1e} \otimes S_{2e} \cos\frac{\alpha}{2} - 2iS_{1x} \otimes S_{2e} \sin\frac{\alpha}{2},\end{aligned} \tag{B.22a}$$

$$\begin{aligned}R_{yS_1} &= \exp(-i\alpha S_{1y}) \\ &= S_{1e} \otimes S_{2e} \cos\frac{\alpha}{2} - 2iS_{1y} \otimes S_{2e} \sin\frac{\alpha}{2},\end{aligned} \tag{B.22b}$$

$$\begin{aligned}R_{zS_1} &= \exp(-i\alpha S_{1z}) \\ &= S_{1e} \otimes S_{2e} \cos\frac{\alpha}{2} - 2iS_{1z} \otimes S_{2e} \sin\frac{\alpha}{2},\end{aligned} \tag{B.22c}$$

$$\begin{aligned}R_{xS_2} &= \exp(-i\alpha S_{2x}) \\ &= S_{1e} \otimes S_{2e} \cos\frac{\alpha}{2} - 2iS_{1e} \otimes S_{2x} \sin\frac{\alpha}{2},\end{aligned} \tag{B.23a}$$

$$\begin{aligned}R_{yS_2} &= \exp(-i\alpha S_{2y}) \\ &= S_{1e} \otimes S_{2e} \cos\frac{\alpha}{2} - 2iS_{1e} \otimes S_{2y} \sin\frac{\alpha}{2},\end{aligned} \tag{B.23b}$$

$$\begin{aligned}R_{zS_2} &= \exp(-i\alpha S_{2z}) \\ &= S_{1e} \otimes S_{2e} \cos\frac{\alpha}{2} - 2iS_{1e} \otimes S_{2z} \sin\frac{\alpha}{2}.\end{aligned} \tag{B.23c}$$

The rotation operators for the single-transition between $|i\rangle$ and $|j\rangle$ states are also given by

$$R_x^{i\text{-}j}(\alpha) = \exp(-i\alpha S_x^{i\text{-}j}), \tag{B.24a}$$
$$R_y^{i\text{-}j}(\alpha) = \exp(-i\alpha S_y^{i\text{-}j}), \tag{B.24b}$$
$$R_z^{i\text{-}j}(\alpha) = \exp(-i\alpha S_z^{i\text{-}j}). \tag{B.24c}$$

The matrix representations of the rotation operators for the single-transition between $|1\rangle(\equiv |1/2, 1/2\rangle)$ and $|3\rangle(\equiv |-1/2, 1/2\rangle)$ states are explicitly given by

$$R_x^{1\text{-}3}(\alpha) = \begin{pmatrix} \cos(\alpha/2) & 0 & -i\sin(\alpha/2) & 0 \\ 0 & 1 & 0 & 0 \\ -i\sin(\alpha/2) & 0 & \cos(\alpha/2) & 0 \\ 0 & 0 & 0 & 1 \end{pmatrix}, \quad \text{(B.25a)}$$

$$R_y^{1\text{-}3}(\alpha) = \begin{pmatrix} \cos(\alpha/2) & 0 & -\sin(\alpha/2) & 0 \\ 0 & 1 & 0 & 0 \\ \sin(\alpha/2) & 0 & \cos(\alpha/2) & 0 \\ 0 & 0 & 0 & 1 \end{pmatrix}, \quad \text{(B.25b)}$$

$$R_z^{1\text{-}3}(\alpha) = \begin{pmatrix} \exp(-i\alpha/2) & 0 & 0 & 0 \\ 0 & 1 & 0 & 0 \\ 0 & 0 & \exp(i\alpha/2) & 0 \\ 0 & 0 & 0 & 1 \end{pmatrix}. \quad \text{(B.25c)}$$

Appendix B.3. *A Single Spin System (S = 1/2) in the Presence of a Static Magnetic Field*

The spin Hamiltonian of a spin-1/2 in a static magnetic field $\boldsymbol{B}_0 = (0, 0, B_0)$ is given by

$$H = \beta \tilde{\boldsymbol{S}} \cdot \boldsymbol{g} \cdot \boldsymbol{B}_0 = g\beta B_0 S_z = \omega_S S_z, \quad \text{(B.26)}$$

where g and β are electron g factor and the Bohr magneton, respectively, as previously given. In Eq. (B.26), we introduce the Larmor frequency ω_S equal to $g\beta$. When an irradiation field $\boldsymbol{B}_1$ with frequency ω_{MF} is applied in the xy plane, the spin Hamiltonian is

$$\begin{aligned} H(t) &= H_0 + H_1(t) \\ &= \beta \tilde{\boldsymbol{S}} \cdot \boldsymbol{g} \cdot (\boldsymbol{B}_0 + \boldsymbol{B}_1(t)) \\ &= \omega_S S_z + \omega_1 \left[S_x \cos(\omega_{\text{MF}} t + \phi) + S_y \sin(\omega_{\text{MF}} t + \phi) \right], \end{aligned} \quad \text{(B.27)}$$

where $\omega_1 (= g\beta B_1)$ and ϕ are the strength and phase of the irradiation field. Transforming the laboratory frame to a rotating frame that rotates about the z axis with frequency ω_{MF}, we can make a time-independent rotating-frame Hamiltonian H^r. The density matrix in the rotating-frame is then given by

$$\rho^r(t) = \exp(i\omega_{\text{MF}} S_z t)\rho(t) \exp(-i\omega_{\text{MF}} S_z t). \quad \text{(B.28)}$$

From Eqs. (B.1), (B.27), and (B.28), he Liouville-von Neumann equation in the rotating frame becomes

$$\frac{\partial}{\partial t}\rho^r(t) = i[H_{\text{eff}}^r, \rho^r(t)], \tag{B.29}$$

where $H_{\text{eff}}^r = (\omega_S - \omega_{\text{MF}})S_z + \omega_1(S_x \cos\phi + S_y \sin\phi)$.

Here, we consider a two-pulse Hahn echo experiment of the single spin system. In the pulsed EMR experiments, we apply a strong irradiation pulse to the spin system on resonance condition, $\Delta\omega = \omega_S - \omega_{\text{MW}} = 0$. The rotating frame Hamiltonian during the pulse irradiation is described as

$$H_p^r = \omega_1(S_x \cos\phi + S_y \sin\phi). \tag{B.30}$$

The propagator defined by Eq. (B.5) is given by

$$U_p(t) = \exp(-iH_p^r t)$$
$$= \exp[-i\omega_1 t(S_x \cos\phi + S_y \sin\phi)]$$
$$= \begin{pmatrix} \cos\frac{\omega_1 t}{2} & \sin\frac{\omega_1 t}{2}(-i\cos\phi - \sin\phi) \\ \sin\frac{\omega_1 t}{2}(-i\cos\phi + \sin\phi) & \cos\frac{\omega_1 t}{2} \end{pmatrix}, \tag{B.31}$$

for the irradiation pulse. The rotation angle of spin depends on both the strength ω_1 and irradiation time t of the pulse. Using a relationship of

$$R_\phi(\alpha) = R_z(\phi)R_x(\alpha)R_z^{-1}(\phi)$$
$$= \exp(-i\phi S_z)\left(S_e \cos\frac{\alpha}{2} - 2iS_x \sin\frac{\alpha}{2}\right)\exp(i\phi S_z)$$
$$= S_e \cos\frac{\alpha}{2} - 2i(S_x \cos\phi + S_y \sin\phi)\sin\frac{\alpha}{2}.$$

The propagator $U_p(t)$ for the pulse with the phase ϕ is rewritten as

$$U_p(t) = R_z(\phi)R_x(\omega_1 t)R_z^{-1}(\phi). \tag{B.32}$$

In particular, $R_{\phi=0}(\alpha)$ and $R_{\phi=\pi/2}(\alpha)$ generate rotations by the angle $\alpha = \omega_1 t$ about the x and y axes, respectively.

In the absence of the irradiation field, the spin system develops in the static magnetic field. The rotating frame Hamiltonian during the free evolution period is described as

$$H_{\text{eff}}^r = (\omega_S - \omega_{\text{MF}})S_z = \Delta\omega S_z. \tag{B.33}$$

Then the evolution propagator in the rotating frame is

$$U(t) = \exp(-iH_{\text{eff}}^r t) = \exp(-i\Delta\omega t). \tag{B.34}$$

Hahn spin-echo experiment is performed using a pulse sequence of

$$\rho_0 \xrightarrow{R_x(\pi/2)} \rho_1 \xrightarrow{U(t)} \rho_2 \xrightarrow{R_x(\pi)} \rho_3 \xrightarrow{U(t)} \rho_{\text{echo}}. \tag{B.35}$$

For simplicity, we apply the delta pulse approximation to the mathematical derivations. The density matrix of the refocused echo is calculated by

$$\rho_{\text{echo}} = U(t)R_x(\pi)U(t)R_x(\pi/2)\rho_0 R_x^{-1}(\pi/2)U^{-1}(t)R_x^{-1}(\pi)U^{-1}(t). \quad (B.36)$$

Here the density matrix of the initial state is given by

$$\begin{aligned}
\rho_0 &= \frac{\exp(-\omega_S S_z/k_B T)}{\text{Tr}\left[\exp(-\omega_S S_z/k_B T)\right]} \\
&= \frac{S_e - (\omega_S S_z/k_B T) + (\omega_S S_z/k_B T)^2/2! - \cdots}{\text{Tr}\left[S_e - (\omega_S S_z/k_B T) + (\omega_S S_z/k_B T)^2/2! - \cdots\right]} \\
&= \frac{S_e - (\omega_S S_z/k_B T)}{\text{Tr}\, S_e} = \frac{1}{2}\left(S_e - \frac{\omega_S}{k_B T}S_z\right) \quad (B.37)
\end{aligned}$$

in the high-temperature approximation. Since the first term S_e in Eq. (B.37) is irrelevant in the pulsed experiments, we may simply write the initial density matrix as

$$\rho_0 \approx -\frac{\omega_S}{k_B T}S_z \propto -S_z. \quad (B.38)$$

As the proportional factor is not important in order to consider the expression for the pulsed experiments, we approximate the initial density matrix as

$$\rho_0 = -S_z. \quad (B.39)$$

Then, we have the corresponding density matrix in the Hahn echo experiment from Eq. (B.34).

$$\begin{aligned}
\rho_{\text{echo}} &= U(t)R_x(\pi)U(t)R_x(\pi/2)(-S_z) \\
&\quad \times R_x^{-1}(\pi/2)U^{-1}(t)R_x^{-1}(\pi)U^{-1}(t) \\
&= -S_y. \quad (B.40)
\end{aligned}$$

The expectation value of the operator S_y at time $2t$ is

$$I^{\text{echo}} = \langle S_y \rangle = \text{Tr}\,(S_y \rho_{\text{echo}}) = -\frac{1}{2}. \quad (B.41)$$

When we apply the second π pulse with general phase ϕ instead of the x-pulse, we have as the density matrix and the expectation value of the operator S_y:

$$\begin{aligned}\rho_{\text{echo}} &= U(t)R_\phi(\pi)U(t)R_x(\pi/2)(-S_z) \\ &\quad \times R_x^{-1}(\pi/2)U^{-1}(t)R_\phi^{-1}(\pi)U^{-1}(t) \\ &= U(t)\exp(-i\phi S_z)\exp(-i\pi S_x)\exp(-i\phi S_z)U(t) \\ &\quad \times \exp(-i\pi S_x/2)(-S_z)[\exp(-i\pi S_x/2)]^{-1}U^{-1}(t) \\ &\quad \times [\exp(-i\phi S_z)]^{-1}[\exp(-i\pi S_x)]^{-1}[\exp(i\phi S_z)]^{-1}U^{-1}(t) \\ &= \begin{pmatrix} 0 & \frac{i}{2}\exp(2i\phi) \\ -\frac{i}{2}\exp(2i\phi) & 0 \end{pmatrix},\end{aligned}$$

and

$$I^{\text{echo}} = \langle S_y \rangle = \text{Tr}\,(S_y \rho_{\text{echo}}) = -\frac{1}{2}\cos 2\phi. \tag{B.42}$$

Appendix B.4. *The Electron and Nuclear Spins-System Manipulated by Pulsed ENDOR ($S = I = 1/2$)*

The spin Hamiltonian of the electron and nuclear spins system in a static magnetic field $\boldsymbol{B}_0 = (0,0,B_0)$ is given by

$$\begin{aligned}H_0 &= \beta_e \tilde{\boldsymbol{S}} \cdot \boldsymbol{g}_e \cdot \boldsymbol{B}_0 - \beta_n \tilde{\boldsymbol{I}} \cdot \boldsymbol{g}_n \cdot \boldsymbol{B}_0 + \tilde{\boldsymbol{S}} \cdot \boldsymbol{A} \cdot \boldsymbol{I} \\ &= g_e\beta_e B_0 S_z - g_n\beta_n B_0 I_z + A_{zz}S_z I_z + A_{zx}S_z I_x + A_{zy}S_z I_y \\ &= \omega_S S_z - \omega_I I_z + A_{zz}S_z I_z + A_{zx}S_z I_x + A_{zy}S_z I_y, \end{aligned}\tag{B.43}$$

where g_e denotes the electron g factor, β_e Bohr magneton, g_n nuclear g factor and β_n nuclear magneton, respectively. The Hamiltonian consists of the electron-Zeeman, nuclear-Zeeman and hyperfine coupling terms. In Eq. (B.43), we introduced the Larmor frequencies ω_S equal to $g_e\beta_e$ and ω_I to $g_n\beta_n$. In the high-filed approximation, we may simply write the spin Hamiltonian as

$$H_0 \approx \omega_S S_z - \omega_I I_z + A_{zz}S_z I_z. \tag{B.44}$$

Then the energy eigenvalues of the Hamiltonian in the direct product basis of the electron Zeeman and nuclear Zeeman bases, $|M_S, M_I\rangle$, are described as

$$E = \omega_S M_S - \omega_I M_I + A_{zz} M_S M_I. \tag{B.45}$$

The typical energy diagram of the spin state is shown in Fig. B1. When, $\omega_S \gg \omega_I simeq A_{zz}$, the density matrix in the thermal equilibrium state is approximated to be

$$\rho_0 = -S_z \otimes I_e. \tag{B.46}$$

When an irradiation field $\boldsymbol{B}_1$ with frequency $\omega_{\rm MF}$ or an irradiation field $\boldsymbol{B}_2$ with frequency $\omega_{\rm RF}$ is applied in the xy plane, the time-dependent spin Hamiltonian is

$$\begin{aligned}
H^{\rm MW}(t) &= H_0 + H_1(t) \\
&= \beta_{\rm e}\tilde{\boldsymbol{S}} \cdot \boldsymbol{g}_{\rm e} \cdot (\boldsymbol{B}_0 + \boldsymbol{B}_1(t)) + \beta_{\rm n}\tilde{\boldsymbol{I}} \cdot \boldsymbol{g}_{\rm n} \cdot \boldsymbol{B}_0 + \tilde{\boldsymbol{S}} \cdot \boldsymbol{A} \cdot \boldsymbol{I} \\
&= \omega_{\rm S} S_z + \omega_{\rm I} I_z + A_{zz} S_z I_z \\
&\quad + \omega_1 \left[S_x \cos(\omega_{\rm MW} t + \phi) + I_y \sin(\omega_{\rm MW} t + \phi) \right], \quad {\rm (B.47)}
\end{aligned}$$

or

$$\begin{aligned}
H^{\rm RF}(t) &= H_0 + H_2(t) \\
&= \beta_{\rm e}\tilde{\boldsymbol{S}} \cdot \boldsymbol{g}_{\rm e} \cdot \boldsymbol{B}_0 + \beta_{\rm n}\tilde{\boldsymbol{I}} \cdot \boldsymbol{g}_{\rm n} \cdot (\boldsymbol{B}_0 + \boldsymbol{B}_2(t)) + \tilde{\boldsymbol{S}} \cdot \boldsymbol{A} \cdot \boldsymbol{I} \\
&= \omega_{\rm S} S_z + \omega_{\rm I} I_z + A_{zz} S_z I_z \\
&\quad + \omega_2 \left[I_x \cos(\omega_{\rm RF} t + \phi') + S_y \sin(\omega_{\rm RF} t + \phi') \right], \quad {\rm (B.48)}
\end{aligned}$$

where $\omega_1 (= g_{\rm e}\beta_{\rm e} B_1)$ and ϕ are the strength and phase of the microwave irradiation field and $\omega_2 (= g_{\rm n}\beta_{\rm n} B_2)$ and ϕ' are the strength and phase of the RF irradiation field. Transforming the laboratory frame to a doubly rotating frame that rotates about the z axis with frequencies $\omega_{\rm MW}$ and

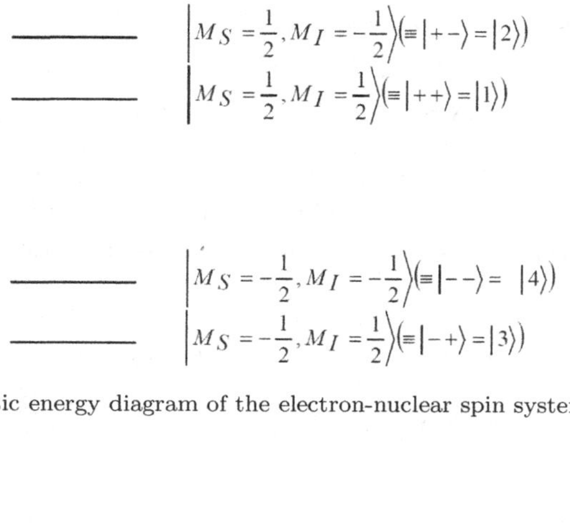

$\left| M_S = \frac{1}{2}, M_I = -\frac{1}{2} \right\rangle (\equiv | + - \rangle = |2\rangle)$

$\left| M_S = \frac{1}{2}, M_I = \frac{1}{2} \right\rangle (\equiv | + + \rangle = |1\rangle)$

$\left| M_S = -\frac{1}{2}, M_I = -\frac{1}{2} \right\rangle (\equiv | - - \rangle = |4\rangle)$

$\left| M_S = -\frac{1}{2}, M_I = \frac{1}{2} \right\rangle (\equiv | - + \rangle = |3\rangle)$

Fig. B1. Schematic energy diagram of the electron-nuclear spin system ($S = I = 1/2$). $2\omega_{\rm I} > A_{zz} > 0$.

ω_{RF}, we can make a rotating-frame Hamiltonians $H^{\text{r\,MW}}$ and $H^{\text{r\,RF}}$:

$$\begin{aligned}
H^{\text{r\,MW}} &= \exp(i\omega_{\text{RF}} I_z t)\exp(i\omega_{\text{MW}} S_z t) H^{\text{MW}}(t) \\
&\quad \times \exp(-i\omega_{\text{MW}} S_z t)\exp(-i\omega_{\text{RF}} I_z t) \\
&= \omega_S S_z + \omega_I I_z + A_{zz} S_z I_z + \omega_1(S_x \cos\phi + S_y \sin\phi), \quad \text{(B.49)} \\
H^{\text{r\,RF}} &= \exp(i\omega_{\text{RF}} I_z t)\exp(i\omega_{\text{MW}} S_z t) H^{\text{RF}}(t) \\
&\quad \times \exp(-i\omega_{\text{MW}} S_z t)\exp(-i\omega_{\text{RF}} I_z t) \\
&= \omega_S S_z + \omega_I I_z + A_{zz} S_z I_z + \omega_2(I_x \cos\phi' + I_y \sin\phi'). \quad \text{(B.50)}
\end{aligned}$$

The density matrix in the doubly rotating-frame is then given by

$$\begin{aligned}
\rho^{\text{r}}(t) &= \exp(i\omega_{\text{RF}} I_z t)\exp(i\omega_{\text{MW}} S_z t)\rho(t) \\
&\quad \times \exp(-i\omega_{\text{MW}} S_z t)\exp(-i\omega_{\text{RF}} I_z t). \quad \text{(B.51)}
\end{aligned}$$

From Eqs. (B.1), (B.22), and (B.23), the Liouville-von Neumann equation in the doubly rotating frame becomes

$$\frac{\partial}{\partial t}\rho^{\text{r}}(t) = i[H^{\text{r}}_{\text{eff}}, \rho^{\text{r}}(t)]. \quad \text{(B.52)}$$

The effective Hamiltonian with microwave irradiation in the doubly rotating frame is

$$\begin{aligned}
H_0 &= (\omega_S - \omega_{\text{MW}})S_z + (\omega_I - \omega_{\text{RF}})I_z + A_{zz}S_z I_z \\
&\quad + \omega_1(S_x \cos\phi + S_y \sin\phi) \\
&= \Delta\omega_S S_z + \Delta\omega_I I_z + A_{zz}S_z I_z + \omega_1(S_x \cos\phi + S_y \sin\phi). \quad \text{(B.53)}
\end{aligned}$$

The effective Hamiltonian with RF irradiation is

$$\begin{aligned}
H_0 &= (\omega_S - \omega_{\text{MW}})S_z + (\omega_I - \omega_{\text{RF}})I_z + A_{zz}S_z I_z \\
&\quad + \omega_2(I_x \cos\phi' + I_y \sin\phi') \\
&= \Delta\omega_S S_z + \Delta\omega_I I_z + A_{zz}S_z I_z + \omega_2(I_x \cos\phi' + I_y \sin\phi'). \quad \text{(B.54)}
\end{aligned}$$

When we can selectively excite an allowed electron-spin transition, we may focus on the transition only excited by the microwave irradiation. The effective Hamiltonians for the selective excitation of the transitions are described as

$$H^{\text{r},1\text{-}3}_{\text{eff}} = \omega_1\left(S^{1\text{-}3}_x \cos\phi + S^{1\text{-}3}_y \sin\phi\right) \text{ for } \omega_{\text{MW}} = \omega_S + \frac{A_{zz}}{2}, \quad \text{(B.55a)}$$

$$H^{\text{r},2\text{-}4}_{\text{eff}} = \omega_1\left(S^{2\text{-}4}_x \cos\phi + S^{2\text{-}4}_y \sin\phi\right) \text{ for } \omega_{\text{MW}} = \omega_S - \frac{A_{zz}}{2}. \quad \text{(B.55b)}$$

The effective Hamiltonians for the selective nuclear-spin transitions are described as

$$H_{\text{eff}}^{\text{r},1\text{-}2} = \omega_2 \left(S_x^{1\text{-}2} \cos\phi + S_y^{1\text{-}2} \sin\phi \right) \text{ for } \omega_{\text{RF}} = \omega_{\text{I}} + \frac{A_{zz}}{2}, \quad \text{(B.56a)}$$

$$H_{\text{eff}}^{\text{r},3\text{-}4} = \omega_2 \left(S_x^{3\text{-}4} \cos\phi + S_y^{3\text{-}4} \sin\phi \right) \text{ for } \omega_{\text{RF}} = \omega_{\text{I}} - \frac{A_{zz}}{2}. \quad \text{(B.56b)}$$

In Eqs. (B.55) and (B.56), the Hamiltonians are given using the single transition operators defined in Appendix B.2.

Appendix B.4.1. *Initialization; Generation of the Pseudo-Pure State*

A pseudo-pure state is prepared in the selective excitation arguments by application of a pulse sequence a microwave pulse with $\theta = \arccos(-1/3) \sim 109.471$ degrees and RF $\pi/2$ pulse as follows:

$$\rho_0 \xrightarrow{R^{2\text{-}4}(\theta = \arccos(-1/3))} \rho_1 \xrightarrow{L(t_1)} \rho_2 \xrightarrow{R_x^{1\text{-}2}(\pi/2)} \rho_3 \xrightarrow{L(t_2)} \rho_{\text{pp}}. \quad \text{(B.57)}$$

Then waiting times t_1 and t_2 are taken after the first and second pulses in order to decay off-diagonal elements in the density matrix. When MW pulse for a $|2\rangle \leftrightarrow |4\rangle$ electron spin transition and RF1 pulse for $|1\rangle \leftrightarrow |2\rangle$ nuclear spin transition are selectively applied to the thermal equilibrium state that is given by $\rho_0 \approx -S_z$, the following pseudo-pure state is generated.

$$\rho_{\text{pp}} = L(t_2) R_x^{1\text{-}2}(\pi/2) L(t_1) R_x^{2\text{-}4}(\theta)(-S_z)$$
$$\times [R_x^{2\text{-}4}(\theta)]^{-1} L(t_1) [R_x^{1\text{-}2}(\pi/2)]^{-1} L(t_2), \quad \text{(B.58)}$$

where $L(t_1)$ and $L(t_2)$ mean the waiting times to decay the coherence. $L(t_1)$ is defined by

$$L(t) = \exp(-iH_{\text{eff}}^{\text{r}} t) = \exp\left[-i \left(\Delta\omega_S S_z + \Delta\omega_I I_z + A_{zz} S_z I_z \right) \right]. \quad \text{(B.59)}$$

After the waiting time t, the density matrix of the system may be considered as a diagonal matrix. The matrix representation of the generated spin state is given by

$$\rho_{\text{pp}} = \begin{pmatrix} -\frac{1}{2}\cos^2\frac{\theta}{2} & 0 & 0 & 0 \\ 0 & -\frac{1}{2}\cos^2\frac{\theta}{2} & 0 & 0 \\ 0 & 0 & \frac{1}{2} & 0 \\ 0 & 0 & 0 & \frac{1}{2}\cos\theta \end{pmatrix}. \quad \text{(B.60)}$$

In order that the spin state described by ρ_{pp} is a pseudo-pure state, the following relation must be satisfied.

$$-\frac{1}{2}\cos^2\frac{\theta}{2} = \frac{1}{2}\cos\theta \iff \cos\theta = -\frac{1}{3}. \quad \text{(B.61)}$$

Then, we have

$$\rho_{pp} = \begin{pmatrix} -1/6 & 0 & 0 & 0 \\ 0 & -1/6 & 0 & 0 \\ 0 & 0 & 1/2 & 0 \\ 0 & 0 & 0 & -1/6 \end{pmatrix} = \rho_{|3\rangle}. \tag{B.62}$$

When we apply the selective excitation pulse of the $|1\rangle \leftrightarrow |3\rangle$ electron spin transition instead of that for the $|2\rangle \leftrightarrow |3\rangle$ transition, the pseudo-pure state of $\rho_{|2\rangle} \equiv |+-\rangle\langle +-|$ and $\rho_{|4\rangle} \equiv |--\rangle\langle --|$ are generated in the same manner.

$$\rho_0 \xrightarrow{R_x^{1\text{-}3}(\theta=\arccos(-1/3))} \rho_1 \xrightarrow{L(t_1)} \rho_2 \xrightarrow{R_x^{3\text{-}4}(\pi/2)} \rho_3 \xrightarrow{L(t_2)} \rho_{|2\rangle}.$$

$$\rho_{|2\rangle} = L(t_2) R_x^{3\text{-}4}(\pi/2) L(t_1) R_x^{2\text{-}4}(\theta)(-S_z)[R_x^{2\text{-}4}(\theta)]^{-1} L(t_1)[R_x^{3\text{-}4}(\pi/2)]^{-1} L(t_2)$$

$$= \begin{pmatrix} -\tfrac{1}{2}\cos\theta & 0 & 0 & 0 \\ 0 & -\tfrac{1}{2} & 0 & 0 \\ 0 & 0 & \tfrac{1}{2}\cos^2\tfrac{\theta}{2} & 0 \\ 0 & 0 & 0 & \tfrac{1}{2}\cos^2\tfrac{\theta}{2} \end{pmatrix}. \tag{B.63}$$

$$\rho_0 \xrightarrow{R_x^{1\text{-}3}(\theta=\arccos(-1/3))} \rho_1 \xrightarrow{L(t_1)} \rho_2 \xrightarrow{R_x^{1\text{-}2}(\pi/2)} \rho_3 \xrightarrow{L(t_2)} \rho_{|4\rangle},$$

$$\rho_{|4\rangle} = L(t_2) R_x^{3\text{-}4}(\pi/2) L(t_1) R_x^{2\text{-}4}(\theta)(-S_z)[R_x^{2\text{-}4}(\theta)]^{-1} L(t_1)[R_x^{3\text{-}4}(\pi/2)]^{-1} L(t_2)$$

$$= \begin{pmatrix} -\tfrac{1}{2}\cos^2\tfrac{\theta}{2} & 0 & 0 & 0 \\ 0 & -\tfrac{1}{2}\cos^2\tfrac{\theta}{2} & 0 & 0 \\ 0 & 0 & \tfrac{1}{2}\cos\theta & 0 \\ 0 & 0 & 0 & \tfrac{1}{2} \end{pmatrix}. \tag{B.64}$$

Appendix B.4.2. *Entangling the Pure States*

The entangled states of the electron-nuclear spin system are generated by application of the $\pi/2$-RF pulse and π-MW pulse. The density matrices of the entangled states generated are shown as follows:

$$\rho_{|3\rangle} \xrightarrow{R_x^{3\text{-}4}(\pi/2)} \rho_1 \xrightarrow{R_{\pm x}^{2\text{-}4}(\pi)} \rho_{\Psi\mp},$$

$$\rho_{\Psi\mp} = R_{\pm x}^{2\text{-}4}(\pi) R_x^{3\text{-}4}(\pi/2) \rho_{|3\rangle} [R_x^{3\text{-}4}(\pi/2)]^{-1} [R_{\pm x}^{2\text{-}4}(\pi)]^{-1}$$

$$= \frac{1}{2}\begin{pmatrix} 0 & 0 & 0 & 0 \\ 0 & 1 & \mp 1 & 0 \\ 0 & \mp 1 & 1 & 0 \\ 0 & 0 & 0 & 0 \end{pmatrix}, \tag{B.65a}$$

$$\rho_{|1\rangle} \xrightarrow{R_x^{1\text{-}2}(\pi/2)} \rho_1 \xrightarrow{R_{\pm x}^{2\text{-}4}(\pi)} \rho_{\Phi\mp},$$

$$\rho_{\Phi\mp} = R_{\pm x}^{2\text{-}4}(\pi)R_x^{1\text{-}2}(\pi/2)\rho_{|1\rangle}[R_x^{1\text{-}2}(\pi/2)]^{-1}[R_{\pm x}^{2\text{-}4}(\pi)]^{-1}$$

$$= \frac{1}{2}\begin{pmatrix} 1 & 0 & 0 & \mp 1 \\ 0 & 0 & 0 & 0 \\ 0 & 0 & 0 & 0 \\ \mp 1 & 0 & 0 & 1 \end{pmatrix}, \quad (B.65b)$$

$$\rho_{|2\rangle} \xrightarrow{R_x^{1\text{-}2}(\pi/2)} \rho_1 \xrightarrow{R_{\pm x}^{1\text{-}3}(\pi)} \rho_{\Psi\mp},$$

$$\rho_{\Psi\mp} = R_{\pm x}^{1\text{-}3}(\pi)R_x^{1\text{-}2}(\pi/2)\rho_{|2\rangle}[R_x^{1\text{-}2}(\pi/2)]^{-1}[R_{\pm x}^{1\text{-}3}(\pi)]^{-1}$$

$$= \frac{1}{2}\begin{pmatrix} 1 & 0 & 0 & 0 \\ 0 & 1 & \mp 1 & 0 \\ 0 & \mp 1 & 1 & 0 \\ 0 & 0 & 0 & 0 \end{pmatrix}, \quad (B.65c)$$

$$\rho_{|4\rangle} \xrightarrow{R_x^{3\text{-}4}(\pi/2)} \rho_1 \xrightarrow{R_{\pm x}^{1\text{-}3}(\pi)} \rho_{\Phi\mp},$$

$$\rho_{\Phi\mp} = R_{\pm x}^{1\text{-}3}(\pi)R_x^{3\text{-}4}(\pi/2)\rho_{|4\rangle}[R_x^{3\text{-}4}(\pi/2)]^{-1}[R_{\pm x}^{1\text{-}3}(\pi)]^{-1}$$

$$= \frac{1}{2}\begin{pmatrix} 1 & 0 & 0 & \mp 1 \\ 0 & 0 & 0 & 0 \\ 0 & 0 & 0 & 0 \\ \mp 1 & 0 & 0 & 1 \end{pmatrix}, \quad (B.65d)$$

where

$$|\Psi^{\pm}\rangle = \frac{1}{\sqrt{2}}(|+-\rangle \pm |-+\rangle), \quad |\Phi^{\pm}\rangle = \frac{1}{\sqrt{2}}(|++\rangle \pm |--\rangle). \quad (B.66)$$

Appendix B.4.3. *Inter-Conversion between the Entangled States*

Here we demonstrate the inter-conversion between the entangled states applying the unitary operations in the scenario of the pulsed ENDOR-QC experiments based on the selective excitation. When we define the unitary operators X^S, Y^S, and Z^S as the rotation operators for the electron spin by angle π about the x, y and z-axes, respectively, the unitary operators convert the entangled state to the others. For instance,

$$|\Phi^{\pm}\rangle = \frac{1}{\sqrt{2}}(|++\rangle \pm |--\rangle)$$
$$\xrightarrow{X^S \equiv \exp(-i\pi S_x)} |\Psi^{\pm}\rangle = \frac{1}{\sqrt{2}}(|+-\rangle \pm |-+\rangle), \quad \text{(B.67a)}$$

$$|\Phi^{\pm}\rangle = \frac{1}{\sqrt{2}}(|++\rangle \pm |--\rangle)$$
$$\xrightarrow{Y^S \equiv \exp(-i\pi S_y)} |\Psi^{\mp}\rangle = \frac{1}{\sqrt{2}}(|+-\rangle \mp |-+\rangle), \quad \text{(B.67b)}$$

$$|\Phi^{\pm}\rangle = \frac{1}{\sqrt{2}}(|++\rangle \pm |--\rangle)$$
$$\xrightarrow{Z^S \equiv \exp(-i\pi S_z)} |\Phi^{\mp}\rangle = \frac{1}{\sqrt{2}}(|++\rangle \mp |--\rangle). \quad \text{(B.67c)}$$

As shown in Eq. (B.67), all the unitary operators are implemented by using a combination of selective excitation pulses for the electron spin transitions.

$$R_x^{2\text{-}4}(\pi)R_x^{1\text{-}3}(\pi) \quad \text{for} \quad X^S, \quad \text{(B.68a)}$$
$$R_x^{2\text{-}4}(2\pi)R_x^{2\text{-}4}(\pi)R_x^{1\text{-}3}(\pi) \quad \text{for} \quad Y^S, \quad \text{(B.68b)}$$
$$R_x^{2\text{-}4}(2\pi) \quad \text{for} \quad Z^S. \quad \text{(B.68c)}$$

For the unitary operators X^I, Y^I, and Z^I defined as the rotation operators for the nuclear spin by angle π, we can construct the operators applied to the nuclear spin:

$$|\Phi^{\pm}\rangle = \frac{1}{\sqrt{2}}(|++\rangle \pm |--\rangle)$$
$$\xrightarrow{X^I \equiv \exp(-i\pi I_x)} |\Psi^{\pm}\rangle = \frac{1}{\sqrt{2}}(|+-\rangle \pm |-+\rangle), \quad \text{(B.69a)}$$

$$|\Phi^{\pm}\rangle = \frac{1}{\sqrt{2}}(|++\rangle \pm |--\rangle)$$
$$\xrightarrow{Y^I \equiv \exp(-i\pi I_y)} |\Psi^{\mp}\rangle = \frac{1}{\sqrt{2}}(|+-\rangle \mp |-+\rangle), \quad \text{(B.69b)}$$

$$|\Phi^{\pm}\rangle = \frac{1}{\sqrt{2}}(|++\rangle \pm |--\rangle)$$
$$\xrightarrow{Z^I \equiv \exp(-i\pi I_z)} |\Phi^{\mp}\rangle = \frac{1}{\sqrt{2}}(|++\rangle \mp |--\rangle). \quad \text{(B.69c)}$$

A combination of selective excitation pulses for manipulating the nuclear spin are also given as

$$R_x^{3\text{-}4}(\pi)R_x^{1\text{-}2}(\pi) \quad \text{for} \quad X^I, \quad \text{(B.70a)}$$
$$R_x^{3\text{-}4}(2\pi)R_x^{3\text{-}4}(\pi)R_x^{1\text{-}2}(\pi) \quad \text{for} \quad Y^I, \quad \text{(B.70b)}$$
$$R_x^{3\text{-}4}(2\pi) \quad \text{for} \quad Z^I. \quad \text{(B.70c)}$$

Appendix B.4.4. *Characterization of the Entangled States and Readout*

In order to characterize the entangled states by monitoring the interference between the phases of the MW and RF pulses, we apply the MW π-pulse and RF $\pi/2$-pulse with general phases ϕ_{MF} and ϕ_{RF}, measuring the phase dependence on the Hahn echo signals after the manipulation. We consider the entangled state given by the form of

$$\rho = \begin{pmatrix} 0 & 0 & 0 & 0 \\ 0 & 1/2 & x & 0 \\ 0 & x & 1/2 & 0 \\ 0 & 0 & 0 & 0 \end{pmatrix}. \tag{B.71}$$

When $x = \pm 1/2$, the density matrix ρ corresponds to that of $|\Psi^\pm\rangle$. $x = 0$ means that the spin system does not have any coherence between the different sublevels. The system is considered to be a mixed spin system. Applying the MW π-pulse with phase ϕ_{MW} and RF $\pi/2$-pulse with phase ϕ_{MW} and ϕ_{RF} to the entangled state described in Eq.(B.71), transformation of the spin state is described as

$$\rho \xrightarrow{R^{2\text{-}4}_{\phi_{\text{MW}}}(\pi)} \rho_1 \xrightarrow{R^{1\text{-}2}_{\phi_{\text{RF}}}(\pi/2)} \rho_2.$$

We observe the spin state of by the Hahn echo sequence ($\pi/2$-π-echo). Then the time evolution of the spin system is calculated by

$$\rho_2 = R^{3\text{-}4}_{\phi_{\text{RF}}}(\pi/2) R^{2\text{-}4}_{\phi_{\text{MW}}}(\pi) \rho [R^{2\text{-}4}_{\phi_{\text{MW}}}(\pi)]^{-1} [R^{3\text{-}4}_{\phi_{\text{RF}}}(\pi/2)]^{-1},$$

$$\rho_{\text{echo}} = U(\tau) R^{2\text{-}4}_X(\pi) U(\tau) R^{2\text{-}4}_X(\pi/2) \rho_2$$
$$\times [R^{2\text{-}4}_X(\pi/2)]^{-1} U^{-1}(\tau) [R^{2\text{-}4}_X(\pi)]^{-1} U^{-1}(\tau). \tag{B.72}$$

The observed echo intensity is given by

$$I^{\text{echo}} = \text{Tr}(\rho_{\text{echo}} S^{2\text{-}4}_y) = -\frac{1}{4} [1 + 2x \cos(\phi_{\text{MW}} - \phi_{\text{RF}})]. \tag{B.73}$$

Considering the another entangled state which is a type of $|\Phi^\pm\rangle$, we have as the echo intensity

$$I^{\text{echo}} = \text{Tr}(\rho_{\text{echo}} S^{2\text{-}4}_y) = -\frac{1}{4} [1 + 2x \cos(\phi_{\text{MW}} + \phi_{\text{RF}})], \tag{B.74}$$

where

$$\rho = \begin{pmatrix} 1/2 & 0 & 0 & x \\ 0 & 0 & 0 & 0 \\ 0 & 0 & 0 & 0 \\ x & 0 & 0 & 1/2 \end{pmatrix}, \tag{B.75}$$

and

$$\rho_2 = R^{1\text{-}2}_{\phi_{\text{RF}}}(\pi/2) R^{2\text{-}4}_{\phi_{\text{MW}}}(\pi) \rho \, [R^{2\text{-}4}_{\phi_{\text{MW}}}(\pi)]^{-1} [R^{1\text{-}2}_{\phi_{\text{RF}}}(\pi/2)]^{-1},$$
$$\rho_{\text{echo}} = U(\tau) R^{2\text{-}4}_X(\pi) U(\tau) R^{2\text{-}4}_X(\pi/2) \rho_2$$
$$\times [R^{2\text{-}4}_X(\pi/2)]^{-1} U^{-1}(\tau) [R^{2\text{-}4}_X(\pi)]^{-1} U^{-1}(\tau). \quad \text{(B.76)}$$

Appendix B.4.5. *Mathematical Features of the Super Dense Coding by Pulsed ENDOR Technique*

We consider the echo signal intensities when applying the quantum operation in Fig. 30. The density matrix starting from the pure state $\rho_{|i\rangle}$ is evolved by a series of the MW and RF pulses. The evolution from $\rho_{|4\rangle}$ is demonstrated below:

$$\rho_{|4\rangle} \xrightarrow{R^{3\text{-}4}_x(\pi/2)} \rho_1 \xrightarrow{R^{1\text{-}3}_{-x}(\pi)} \rho_{\Phi^+} \xrightarrow{Op} \rho_{\Psi^\pm} \text{ or } \Phi^\mp$$
$$\xrightarrow{R^{1\text{-}3}_{\phi_{\text{MW}}}(\pi)} \rho_2 \xrightarrow{R^{3\text{-}4}_{\phi_{\text{RF}}}(\pi/2)} \rho_3 \xrightarrow{R^{1\text{-}3}_{\phi_x}(\pi/2)} \rho_4 \xrightarrow{U(t)} \rho_5$$
$$\xrightarrow{R^{1\text{-}3}_x(\pi)} \rho_6 \xrightarrow{U(t)} \rho_{\text{echo}} \quad \text{(B.77)}$$

where Op stands for the unitary operator indicating the quantum operation to the entangled state.

For $Op = I$ (No operation),

$$\rho_{\text{echo}} = \begin{pmatrix} \frac{1+\cos(\phi_{\text{MW}}+\phi_{\text{RF}})}{4} & 0 & i\frac{1+\cos(\phi_{\text{MW}}+\phi_{\text{RF}})}{4} & \frac{i e^{i\phi_{\text{RF}}} \sin(\phi_{\text{MW}}+\phi_{\text{RF}})}{2\sqrt{2}} \\ 0 & 0 & 0 & 0 \\ \frac{1+\cos(\phi_{\text{MW}}+\phi_{\text{RF}})}{4} & 0 & \frac{1+\cos(\phi_{\text{MW}}+\phi_{\text{RF}})}{4} & \frac{e^{i\phi_{\text{RF}}} \sin(\phi_{\text{MW}}+\phi_{\text{RF}})}{2\sqrt{2}} \\ \frac{-i e^{i\phi_{\text{RF}}} \sin(\phi_{\text{MW}}+\phi_{\text{RF}})}{2\sqrt{2}} & 0 & \frac{e^{i\phi_{\text{RF}}} \sin(\phi_{\text{MW}}+\phi_{\text{RF}})}{2\sqrt{2}} & \sin^2 \frac{\phi_{\text{MW}}+\phi_{\text{RF}}}{2} \end{pmatrix},$$
(B.78)

$$I^{\text{echo}} = -\frac{1}{4}[1 + \cos(\phi_{\text{MW}} + \phi_{\text{RF}})]. \quad \text{(B.79)}$$

For $R_x^{3\text{-}4}(\pi)R_x^{1\text{-}2}(\pi)$ (X^I operation),

$$\rho_{\text{echo}} = \begin{pmatrix} \frac{1}{4} & \frac{ie^{i\phi_{\text{MW}}}}{2\sqrt{2}} & -\frac{i}{4} & 0 \\ -\frac{ie^{-i\phi_{\text{MW}}}}{2\sqrt{2}} & \frac{1}{2} & \frac{e^{-i\phi_{\text{MW}}}}{2\sqrt{2}} & 0 \\ -\frac{i}{4} & -\frac{e^{i\phi_{\text{MW}}}}{2\sqrt{2}} & \frac{1}{4} & 0 \\ 0 & 0 & 0 & 0 \end{pmatrix}, \quad (B.80)$$

$$I^{\text{echo}} = \frac{1}{4}. \quad (B.81)$$

For $R_x^{3\text{-}4}(2\pi)R_x^{3\text{-}4}(\pi)R_x^{1\text{-}2}(\pi)$ (Y^I operation),

$$\rho_{\text{echo}} = \begin{pmatrix} \frac{1}{4} & -\frac{ie^{i\phi_{\text{MW}}}}{2\sqrt{2}} & -\frac{i}{4} & 0 \\ \frac{ie^{-i\phi_{\text{MW}}}}{2\sqrt{2}} & \frac{1}{2} & \frac{e^{-i\phi_{\text{MW}}}}{2\sqrt{2}} & 0 \\ -\frac{i}{4} & \frac{e^{i\phi_{\text{MW}}}}{2\sqrt{2}} & \frac{1}{4} & 0 \\ 0 & 0 & 0 & 0 \end{pmatrix}, \quad (B.82)$$

$$I^{\text{echo}} = \frac{1}{4}. \quad (B.83)$$

For $Op = R^{3\text{-}4}(2\pi)$ (Z^I operation),

ρ_{echo}

$$= \begin{pmatrix} \frac{1-\cos(\phi_{\text{MW}}+\phi_{\text{RF}})}{4} & 0 & -i\frac{-1+\cos(\phi_{\text{MW}}+\phi_{\text{RF}})}{4} & -\frac{ie^{-i\phi_{\text{RF}}}\sin(\phi_{\text{MW}}+\phi_{\text{RF}})}{2\sqrt{2}} \\ 0 & 0 & 0 & 0 \\ i\frac{-1+\cos(\phi_{\text{MW}}+\phi_{\text{RF}})}{4} & 0 & \frac{1-\cos(\phi_{\text{MW}}+\phi_{\text{RF}})}{4} & \frac{e^{i\phi_{\text{RF}}}\sin(\phi_{\text{MW}}+\phi_{\text{RF}})}{2\sqrt{2}} \\ \frac{ie^{-i\phi_{\text{RF}}}\sin(\phi_{\text{MW}}+\phi_{\text{RF}})}{2\sqrt{2}} & 0 & -\frac{e^{-i\phi_{\text{RF}}}\sin(\phi_{\text{MW}}+\phi_{\text{RF}})}{2\sqrt{2}} & \cos^2\frac{\phi_{\text{MW}}+\phi_{\text{RF}}}{2} \end{pmatrix},$$

(B.84)

$$I^{\text{echo}} = \frac{1}{4}\left[-1 + \cos(\phi_{\text{MW}} + \phi_{\text{RF}})\right]. \quad (B.85)$$

More generally, applying an operator $Op = R_x^{3\text{-}4}(\theta)R_x^{1\text{-}4}(\phi)$ with variables θ and ϕ to $|\Phi^+\rangle$, we have

$$I^{\text{echo}}(\theta,\phi) = \frac{1}{16}\left[-3\cos\theta - \cos\phi - 4\cos\frac{\theta}{2}\cos\frac{\phi}{2}\cos(\phi_{\text{MW}}+\phi_{\text{RF}})\right]. \quad (B.86)$$

When the operator $Op = R_x^{3\text{-}4}(\theta)R_x^{1\text{-}2}(\phi)$ is applied to the entangled state of $|\Psi^+\rangle(|+-\rangle+|-+\rangle)/\sqrt{2}$, the electron spin echo intensity is given below.

$$\rho_{\text{echo}} = U(\tau)R_X^{1\text{-}3}(\pi)U(\tau)R_X^{1\text{-}3}(\pi/2)R_{\phi_{\text{RF}}}^{1\text{-}2}(\pi/2)R_{\phi_{\text{MF}}}^{1\text{-}3}(\pi)R_X^{1\text{-}2}(\phi)R_X^{3\text{-}4}(\theta)$$
$$\times \rho(\Psi^+)[R_X^{3\text{-}4}(\theta)]^{-1}[R_X^{1\text{-}2}(\phi)]^{-1}[R_{\phi_{\text{MF}}}^{1\text{-}3}(\pi)]^{-1}[R_{\phi_{\text{RF}}}^{1\text{-}2}(\pi/2)]^{-1}$$
$$\times [R_X^{1\text{-}3}(\pi/2)]^{-1}U^{-1}(\tau)[R_X^{1\text{-}3}(\pi)]^{-1}U^{-1}(\tau),$$

$$I^{\text{echo}} = \text{Tr}(\rho_{\text{echo}} S_y^{1\text{-}3})$$
$$= \frac{1}{16}\left[\cos\theta + 3\cos\phi + 4\cos\frac{\theta}{2}\cos\frac{\phi}{2}\cos(\phi_{\text{MW}} + \phi_{\text{RF}})\right]. \quad \text{(B.87)}$$

In Eqs. (B.77)-(B.87), the selective RF pulses are applied to the quantum operation $Op = r_x^{3\text{-}4}(\theta)$. In the coherent dual ELDOR technique, the operation is able to replace with the selective MW pulses, $Op = R_x^{2\text{-}4}(\theta)R_x^{1\text{-}3}(\phi)$. Applying Op to Φ^+ and Ψ^+, we have the angular dependences of the echo intensities as follows:

$$\rho_{\text{echo}} = U(\tau)R_X^{1\text{-}3}(\pi)U(\tau)R_X^{1\text{-}3}(\pi/2)R_{\phi_{\text{RF}}}^{1\text{-}2}(\pi/2)R_{\phi_{\text{MF}}}^{1\text{-}3}(\pi)R_X^{1\text{-}3}(\theta)R_X^{2\text{-}4}(\phi)$$
$$\times \rho(\Psi^+)[R_X^{2\text{-}4}(\phi)]^{-1}[R_X^{1\text{-}3}(\theta)]^{-1}[R_{\phi_{\text{MF}}}^{1\text{-}3}(\pi)]^{-1}[R_{\phi_{\text{RF}}}^{1\text{-}2}(\pi/2)]^{-1}$$
$$\times [R_X^{1\text{-}3}(\pi/2)]^{-1}U^{-1}(\tau)[R_X^{1\text{-}3}(\pi)]^{-1}U^{-1}(\tau),$$

$$I^{\text{echo}} = \text{Tr}(\rho_{\text{echo}} S_y^{1\text{-}3})$$
$$= \frac{1}{16}\left[3\cos\theta + \cos\phi + 4\cos\frac{\theta}{2}\cos\frac{\phi}{2}\cos(\phi_{\text{MW}} - \phi_{\text{RF}})\right], \quad \text{(B.88)}$$

$$\rho_{\text{echo}} = U(\tau)R_X^{1\text{-}3}(\pi)U(\tau)R_X^{1\text{-}3}(\pi/2)R_{\phi_{\text{RF}}}^{1\text{-}2}(\pi/2)R_{\phi_{\text{MF}}}^{1\text{-}3}(\pi)R_X^{1\text{-}3}(\theta)R_X^{2\text{-}4}(\phi)$$
$$\times \rho(\Phi^+)[R_X^{2\text{-}4}(\phi)]^{-1}[R_X^{1\text{-}3}(\theta)]^{-1}[R_{\phi_{\text{MF}}}^{1\text{-}3}(\pi)]^{-1}[R_{\phi_{\text{RF}}}^{1\text{-}2}(\pi/2)]^{-1}$$
$$\times [R_X^{1\text{-}3}(\pi/2)]^{-1}U^{-1}(\tau)[R_X^{1\text{-}3}(\pi)]^{-1}U^{-1}(\tau),$$

$$I^{\text{echo}} = \text{Tr}(\rho_{\text{echo}} S_y^{1\text{-}3})$$
$$= \frac{1}{16}\left[-3\cos\theta + \cos\phi - 4\cos\frac{\theta}{2}\cos\frac{\phi}{2}\cos(\phi_{\text{MW}} - \phi_{\text{RF}})\right]. \quad \text{(B.89)}$$

References

1. R. P. Feynman, *Int. J. Theor. Phys.* **21** (1982) 467; *ibid.*, *Opt. News* **11** (1985) 11.
2. M. A. Nielsen and I. L. Chuang, *Quantum Computation and Quantum Information* (Cambridge University Press, 2000).
3. P. W. Shor, *In Proceedings, 35th Annual Symposium on Foundations of Computer Science* (IEEE Press, LosAlamitos, CA, 1994).
4. L. K. Grover, *Phys. Rev. Lett.* **79** (1997) 325.
5. C. H. Bennett, G. Brassard, C. Crepeau, R. Jozsa, A. Peres and W. Wootters, *Phys. Rev. Lett.* **70** (1993) 1895.
6. C. H. Bennett and S. J. Wiesner, *Phys. Rev. Lett.* **69** (1992) 2881.
7. M. Mehring, J. Mende and W. Scherer, *Phys. Rev. Lett.* **90** (2003) 153001.
8. M. Mehring, W. Scherer and A. Weidinger, *Phys. Rev. Lett.* **93** (2004) 206603.
9. R. Rahimi, K. Sato, K. Furukawa, K. Toyota, D. Shiomi, T. Nakamura, M. Kitagawa and T. Takui, *Proc. the 1st Asia-Pacific Conference on Quantum Information Science* (Dec. 10-13, 2004) 197.
10. R. Rahimi, K. Sato, K. Furukawa, K. Toyota, D. Shiomi, T. Nakamura, M. Kitagawa and T. Takui, *Int. J. Quantum Inf.* **3** (2005) 197.
11. J. J. L. Morton, A. Tyryshkin, A. Ardavan, S. Benjamin, K. Porfyrakis, S. A. Lyon and G. A. D. Briggs, *Nature Physics* **2** (2006) 40.
12. K. Sato, R. Rahimi, S. Nishida, K. Toyota, D. Shiomi, Y. Morita, A. Ueda, S. Suzuki, K. Furukawa, T. Nakamura, M. Kitagawa, K. Makasuji, M. Nakahara, H. Hara, P. Carl, P. Höefer and T. Takui, *Physica E* **40** (2007) 363.
13. W. Sherer and M. Mehring, *J. Chem. Phys.* **128** (2008) 052305.
14. J. A. Jones, M. Mosca and R. H. Hansen, *Nature* **393** (1998) 344.
15. I. L. Chuang, L. M. K.Vandersypen, X. Zhou, D. W. Leung and S. Lloyd, *Nature* **393** (1998) 143.
16. I. L. Chuang and Y. Yamamoto, *Phys. Rev. A* **52** (1995) 3489.
17. J. I. Cirac and P. Zoller, *Phys. Rev. Lett.* **74** (1995) 4091.
18. D. P. Divincenzo, *in Mesoscopic Electron Transport*, edited by L. Sohn, L. Kouwenhoven and G. Schon (vol. 345, NATO ASI Series E, Kluwer, 1997) 657; cond-mat/9612126.
19. S. Nakazawa, K. Sato, T. Takui, *et al.* (unpublished)
20. M. Mehring, P. Höefer and A. Grupp, *Phys. Rev. A* **33** (1986) 3523.
21. I. L. Chuang, N. Gershenfeld and M. Kubinec, *Phys. Rev. Lett.* **18** (1998) 3408.
22. M. A. Nielsen, E. Knill and R. Laflamme, *Nature* **396** (1998) 52.
23. L. M. K. Vandersypen, M. Steffen, G. Breytra, C. S. Yannoni, M. H. Sherwood and I. L. Chuang, *Nature* **414** (2001) 883.
24. X. Fang, X. Zhu, M. Feng, X. Mao and F. Du, *Phys. Rev. A* **61** (2000) 022307.
25. D. G. Cory, A. F. Fahmy and T. F. Havel, *Proc. Natl. Acad. Sci. USA* **94** (1997) 1634.
26. N. Gershenfeld and I. L. Chuang, *Science* **275** (1997) 350.
27. A. Ekert and R. Jozsa, *Phil. Trans. R. Soc. Lond. A* **356** (1998) 1769.
28. N. Linden and S. Popescu, *Phys. Rev. Lett.* **87** (2001) 047901.
29. K. Życzkowski, P. Horodecki, A. Sanpera and M. Lewenstein, *Phys. Rev. A*

58 (1998) 883.
30. S. L. Braunstein, S. M. Caves, R. Jozsa, N. Linden, S. Popescu and R. Schack, *Phys. Rev. Lett.* **83** (1999) 1054.
31. M. S. Anwar, D. Blazina, H. Carteret, S. B. Duckett, T. K. Halstead, J. A. Jones, C. M. Kozak and R. J. K. Taylor, *Phys. Rev. Lett.* **93** (2004) 040501.
32. M. S. Anwar, J. A. Jones, D. Blazina, S. B. Duckett and H. A. Carteret, *Phys. Rev. A* **70** (2004) 032324.
33. G. Feher, *Phys. Rev.* **103** (1956) 834; W. B. Mims, *Proc. Roy. Soc. London* **283** (1965); A. Grupp and M. Mehring, *In Modern pulsed and continuous-wave electron spin resonance*, editted by L. Kevan and M. K. Bowman (Wiley, New York, 1990) 195.
34. Y. Morita, Y. Yakiyama, T. Murata, T. Ise, D. Hashizume, D. Shiomi, K. Sato, M. Kitagawa, T. Takui and K. Nakasuji (unpublished).
35. A. Peres, *Phys. Rev. Lett.* **77** (1996) 1413.
36. M. Horodecki, P. Horodecki and R. Horodecki, *Phys. Lett. A* **223** (1996) 1.
37. H. M. McConnell, C. Heller, T. Cole and R. W. Fessenden, *J. Am. Chem. Soc.* **82** (1960) 766.
38. E. R. Davies, *Phys. Lett. A* **47** (1974) 1.
39. R. Rahimi, K. Sato, S. Nishida, Y. Morita, M. Kitagawa, T. Takui, *et al.* (unpublished).
40. T. Yoshino, S. Nishida, S. Nakazawa, R. Rahimi, K. Sato, Y. Morita, M. Kitagawa, T. Takui, *et al.*, to be published.
41. S. Nakazawa, T. Ise, K. Sato, T. Yoshino, Y. Morita, M. Kitagawa, T. Takui, *et al.* (unpublished).
42. S. Lloyd, *Scientific American*, **273** (1995) 140.
43. S. Lloyd, *Science*, **261** (1993) 1569.

FULLERENE C_{60}: A POSSIBLE MOLECULAR QUANTUM COMPUTER

TOMONARI WAKABAYASHI

Department of Chemistry, School of Science and Engineering, Kinki University,
Kowakae 3-4-1, Higashiosaka, Osaka 577-8502, Japan
E-mail: wakaba@chem.kindai.ac.jp

Fullerene C_{60} is a molecule with a hollow space in its closed cage of sixty carbon atoms and able to accommodate atoms or molecules inside. Recently, the system of an electron spin coupled with a nuclear spin of a nitrogen atom trapped inside C_{60} (N@C_{60}) has been demonstrated to be a possible qubit system for the implementation of quantum computation (QC) and quantum information processing (QIP). In the last decade, there has appeared an increasing number of reports for this molecule, on its magnetic properties, ideas for physical realization as quantum computers, and experimental approaches based on the magnetic resonance techniques. In this Chapter, the research areas on the topic on the fullerene-based QC/QIP are reviewed together with a relevant part of the fullerene stories. The production of N@C_{60} in our research group is introduced.

Keywords: Fullerenes; Endohedral Fullerenes; N@C_{60}

1. Introduction

Quantum computation (QC) and quantum information processing (QIP) are based on simultaneous operations of unitary transformations in multi-dimensional quantum systems.[1,2] The algorithm using the quantum nature is, in some cases, shown to be superior to the classical algorithm when the number of quantum bits (qubits) is sufficiently large. So far, the properties for the system required for QC/QIP have been thoroughly investigated and the implementations have been performed experimentally on a variety of systems, such as nuclear magnetic resonance (NMR) at room temperature, Doppler-cooled ions in a Paul trap, superconducting Josephson junctions at low temperatures, and so on. The NMR is one of the choices for the implementation of QC.[5-9] The spin echo emvelope modulation observed in pulse NMR experiments is a basic constituent of the manipulation of the spin states.[85] The sensitivity for the detection of NMR transitions can be enhanced by polarization transfer from the nuclear spins to the electron spins, followed by the detection of transitions for the electron spin resonance (ESR), namely electron/nuclear double resonance (ENDOR).[17] The experimental techniques based on the electron spin echo envelope modulation (ESEEM) and ENDOR, originally developed for the measurement of magnetic properties of bulk materials, are opened for manipulation of spin dynamics for QC/QIP.

This Chapter deals with physical implementation of QIP using a specific endohedral fullerene, i.e., a spherical shell molecule of sixty carbon atoms, C_{60}, in a dimension of 0.7 nm in daimeter posessing a nitrogen atom, N, inside. Trapped inside a hollow closed cage of C_{60}, the nitrogen atom can maintain its unpaired electron in a spin quartet state. The finding that a nitrogen atom having both the electron and nuclear spins in its ground state is encapsulated inside a cage of C_{60} opened the way for a number of research activities on physical realization of quantum computer using this material, $N@C_{60}$.

In the following, Section 2 describes historical overview staring from the discovery of C_{60} to the finding of a free nitrogen atom trapped inside a C_{60} cage. Section 3 provides basic ideas necessary for the discussion on quantum information processing and recent activities in the field of fullerene-based QC/QIP. Section 4 explains a progress on the production of $N@C_{60}$ in our group.

2. Fullerenes and Endohedral Fullerenes

2.1. *Historical Overview*

In 1985, Kroto, Curl, Smalley, and co-workers found the evidence for the stability of carbon cluster C_{60} under the condition of laser vaporization of graphite in a helium gas flow.[18] Despite the tiny amount of ions in the mass spectrometer, the signals of C_{60}^+ and C_{70}^+ exhibited their relative abundance and stability compared to the other clusters C_{2n}^+. The molecule having a polygonal shape with 12 pentagons and 20 hexagons was coined Buckminsterfullerene, or Fullerene in short, after Buckminster Fuller who designed many types of frameworks for the dome architectures. In 1990, the breakthrough was brought by Krätschmer and Huffmann reporting the production of C_{60} in macroscopic quantities by evaporation of graphite rods under the suitable conditions in helium buffer gas.[19] Once having a solid form of C_{60}, researchers in the diverse areas of physics and chemistry

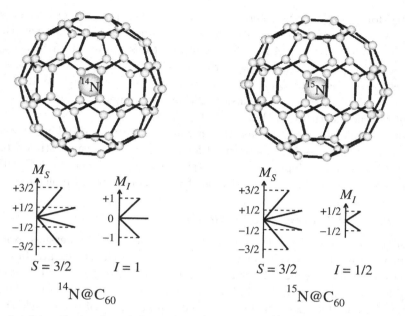

Fig. 1. Schemtic drawing of the endohedral fullerenes, ^{14}N@C$_{60}$ and ^{15}N@C$_{60}$. Depending on the nuclear spin of nitrogen, $I = 1$ for ^{14}N or $I = 1/2$ for ^{15}N, two types of molecules (isotopomers) are conceivable. The natural abundance is 99.6% for ^{14}N and 0.4% for ^{15}N.

started investigations of the properties for the new form of carbon.

The soccer-ball-shaped molecule, C$_{60}$, had been proposed earlier, e.g., in 1970 by Osawa and Yoshida.[20,21] They quoted its aromatic stability due to delocalized π electrons along both sides of the surface of the cage.[21] Inspired by the molecular shape of corannulene C$_{20}$H$_{10}$, a bowl-shaped molecule consisted of a flat but slightly tapered ring of five fused-hexagons with a pentagon at the center, they thought up an arrangement of the polygonal patches to have a truncated icosahedron with sixty vertices. However, the experimental evidence had to be awaited for recognition of the significance.[18,19]

The stability of C$_{60}$ under the mass spectroscopic conditions was accounted for by the closed-cage structure of the molecule,[18] leading the authors to conceive of a perspective in the research stream at that time, *"We also have evidence that an atom can be placed in the interior, producing molecules which may exhibit unusual properties."*[18] The idea of "endohedral fullerene" that contains atoms or molecules inside meant an additional variety for a single-cage fullerene species and was put forward to grow into a new class of chemical substances. In a concrete sense, however, the molecular QC/QIP exploiting the quantum states of the atom in a C_{60} cage would be beyond the scope at that time. The connection of the ideas between fullerenes and QC/QIP was inspired from the finding in 1996 of a free-standing nitrogen atom inside C_{60} by Weidinger and co-workers at Hahn-Meitner Institute in Berlin (see Fig. 1).[73]

In the following subsections, the intrinsic properties of fullerenes and endohedral fullerenes are summarized for showing the research stream towards the recent developments in the experimental approaches for QC/QIP using electron/nuclear-spin transitions of the fullerene-based molecules.

2.2. Physical Properties of Fullerenes

Prior to the discussion of the QC/QIP of the endohedral fullerenes, some intrinsic properties of the molecular C_{60} and its solid states are briefly reviewed. Krätschmer and Huffman reported the X-ray powder diffraction of solid C_{60}/C_{70} and deduced that the molecules were packed in a face-centered-cubic (fcc) manner, but could not find a signal of large scattering angle to obtain a molecular structure.[19] The spherical molecule is rotating about the lattice point in the fcc crystal at room temperature.

The molecular symmetry of C_{60} was confirmed by ^{13}C-NMR spectra in solutions.[22,23] Representing all the sixty carbon nuclei in the molecule being identical under the icosahedral symmetry, a single peak was observed at the chemical shift of 143 ppm. For C_{70} with a slightly elongated cage structure, five lines of absorption were observed between 130-150 ppm with relative intensities of about 1:2:1:2:1.[23] The 2D-NMR spectra clarified the connections between the atoms associated with these lines.[24] Using the modified C_{60} compounds having a chemically attached "handle" outside the cage, which was evidenced by 2D-NMR in solutions, the molecular rotation in the crystal was frozen to analyze the atomic structure within the fullerene cage directly from the X-ray diffraction pattern.[25,26]

In the NMR spectroscopy of solid C_{60}, anisotropic lineshape was observed at low temperatures.[27,28] Tycko and his colleagues found an anomalous behavior in the ^{13}C nuclear spin-lattice relaxation time T_1 at $\sim$ 260 K.[29] In addition to the sharp line exhibiting rotational narrowing, a broad band due to the anisotropic component was superposed below $\sim$ 150 K.[29] The rotational ordering transition in solid C_{60} was observed also by low temperature X-ray diffraction.[30] The diffractions for short-range orders ($Q > 2.7\,\text{Å}^{-1}$) appear in a complex pattern at low temperatures. Upon warming, the intensity for a selected one among these lines decreased to zero and the differential scanning calorimetry (DSC) exhibited an exothermic behavior at 260 K. This observation indicates that the rotational ordering trasnsition at 260 K for solid C_{60} is a first-order phase transition.

Another phase transition was found at around ~ 90 K. This is called glass transition.[31,32] Bellow the temperature, the rotational motion of the molecules is frozen for the experimental time scales. In a low temperature phase of solid C_{60}, the energetically favorable orientation of the molecule with respect to its nearest neighbors is such that the center of the C-C bond of a molecule faces to a hexagonal or a pentagonal face of the adjacent molecule.[4,32] At temperatures between those of the two phase transitions, ~ 90 and 260 K, the molecule in the crystal librates within a potential well for one of the stable orientations and some times rotates or jumps to another by, namely, a ratchet motion.

The solid state properties of C_{60} dramatically change by a different way of doping, namely, doping atoms or molecules at interstitial sites. The most striking system is alkali-doped C_{60} which exhibits superconductivity. The compound of K_3C_{60}, a potassium ion at an octahedral site and two potassium ions at tetrahedral sites per one C_{60} molecule in the fcc lattice, exhibits superconductivity below the critical temperature, $T_c = 18$ K. Three electrons transfer from potassium to fullerene so that the conduction band is half-filled to realize non-zero density of states (DOS) at the Fermi level, E_F. With a variety of alikali metals from Li to Cs, the different elements and their combinations were examined for doping to plot the critical temperature, T_c, as a function of the lattice constant, c.f., the unit volume in solid C_{60} increases upon doping. The system of $RbCs_2C_{60}$ recorded $T_c = 33$ K and the T_c increased systematically to the lattice constant.[34] Recently, the largest alkali-doped C_{60}, namely Cs_3C_{60}, was revisited to show superconductivity under high pressure, > 3 kbar, with $T_c = 38$ K, the highest-temperature record in T_c for molecular materials.[35]

The electronic structure of the C_{60} molecule is responsible in many cases for the primary properties of the related materials. The sixty p electrons constitute the valence system of delocalized π electrons. In the one-particle picture in a spherical shell model, the energy levels are labeled by the orbital angular momentum quantum number, $l = 0, 1, 2, \ldots$, and each of the levels has $2l + 1$-fold degeneracy according to the magnetic quantum number, $m = -l, \ldots, +l$. The introduction of icosahedral symmetry results in splitting of the energy levels into those with the irreducible representation of $A(1)$, $T(3)$, $G(4)$, or $H(5)$ in the point group symmetry (the numbers in parenthesis are the degeneracy). For neutral C_{60}, the highest occupied level of the 5-fold H_u symmetry (one of the $l=5$ states) are filled with ten electrons, thus the C_{60} molecule has the spin-singlet totally-symmetric ground state of $^1A_{1g}$ in the point group of I_h. Since the lowest unoccupied level has T_{1u} symmetry (also from $l=5$), the lowest-energy electric-dipole transition ($H_u \rightarrow T_{1u}$) of C_{60} is symmetry forbidden. The electronic transition occurs from the H_u level to the upper empty level of g symmetry ($H_u \rightarrow T_{1g}$) and from the lower occupied level of H_g to the T_{1u} level ($H_g \rightarrow T_{1u}$) to show a group of intense absorption bands in the ultraviolet (UV) region. The weak absorption features are known in the visible region, which correspond to the vibronic bands of the forbidden transition ($H_u \rightarrow T_{1u}$) of C_{60}.

The spin-triplet state of C_{60} is generated during the relaxation processes following the photoexcitation of the molecule via the allowed transitions in the UV. The lifetimes of the triplet C_{60} have been measured in solutions by transient absorption,[36] phosphorescence and transient EPR absorption,[37] photoluminescence-detected ORMR and light-induced ESR,[38] pulsed ENDOR experiments,[40] and phosphorescence in rare gas matrices.[41] The EPR of C_{60} in liquid crystal[42] and phosphorescence of C_{60} in rare gas matrices[41] also reported on the triplet state of neutral C_{60}. The electron-doped C_{60}, the radical anion, was also detected by the electronic absorption and fluorescence spectra in the near infrared regions and by the ESR spectra.[43–46] The C_{60} anion is of basic importance, because, when a metal atom is doped inside the fullerene cage, the electron of the metal atom often moves to the cage to form anionic C_{60}.

2.3. *Putting Atoms inside Forming Endohedral Fullerenes*

The fullerene molecule has an empty space inside, and a potential to accommodate atoms and molecules in the cage. Reactive species such as spin multiplet species of atoms and molecules can be stabilized. The high-spin atom or molecule has potential applications for QC/QIP by magnetic resonance. Here, the methods for encapsulating atoms and molecules into fullerene cages are discussed in view of the difference in products and efficiency.

Soon after the large-scale production of fullerenes by resistive heating of graphite,[19] another efficient way for producing fullerenes, namely, contact arc, was reported by which fullerene molecules with a lanthanum atom inside were also produced and extracted by solvents from the soot.[47] Using the arc discharge method, in addition to C_{60} and C_{70}, higher fullerenes of C_n ($n=76,78,82,84,\ldots$) were detected and isolated to reveal their cage structures of lower symmetries.[49–52] The endohedral fullerene, La@C_{82}, was remarkably abundant in the soot relative to the others of La@C_{2n}, where the symbol "A@B" represents "A inside B".[48] The preference for larger cages for the encapsulation, e.g., La atom inside C_{82} rather than C_{60}, has been an issue for discussion and explained in relation to the cage structure of the empty fullerenes, the degree of charge transfer from the atom to the cage, and the mechanism by which the endohedral molecules are formed. As a new branch of fullerene-related meterials, a variety of metallofullerenes M@C_{2n} have been produced, isolated, and characterized so far.[53] Under the typical conditions of arc discharge using metal-containing carbon rods as electrodes, however, the metals are mainly trapped in the higher fullerenes (C_{2n} of $2n > 70$) of lower symmetries (except for I_h-C_{80}) and the amount of M@C_{60} is less than the detection limit. The propensity rule for the preference of lanthanides to be encapsulated has been also discussed along the the stability of the intermediate clusters and/or the difference in the binding energies between the metal and the carbon clusters. In the arc discharge, metallofullerenes are formed during the vaporization of the metal containing carbon rods and the condensation of the carbon vapor molecules in a buffer gas.

From the soot that contains mixtures of various sizes of fullerenes and metallofullerenes, a desired specie is isolated by using high performance liquid chromatography (HPLC), e.g., Sc_3@C_{82}[61] and La@C_{82}.[62] The elements that have been encapsulated by arc discharge are distributed on the periodic table, from the group III (Sc, Y, La), lanthanides (Ce, Pr, Nd, Sm, Eu, Gd, Tb, Dy, Ho, Er, Tm, Yb, Lu), group II (Ca, Sr, Ba) to transient metals (Ti, Fe).[53] Concerning the nitrogen-contaning endohedrals, it is noted that Sc_3N@C_{80} was isolated from the soot,[63] and characterised by 22-fold hyperfine lines in the ESR spectra.[64] For trapping solely the atoms of non-metal elements, the other methods have been examined.

The approaches for encapsulating atoms after the formation of C_{60} have been also studied. Saunders *et al.* reported that helium atoms were encapsulated in fullerene cages by pressurizing at $\sim$ 2500 atm at 600 °C to form ^{3}He@C_{60} and ^{3}He@C_{70}.[54] Komatsu *et al.* succeeded in encapsulation of molecular hydrogen H_2 to form H_2@C_{60} by an elaborating fashion, so called "molecular surgery".[56] The empty C_{60} was opened to make a hole of an 11-membered ring on the cage by step-by-step reactions in organic chemistry, to which the gaseous H_2 was adsorbed under moderate conditions, and finally the molecular moieties attached as scissors for keeping the hole open was eliminated to close the hole on the cage. Ohtsuki *et al.* exploited recoil energy of a hot atom upon a nuclear reaction to penetrate it into the carbon shell of C_{60} to form $^7Be@C_{60}$.[59] The mixed powder of lithium carbonate with C_{60} was irradiated with a 16 MeV proton beam to promote the nuclear reaction, $^7Li(p,n)\,^7Be$. The beryllium atom kicked by the leaving neutron has a substantial kinetic energy with which the hot atom can be caught then bound in a C_{60} cage. The hot-atom chemistry was applied for incorporation of tritium ^3H in C_{60}.[55]

In contrast to the neutral atoms produced in a nuclear reaction, ions are easily accelerated by electric fields in vacuum. The ion implantation is a general approach for encapsulating atoms after the formation of C_{60}. Positive muons μ^+ are collided with C_{60} to form a neutral pair of a muon and an electron, namely muonium μ^+e^-.[57,58] Aiming at novel conducting materials, lithium ions Li^+ was implanted into C_{60} solids to form $Li@C_{60}$.[60] Nitrogen ions N^+ were formed in plasmas and bombarded onto sediment C_{60} to form $N@C_{60}$.[73] In all these experiments, we see beautiful designs of experiments behind the success of finding the endohedral fullerenes. The methods for detection and the yield of production are discussed in the next section.

2.4. *Detection of the Endohedrals*

The procedure for encapsulation must be combined with a suitable detection technique, because in most cases the primary materials are the mixtures of a majority of empty fullerenes and a tiny amount of the desired species of endohedral fullerenes. It is desirable to have a detection method that is sensitive to the endohedrals but silent for the empty fullerenes. It is a better choice to measure the exclusive property for the encapsulated species rather than that commonly observed for both of the endhedral and empty fullerenes. For example, in the case of $^3He@C_{60}$, the NMR absorption of ^3He nuclei with spin 1/2 was used for the detection. Owing to the high resolution of NMR spectroscopy, the chemical shift of $^3He@C_{60}$ was distinguished from that for $^3He@C_{70}$ and free ^3He.[54] The hot atom of ^7Be, formed by absorption of a proton by a ^7Li nucleus to release a neutron, is in an excited state to decay by the electron capture (EC) decay, $^7Be(e,\gamma)\,^7Li$. The γ-ray photon with a specific energy of 478 keV was an exclusive probe for the presence of the excited ^7Be nuclei. The decay rate for the EC decay of ^7Be turned out to be slightly (0.83%) faster for $^7Be@C_{60}$ than that burried in beryllium metal.[59] Electron spin resonance (ESR) or electron paramagnetic resonance (EPR) is a promising probe for the detection of endohedral species with unpaired electrons, e.g., ESR for metallofullerenes such as $La@C_{82}$, $Y@C_{82}$, $Sc@C_{82}$, $Sc_3@C_{82}$, and μSR for muonium fullerenes. Mass spectrometry is a conventional tool applicable to both of the empty and endohedral fullerenes.

2.5. *The Yields of the Endohedrals*

The necessary choice in the detection methods depends largely on the content of the desired species in the crude materials. When the amount of the endohedral fullerenes is comparable to that of the empty fullerenes, mass spectrometry may be sufficient.[62] However, the relative abundance of the endohedrals in a primary crude material is usually low. The typical yield of the endohedral species are ~0.1-1% for metallofullerenes by arc discharge,[48,53,62] ~0.1% for ^{3}He@C$_{60}$ by pressurizing,[54] up to ~30% for Li@C$_{60}$ by ion bombardment,[60] and ~0.01% for N@C$_{60}$ by ion implantation.[73] Under the reliable control of the relative abundance with a proper detection scheme, the endohedrals has to be concentrated, separated, and purified.

2.6. *Concentration and Purification*

The effective separation of the endohedral fullerenes from the empty fullerenes is performed by using a technique of high performance liquid chromatography (HPLC). Since the fullerenes sublime at elevated temperatures of ~300-400°C, the repeated cycles of sublimation and condensation is another choice for concentration.

The N@C$_{60}$ molecule is detectable by ESR in high sensitively without counting the co-existing empty C$_{60}$ molecules. The presence and abundance can be monitored by ESR upon chromatographic separations by HPLC.[79] In HPLC, the solution containing a mixture of the molecules, i.e., N@C$_{60}$ and C$_{60}$, is put through a stainless-steal column packed with silica-gel particles together with a pressurized solvent such as toluene. By going through the column, the different molecules are separated on the bases of the difference in the mobility and each of the species is eluted in a single bunch of the solution out of the column. At the first stage of the separations, however, the separation of the two bunches of the endohedral and empty C$_{60}$ is not sufficient.

Because of the high abundance of empty C$_{60}$ in the mixture with N@C$_{60}$, C$_{60}$/N@C$_{60}$ ~ $10^4 - 10^5$, and of the close resemblance in the retention time in HPLC, the bunch of N@C$_{60}$ is overlapping with that of C$_{60}$ and not recognizable in the detection by conventional UV absorption. Dinse and colleagues separated the bunch of C$_{60}$/N@C$_{60}$ into fractions according to the retention time and measured the ESR intensity of N@C$_{60}$ for each of the fractions to find the fact that the retention time of N@C$_{60}$ is slightly longer than that of C$_{60}$.[79] By the multi-step processing of taking the slower part within the bunch in each step, N@C$_{60}$ was concentrated relative to C$_{60}$ up to N@C$_{60}$/C$_{60}$ > 0.1.

In the final step of separations, a recycling technique is effective where the bunch of the species to be separated is repeatedly passed through the column in a closed circuit. After several cycles, the two bunches of C$_{60}$ and N@C$_{60}$ is clearly separated in the chromatogram as discernible by UV absorption to obtain ~100% pure N@C$_{60}$. Although the ESR measurement does not detect the empty C$_{60}$ directly, the line width of the ESR signals of N@C$_{60}$ in solid C$_{60}$ increases drastically by increasing the concentration of N@C$_{60}$, indicating the indirect observation of C$_{60}$ molecules or other diamagnetic species that separate the paramagnetic centers distributed in the materials.[79] The considerable broadening observed in the ESR signals for purified N@C$_{60}$ indicates that the magnetic coupling of the spin pairs of N@C$_{60}$-N@C$_{60}$ in the solid N@C$_{60}$ plays a crucial role. In this sense, the precise control of the intermolecular spin coupling is a necessary step for designed processing of the spin dynamics for this system.

3. Experimental Approaches to QC/QIP

3.1. *Necessary Properties for QC/QIP*

Thorough understanding of quantum nature of superposition of states shed light on the states in entanglement. It was triggered in 1935 when Einstein, Podolsky, and Rosen posed a question on the description of quantum states by using wavefunctions (EPR paradox).[65] Bohm[66] and Bohm and Aharanov[67] pursued the issue focusing on the observational aspects in quantum mechanics and introduced a hidden variable along with the discussion on nonlocality and predeterminative property for a coupled pair of states. In 1964, John S. Bell presented an inequality relation that does hold, provided that the theoretical framework involving the hidden variable is supposed to be correct.[68,69] Thereby, the theory has to be testified its validity by experiments. Researchers became aware that the description of quantum mechanics using density matrix is suitable for many particle systems rather than using wavefunctions. The density matrix representation includes both the statistical distribution of states in the many particle system and the superposition of states in the constituent single particles, thus including entangled states from the beginning.

One of the key features in quantum mechanics is superposition of states.[2] For a particle with eigenstates $|n\rangle$ for a dynamical valuable P are depicted by the eigenvalue equation $P|n\rangle = p_n|n\rangle$. The superposition of states is described by a linear combination of the eigenstates, $|\phi\rangle = \sum c_i|i\rangle$ which also satisfies the equation as $P|\phi\rangle = p_\phi|\phi\rangle$, where $p_\phi = \sum_i p_i|c_i|^2$. When going from the system of a single particle to that of two particles, the situation changes for which the issues are to be carefully stated. For simplicity, we consider a two-qubit system of each having two eigenvalues, one with eigenvalues 0 and 1, namely A, and the other again with 0 and 1, namely B. Then, we have four possibilities for the state vector as combinations of the qubits, $|0_A 0_B\rangle$, $|0_A 1_B\rangle$, $|1_A 0_B\rangle$, and $|1_A 1_B\rangle$. For each qubit, one can construct a state being normalized, e.g., for qubit A,

$$|\phi_A\rangle = \lambda_A|0_A\rangle + \mu_A|1_A\rangle, \quad |\lambda_A|^2 + |\mu_A|^2 = 1. \tag{1}$$

Also for B,

$$|\phi_B\rangle = \lambda_B|0_B\rangle + \mu_B|1_B\rangle, \quad |\lambda_B|^2 + |\mu_B|^2 = 1. \tag{2}$$

For states of the two qubit system, the tensor product $|\phi_A \otimes \phi_B\rangle \equiv |\phi_A \phi_B\rangle$ is possible.

$$|\phi_A \phi_B\rangle = \lambda_A \lambda_B |0_A 0_B\rangle + \lambda_A \mu_B |0_A 1_B\rangle + \mu_A \lambda_B |1_A 0_B\rangle + \mu_A \mu_B |1_A 1_B\rangle. \tag{3}$$

However, this tensor product does not cover all the states in the Hilbert space $H_A \otimes H_B$. The state has the general form of

$$|\varphi\rangle = \alpha_{00}|0_A 0_B\rangle + \alpha_{01}|0_A 1_B\rangle + \alpha_{10}|1_A 0_B\rangle + \alpha_{11}|1_A 1_B\rangle. \tag{4}$$

For $|\varphi\rangle$ to be the form $|\phi_A \phi_B\rangle$, a necessary (and sufficient) condition is $\alpha_{00}\alpha_{11} = \alpha_{01}\alpha_{10}$, which is not *a priori* to be valid. A two-qubit state which does not have the form $|\phi_A \phi_B\rangle$ is called entangled state. As an example of such an entangled state, we refer to

$$|\theta_1\rangle = \frac{1}{\sqrt{2}}(|0_A 0_B\rangle + |1_A 1_B\rangle), \tag{5}$$

$$\alpha_{00} = \alpha_{11} = \frac{1}{\sqrt{2}}, \quad \alpha_{01} = \alpha_{10} = 0. \tag{6}$$

Apparently, for this state, the relation $\alpha_{00}\alpha_{11} = \alpha_{01}\alpha_{10}$ does not hold, and the sate of the qubit A is no longer described by a definite state vector, $|0_A\rangle$ or $|1_A\rangle$. Instead, it must be taken as a mixture of 50% of $|0_A\rangle$ and 50% of $|1_A\rangle$. Among the many possible entangled states, the above state (5) is one of the specific states which are characterized as maximally entangled states, for which the density operators for the 2-demensional qubit system of A and B are $\rho_A = \rho_B = 1/2$ (generally for N-demensional system, $\rho_A = \rho_B = \cdots = \rho_N = 1/N$). In the two qubit system, the four states in entanglement, the state (5) and the following states (7)-(9), consist of a set of maximally entangled states, namely Bell states.

$$|\theta_2\rangle = \frac{1}{\sqrt{2}}(|0_A 0_B\rangle - |1_A 1_B\rangle), \tag{7}$$

$$|\theta_3\rangle = \frac{1}{\sqrt{2}}(|0_A 1_B\rangle + |1_A 0_B\rangle), \tag{8}$$

$$|\theta_4\rangle = \frac{1}{\sqrt{2}}(|0_A 1_B\rangle - |1_A 0_B\rangle). \tag{9}$$

The tensor product satisfying the conditions (1), (2), and (3) is called a pure state, a state that is represented by a state vector consisted of definite quantum states for the qubit A and B. For each of the single qubit, when the state is a superposition of more than two orthonomal state vectors, it is always possible to transform it into a single state vector by choosing an appropriate orthonormal basis set. In this sense, the superposition of states is base dependent. In contrast, entanglement as well as pure state for a multiple qubit system is a base independent concept. It can be checked whether the given state is a pure state or not by evaluating the density operator, i.e., the state that satisfies $\rho^2 = \rho$ or $\text{Tr}\rho = 1$ is a pure state, while the state of $\text{Tr}\rho < 1$ is a mixed state.

In QIP, a pure state should be prepared at the beginning. Then, the qubits are interacted and subjected to time evolution during which a desired unitary transformation proceeds under entanglement. Finally, the result is read out from the qubit system. The manipulation of the spin system is performed by applying an appropriate sequence of irradiation with electromagnetic pulses having precisely controlled frequency, duration, amplitude, and delay. For NMR transitions, the radiofrequency (rf) is tuned, while, for ESR, the microwave frequency (μw), in order to drive and control the Rabi oscillations.

An inevitable property for realistic quantum systems is decoherence. The loss of phase coherence disturbs the unitary transformation during the time evolution and results in a false of the QIP. In most cases, decoherence is due to interaction with environment such that the energy of the qubit system is dissipated into surrounding media. The physical systems for manipulation of entangled states having been devised so far are a photonic quantum computer based on nonlinear Kerr effect, optical resonant cavities, microwave resonant cavities, ions in an ion trap, nuclear magnetic resonance, superconducting circuits with Josephson junctions, quantum dots, and atoms of a Bose-Einstein condensate trapped in an optical lattice. For the system for QC/QIP, it is prefered to have a sufficiently long time during which the phase coherence is retained. It is better to choose a system whose interaction with the environment is weak enough or can be decoupled.

Concerning a physical realization of quantum computer, the device should also meet the following requirements i)-v), namely Di Vincenzo's criteria. i) Scalability: A number of qubits can be increased to achieve better performance than classical computer. ii) Initialization: A pure state can be prepared from a mixture of states by appropriate procedures. iii) Coherence: Phase coherence should be retained for a sufficiently long time. iv) A set of universal quantum gates: The device should work for such elementary operations as those with which any complex operations or gates can be constructed, e.g., the controlled-NOT (CNOT) gate and the unitary transformation for each of the single qubits. v) Readout: The result after operations can be extracted from the final state.

3.2. *From NMR to ESR*

In the NMR quantum computer,[1,5-9] a qubit is composed of an ensemble of indistinguishable nuclear spins of $\sim 10^{15}$ identical nuclei. The spin is embedded in individual molecules and the interaction with spins in the other molecules are supposed to be negligibly small or averaged under the experimental conditions. When the electromagnetic fields are applied, the spins in molecules interact with the fields, developing the bulk magnetization. Within a molecule, the nuclear spin can couple with another nuclear spin. The entanglement of the two qubits can be realized within a molecule, and practically the same effect can be observed for the bulk magnetization of an ensemble of identical molecules, each of them having the two distinguishable spins. The signals manipulated in NMR measurements are the bulk magnetization for the ensemble of molecules. It is the ensemble property that, also in the NMR QC, can be exploited for the manipulation of qubits, namely, pseudo-entanglement.[10-12] The upper and lower levels of the molecular spins in a sample are almost equally populated but with small difference according to the Boltzman distribution, typically $\sim 10^{-5}$ at ambient temperatures. The NMR QC is based on the measurements of the variation in the population for these levels as a result of the time evolution performed under the phase coherence in Rabi oscillations.

In the NMR computation, the scalability means that the number of nuclear spins in a molecule is increased, each of which is accessible by a distinct Lamor frequency.[13-16] The accessibility to the ensemble of identical nuclear spins is a necessary property to perform unitary transformation of a single qubit. However, only with a small difference in resonance frequencies or chemical shifts, two spins are no longer excited or deactivated individually. This situation occurs, when the number of nuclear spins in a molecule is increased and the resonance frequencies of neighboring signals become close to each other. With these difficulties in scalability for the NMR QC in mind, exploitation of another dimension with another resonance frequency will be a choice for increasing the number of qubits. The electron-spin transitions accessible by microwave frequencies will be a natural choice. A variation of methods based on the ESR techniques developed so far is useful for characterization and manipulation of the molecular spins, i.e., cw-ESR, pulsed-ESR, ENDOR, and ESEEM. For both the NMR for nuclear spins and the ESR for electron spins, measurements are performed under the condition of magnetic resonance.

Endohedral fullerenes, as demonstrated for N@C_{60}, have potentially advantageous features for QC/QIP by magnetic resonance. First, protected inside the stable cage of C_{60}, the electron/nuclear spins of a nitrogen atom are almost free-standing, having relatively weak interactions with the environmental medium as well as with the surrounding cage. The weakness of the interactions guarantees the long coherence time, necessary for manipulation of the Rabi oscillations of the electron/nuclear spins. The discovery of the remarkable properties[73] have promoted theoretical and experimental studies on the exploitation of N@C_{60} for QC/QIP.

3.3. *Nitrogen Atom in a* C_{60} *Cage*

The nitrogen atom has the electron configuration of $(1s)^2(2s)^2(2p)^3$ and the spin quantum number $S = 3/2$ in its ground state ($S = 3/2, L = 0, J = L + S = 3/2$: $^4S_{3/2}$). In addition, the nuclear spin quantum numbers are $I = 1$ for ^{14}N and $I = 1/2$ for ^{15}N, with natural isotopic abundance of 99.6% and 0.4%, respectively (see Fig. 1). In a magnetic field, the degenerated spin levels split into $(2S + 1)(2I + 1)$, i.e., 12 levels for ^{14}N and 8 levels for ^{15}N. These levels are distinctly depicted by a combination of the two magnetic quantum numbers, M_S and M_I ($M_S = -S, -S + 1, \ldots, +S$ and $M_I = -I, -I + 1, \ldots, +I$). The transitions between two levels which differ in $\Delta M_S = \pm 1$ are accessible by absorption or emission of photons with microwave frequencies (ESR transitions), and those which differ only in $\Delta M_I = \pm 1$ are of photons with radio frequencies (NMR transitions).

In 1996, Almeida-Murphy *et al.* found that the nitrogen-implanted solid C_{60} contains a new type of paramagnetic centers as detected by the EPR signal of a triplet for ^{14}N and a doublet for ^{15}N.[73] The triplet indicates the presence of an electron spin whose energy levels are split by the hyperfine interaction due to the nuclear spin, $I = 1$, while the doublet due to the nuclear spin, $I = 1/2$. The isotropic nature found for these lines indicated the angular momentum for the paramagnetic center being $L = 0$. These observations were compatible with a physical picture that an atomic nitrogen having unpaired electron in its ground state was located at a center of symmetry of the molecular solids. A natural consequence is that a neutral nitrogen atom is situated at the center of the spherical cage of C_{60}. The hyperfine interaction constant (hfc) was 15.73 MHz for ^{14}N@C_{60} which was ~50% larger than free nitrogen atom.[73] Pietzak *et al.* refined the production method to increase the fraction of N@C_{60} over C_{60} and confirmed the inertness and stability of N@C_{60}.[74]

The magnetic properties of the electron/nuclear spin system were thoroughly investigated by Dinse and co-workers using the techniques in pulse electron paramagnetic resonance (EPR) and electron-nuclear double resonance (ENDOR).[75] Focusing on the effect of symmetry braking in N@C_{60} in polycrystalline C_{60}, they observed the spectra below and above the temperature at 258 K for the rotational phase transition of solid C_{60}. Even in the solid solution of N@C_{60} in C_{60}, the spins are located in an isotropic space above the temperature, but the zero-field splitting (ZFS) remains. In each of the three hyperfine lines in the EPR spectra, three transitions are overlapping and the manipulation of each transition seemed to be difficult. Aiming at the modification of the splitting in spin levels of the endohedral fullerenes, chemical derivatives are examined.[76] Hirsch with the groups of Weidinger and Dinse produced six derivatives of nitrogen-doped fullerenes, including N@$C_{61}(COOC_2H_5)$ and N@C_{70}, and measured EPR spectra in solutions and in solids. The hyperfine constant was ~5% smaller for N@C_{70} compared to N@C_{60}, but the isotropic nature of the spins of nitrogen atom was conserved in the slightly elongated cage of C_{70} with D_{5h} symmetry which does not provide a site of cubic symmetry. These effects were discussed along non-vanishing ZFS tensor modulated by rotational tumbling, internal motion of the encased atom, or collision induced deformation. The relaxation times, T_1 and T_2, are substantially shortened by the effects of the chemical adducts on the C_{60} cage, which were attributed to the permanent ZFS tensor by rotational tumbling.

Meyer *et al.* reported that the superposed lines in the EPR spectra under isotropic environments can be separated by embedding the spin carrier, N@C_{60} or N@C_{70}, into the nematic phase of the aligned liquid crystal,

4-methoxybenzylidene-4'-n-butylaniline (MBBA).[77] In addition to the hyperfine separations, further splitting was observed for each of the three EPR lines for N@C$_{70}$ below $\sim$320 K. The new splitting is realized by the fine structure term, $\sim SDS$, arising from the spin-spin interaction of the three unpaired electrons of atomic nitrogen. From the temperature dependence of the new splitting, the upper limit for the effective value of $D_{\text{eff}} < 0.418$ MHz was deduced. For N@C$_{60}$ also, the fine-structure splitting was observed, $D_{\text{eff}} < 0.090$ MHz, but smaller than N@C$_{70}$. The alignment of N@C$_{70}$ or N@C$_{60}$ in a liquid crystal which can lead to the discrimination in resonance frequencies for all the nine EPR transitions is important for the access to the individual spin states.

Increasing interests concerned about the effects by the environmental conditions remind again the issues on the purity of N@C$_{60}$. As was mentioned in Section 2.5, the as-prepared sample is a mixture of the endohedral fullerenes and the empty fullerenes. The ratio of N@C$_{60}$/C$_{60}$ is typically 10^{-4}-10^{-5}. By the chromatographic separation, the ratio can be increased by a few orders of magnitude.[78,79] Takagi and the group of Harneit and Weidinger performed a separation of N$_2$@C$_{60}$ and N@C$_{60}$ from C$_{60}$. The diamagnetic specie, N$_2$@C$_{60}$, as well as paramagnetic one, N@C$_{60}$, was detected by laser-desorption time-of-flight mass spectrometry (LD-TOF MS).[78] Dinse and the group of Harneit and Weidinger tried for purification of N@C$_{60}$ by recycling HPLC operations combined with EPR detection.[79] The fraction of N@C$_{60}$ was clearly separated from that of C$_{60}$ after the recycling operation in HPLC. As the concentration of N@C$_{60}$ is increased, the linewidth of the EPR transition increased substantially, indicating that the intermolecular spin-spin interaction can not be negligible under the high concentration. As a pure material, the intrinsic properties of N@C$_{60}$ are still demanded for investigation in more detail. Nevertheless, by focusing only the spin dynamics, it is possible to persuade the way of N@C$_{60}$ for QC/QIP.

3.4. *Proposal and Implementation of Fullerene Quantum Computer*

Inspired by Kane's proposal in 1998 for scalable quantum computers based on nuclear spins of phosphor atoms, ^{31}P, embedded in a solid silicon,[107] Harneit[70] and Suter and Lim[71] proposed fullerene-based electron-spin quantum computers. The endohedral fullerene molecule, N@C$_{60}$ or P@C$_{60}$, carrying qubits in the electron/nuclear spins of the trapped atom inside is aligned on the surface where suitable electronic devises are designed to manipulate individual quibits by using field gradient so as to perform single- and double-qubits operations. Twamley proposed an alternative array of N@C_{60} and P@C_{60} to operate as quantum-cellular-automaton.[72] These approaches are still challenging in a sense that the single spin on an individual molecule is to be distinguished and manipulated, or, at least, even for an ensemble of spins, the subset of spins on the same type of molecules are approached distinctly by adapted environmental differences.

Mehring and Scherer reported the detection of pseudoentanglement for the four-level spin system created in an x-ray irradiated malonic acid single crystal with the electron spin $S = 1/2$ and the proton spin $I = 1/2$.[80] With the collaboration with Weidinger, they applied the same technique of driving nuclear Rabi oscillations by NMR transitions under the detection by ESR echo signals to the four-level subsystems of ^{15}N@C_{60}.[81] By density matrix tomography, they demonstrated to create four Bell states within the selected two levels among the total eight levels of the $S = 3/2$ and $I = 1/2$ system.[82]

The curiosity for QC/QIP by the endohedral fullerene, N@C_{60}, or, in other words, by the trapped spins of otherwise unstable nitrogen atom has been pushed forward after participation of the Oxford research group lead by Briggs into the race. Their major achievements and perspectives were summarized in the article by Benjamin et al.[83] Starting with the synthesis,[84] they performed the experiments on N@C_{60} based on the electron spin echo envelope modulation (ESEEM) techniques.[86] The control of Rabi oscillations leads to high fidelity single qubit operations.[87] Combined with a decoupling technique by irradiation of an μw pulse during the time evolution, the phase control of the nuclear Rabi oscillations was demonstrated to perform controlled gate operations, namely bang-bang control of qubits.[88] In order to achieve a long coherence time, they took care of the homogeneity of the sample as well as the magnetic field and the μw and rf fields.[89] With an advantage of high sensitivity for detection, the manipulation of quantum information in N@C_{60} is designed for reliable QIP.[90] The intrinsic magnetic properties of N@C_{60} were investigated further by Davis ENDOR for higher sensitivity[91] and precise measurement of the relaxation times.[92] The effect of the solvents, such as CS_2, on the relaxation times was thoroughly investigated between 160-300 K, and discussed along with the suggested mechanism of ZFS fluctuations versus the Orbach relaxation mechanism in which some specific vibrational modes of the C_{60} cage is responsible for the temperature dependence in the phase decoherence time, T_2.[93] The solvent effects on the nuclear spins were also examined to identify a dramatic changes in decoherence time, T_2, around 120 K for N@C$_{60}$ in glass matrices of S$_2$Cl$_2$/CS$_2$.[94] The polarization of nuclear spins enhanced by ENDOR (PONSEE) was also applied to N@C$_{60}$ at a relatively high magnetic field of 8.6 T and at a low temperature of 3 K.[97] In relation to the solid state QC/QIP, the nitrogen vacancy in a lattice of diamond also exhibits a long coherence time.[95,96]

Towards a solid state QC/QIP by endohedral fullerenes, the orientation of molecules as spin-carriers, the initialization and read out schemes of spin states, and the modification and stabilization of the electron spin system are crucial. Feng et al. studied a method for read out of the electron spin for a fullerene-based quantum computer by the measurement of variations in the spin-dependent tunneling currents in the scanning tunneling microscope (STM).[98] Neydenov, Harneit and co-workers investigated further the modification of the spin systems of N@C$_{60}$ and P@C$_{60}$ by co-crystallization with 2,4,6-tris-(4-bromophenoxy)-1,3,5-triazine (Br-POT),[99,100] and the morphology upon deposition on a PT/Cr-coated silicon tip.[101] For modification and orientation of the electron spins at desired distances, the dimmers of N@C$_{60}$-C$_{60}$, the endohedrals in other fullerene cavities such as N@C$_{70}$, and chemical derivatives are again noted.[102-104] Corzilius, Dinse, and Hata studied the endohedral fullerenes encapsulated in single-wall carbon nanotubes, N@C$_{60}$/SWNT.[105,106]

3.5. The ESR and ENDOR Experiments

The nitrogen atom has the electron configuration of half-filled three $2p$ orbitals, then the orbital angular momentum vanishes in total, i.e., $L = 0$. Accordingly, relevant contributions in the Hamiltonian are vanishing, i.e., the interaction of the orbital angular momentum with the externally applied magnetic field, $g\beta B_0 \cdot L \to 0$, and the spin-orbit interaction, $\lambda L \cdot S \to 0$, where g denotes the electron g-factor, β the Bohr magneton, B_0 the external magnetic field, L the orbital angular momentum operator, and S the electron spin operator. Thus, for an ensemble of the molecules of N@C$_{60}$, the Hamiltonian H_0 for the spin system can be simply described as follows.

$$H_0 = g\beta B_0 \cdot S - \gamma B_0 \cdot I + aS \cdot I, \qquad (10)$$

$$aS \cdot I = a(S_z I_z + S_x I_x + S_y I_y). \qquad (11)$$

The first term in (10) represents the interaction of the electron spin with the externally applied magnetic field B_0. The second term is the interaction of the nuclear spin with the externally applied magnetic field B_0, where

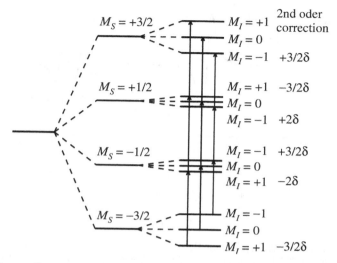

Fig. 2. Energy diagram of the twelve spin levels of ^{14}N@C$_{60}$, reproduced using the experimental parameters by Morton and Benjamin.[83,86] The electron spin Zeeman splitting gives rise to four split levels of $M_S = +3/2, +1/2, -1/2$, and $-3/2$. Each of these levels split further into three levels by hyperfine interaction, $a\hbar^2 M_S M_I$, i.e. the electron spin $S = 3/2$ with the nuclear spin $I = 1$ of ^{14}N. The magnitude of the hyperfine splitting, $a\hbar^2$, is ~15.8 MHz and extended by a factor of 10^2 compared to the electron Zeeman splitting of ~9.4 GHz for a typical X-band ESR measurement at ~335 mT. The magnitude of the nuclear Zeeman splitting, $-\gamma\hbar B_0$, is on the order of ~1 MHz at ~335 mT and not readily discernible. The second-order correction in terms of $\delta = \hbar(a^2\hbar^2/g\beta B_0)$ lifts some of the energy levels by which the observed splitting for the outer signals in the ESR spectra are explained (see Figure 3).

γ denotes the gyromagnetic ratio of the nucleus and I the nuclear spin operator. The third term in (10) corresponds to the hyperfine interaction and given by its components as in (11). When the magnitude of the external magnetic field B_0 is sufficiently large, so that the relation, $2g\beta\hbar B_0 \gg a\hbar^2$, holds, the electron spin magnetic moments are quantized along the axis of the magnetic field B_0, namely z axis. Since the spin operators S and I are diagonalized to $\hbar M_S$ and $\hbar M_I$, the eigenvalues for the Hamiltonian (10) is approximated as,

$$E_0 = (g\beta\hbar M_S - \gamma\hbar M_I)B_0 + a\hbar^2 M_S M_I + \text{S.O.}, \quad (12)$$

where M_S and M_I are the magnetic quantum numbers for the electron spin and the nuclear spin, respectively. The last term is associated with the second-order correction stemming from the off-diagonal elements in the isotropic hyperfine interaction, $S_x I_x + S_y I_y$. The discussion is found on

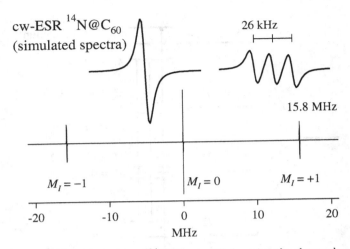

Fig. 3. Simulated cw-ESR spectra of ^{14}N@C$_{60}$ in CS$_2$ by using the observed parameters in the literature.[83,86] The splitting of $\pm a\hbar/2\pi = \pm 15.8$ MHz is due to hyperfine coupling of the electron spin $S = 3/2$ with the nuclear spin $I = 1$ of ^{14}N. The outer two signals are split into three lines separated by $\pm \delta/h = (2\pi)^{-1}(a^2\hbar^2/g\beta B_0) = \pm 26$ kHz due to the second order correction for the isotropic terms of the hyperfine interaction.[83,86]

the effect of the second-order correction along the deviation of the axis of quantization for the nuclear spin from that for the electron spin.[3]

The electron-spin Zeeman splitting in the first term in (12) amounts to $\nu_e = g\beta\hbar B_0/2\pi \sim 9.67$ GHz at $B_0 = 345$ mT, while the nuclear Zeeman splitting in the second term is on the order of ~ 1 MHz, i.e., $\nu_n = |-\gamma B_0/2\pi| \sim 1.06$ MHz for ^{14}N at $B_0 = 345$ mT. The latter term lifts the energy levels of the same nuclear magnetic quantum number M_I to the same amount, downwards for $M_I = +1$ and upwards for $M_I = -1$. Since the ESR transition occurs between two levels of the same nuclear magnetic quantum number M_I, the nuclear Zeeman splitting cannot be measured directly by the cw-ESR measurement. It is measured by ENDOR experiments as described below.

The hyperfine splitting by the isotropic interaction of the form aS_zI_z appears with the coupling constant of $a\hbar/2\pi \sim 15.8$ MHz. The second-order correction in the form of $a(S_xI_x + S_yI_y)$ results in the shifts in six of the twelve energy levels of $M_I = +1$ and -1 on the order of

$$\frac{\delta}{h} = \frac{1}{2\pi}\left(\frac{a^2\hbar^2}{g\beta B_0}\right) \sim 26 \text{ kHz,}$$

i.e., by $-3\delta/2$ for $(M_S, M_I) = (+1/2, +1)$, -2δ for $(-1/2, +1)$, $-3\delta/2$

for $(-3/2, +1)$, $+3\delta/2$ for $(-1/2, -1)$, $+2\delta$ for $(+1/2, -1)$, and $+3\delta/2$ for $(+3\delta/2, -1)$.[83,86]

Figure 2 shows the schematic diagram of the energy level splitting represented by (12) for the spin states of ^{14}N@C$_{60}$.[83,86] The levels are split into four levels by the electron spin $S = 3/2$ (electron Zeeman splitting), and each of which are further split into three by the nuclear spin $I = 1$ (hyperfine splitting). For the case of ^{14}N@C$_{60}$ under the conventional ESR conditions, the nuclear Zeeman splitting is one order of magnitude smaller than the hyperfine splitting, $|-\gamma\hbar B_0| < a\hbar^2/2$, and is not readily discernible in Fig. 2. The hyperfine interaction is an intrinsic property of the molecule, thus it is important for manipulation of the qubits to understand its static and dynamical aspects in the magnetic fields.

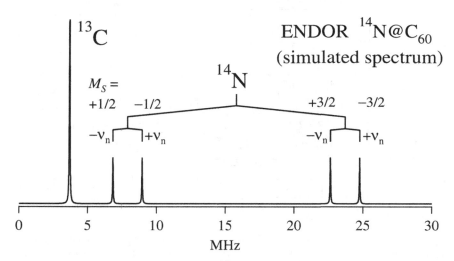

Fig. 4. Simulated ENDOR spectrum of ^{14}N@C$_{60}$ in polycrystalline C$_{60}$ by using the experimental parameters observed by Dinse and co-workers.[75] The signals of ^{14}N are centered at 1/2 and 3/2 of the hyperfine constant $a\hbar/2\pi \sim 15.8$ MHz and further split by twice of the nuclear Zeeman frequency, $2\nu_n$. The signal of ^{13}C is expected for the nuclei situated on the carbon shell of C$_{60}$.

Figure 3 shows simulated ESR spectrum of ^{14}N@C$_{60}$. Three vertical lines in the main panel represent the absorption of microwave of different resonance frequencies (in actual measurements, the microwave frequency is kept fixed and the external magnetic field is swept to vary the Zeemen splitting). The ESR transitions within the $M_I = -1$ manifold resonates at a higher field and $M_I = +1$ at a lower field. These two lines are separated from the central line by ± 15.8 MHz. By using a purified sample of ^{14}N@C$_{60}$ in CS$_2$, further splitting of ± 26 kHz was observed for the outer lines,[83,86] as shown in the inset in Fig. 3.

Morton et al. observed oscillatory behaviors for the decay signals of the outer lines in the electron spin echo envelope modulation (ESEEM) experiments and connected the oscillation with the small splitting of 26 kHz, based on the investigation of the spin Hamiltonian having off-diagonal elements of the isotropic hyperfine interaction. Such a fine analysis became possible because of the narrowness of the signal lines (~ 9 kHz) and the long decoherence time ($\sim 200\,\mu s$).[86]

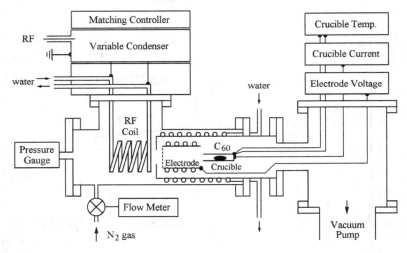

Fig. 5. A schematic view of the apparatus for producing N@C$_{60}$ at Kindai. The C$_{60}$ molecules (powder) in a crucible is sublimed to deposit on the inner surface of the water-cooled cylindrical electrode where the sedimenting C$_{60}$ molecules are bomberded with cationic species that are electrostatically extracted from the plasma of nitrogen gas.

The ENDOR (electron-nuclear double resonance) experiments clarify relation between the hyperfine interaction and the nuclear Zeeman interaction. The ENDOR technique was developed by George Feher for his studies of phosphorus-doped silicon in the '50s.[17] The microwave (μw) for the ESR transition is applied for saturation and the radio frequency (rf) for the NMR transition is swept. When the rf frequency matches to the NMR transition, the balance in the populations between the upper and lower levels of the relevant NMR transition changes thus detected by the difference in the amplitude of the ESR absorption. By monitoring the ESR transition, the NMR transitions are detectable in orders of magnitude higher sensitivity.

Figure 4 shows a simulation of the ENDOR spectrum of ^{14}N@C$_{60}$. The experimental frequencies reported in the literature[75] are used (the amplitudes are supposed to be the same for the series of transitions). As is seen in the level scheme in Fig. 2, the hyperfine interaction, $a\hbar^2 M_S M_I$, is dominant compared to the nuclear Zeeman interaction, $-\gamma\hbar B_0 M_I$. Among the eight NMR transitions, every two for a given M_S have the same transition frequency ($\Delta M_S = 0$, $M_I = -1 \to 0$ and $0 \to +1$). Therefore, depending on M_S, four signals appear in the ENDOR spectrum in different frequencies at

$$\frac{1}{2}\left(\frac{a\hbar}{2\pi}\right) \mp \nu_n \quad \text{and} \quad \frac{3}{2}\left(\frac{a\hbar}{2\pi}\right) \mp \nu_n,$$

for $M_S = \pm 1/2$ and $M_S = \pm 3/2$, respectively, where $\nu_n = |-\gamma B_0/2\pi|$ is the nuclear Larmor frequency at a given magnetic field B_0, e.g., $\nu_n \sim 1.06$ MHz for ^{14}N at ~ 345 mT in this case for Fig. 4. The peak at ~ 3.7 MHz in Fig. 4 depicts the ENDOR signal of ^{13}C nucleus situated on the carbon cage of C$_{60}$, by its natural isotopic abundance of 1.01%.

Dinse and co-workers investigated the ESR and ENDOR spectra of ^{14}N@C$_{60}$ dispersed in polycrystalline C$_{60}$ above and below the temperature of the first-order phase transition of solid C$_{60}$ at 258 K.[75] Below this temperature, the ESR signals were accompanied by shoulders at both sides of the central line, indicating that zero-field splitting (ZFS) tensor components became non-vanishing by the establishment of a long-range order within a crystal to lower the symmetry around the center of the fullerene cage. The ENDOR line of ^{14}N at 80 K, well bellow the temperature of the rotational transition of C$_{60}$, was fitted with a Lorenzian of ~ 4 kHz width and no obvious splitting was observed. Thus, the authors concluded that no quadrupole interaction, $q\mathbf{I} \cdot \mathbf{I}$, was detectable for this system of ^{14}N@C$_{60}$. The ENDOR line of ^{13}C exhibited shoulders, which was attributed to the interaction of the ^{13}C nuclear spin with the electron spin of the nitrogen atom inside the cage.[75]

4. Production of N@C$_{60}$ at Kindai

We report here the ESR detection of N@C$_{60}$ for our ion-bombarded C$_{60}$ sample. Figure 5 illustrates a schematic view of the apparatus for producing N@C$_{60}$ at Kindai (Kinki University). The N$_2$ gas at ~ 2.7 Pa was excited to discharge in an rf coil supplied by 500 W power at 13.56 MHz. The cationic species were extracted through an aperture of 2 cm in diameter and accelerated toward the surface of a water-cooled electrode to which dc voltage of −80 V was applied. The C_{60} molecules were sublimed from a crucible and deposited on the inner surface of the cylindrical electrode where the molecules were bombarded with the cations. Typical ion current under operation was ∼10 mA. The deposited materials were collected in air, sonicated with a small volume of CS_2, and filterated to remove insoluble solid particles and aggregates. The ESR spectra were recorded on JEOL JES-PX1050 spectrometer.

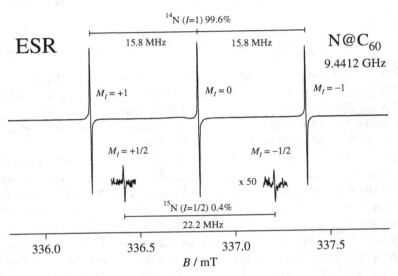

Fig. 6. X-band cw-ESR spectrum of N@C_{60}/C_{60} mixture in CS_2 at room temperature. The relatively sharp, three lines correspond to the electron-spin transitions ($\Delta M_S = +1$) of $M_I = +1, 0, -1$ from the low field, respectively. In each line, three lines of transitions from the electron-spin states of $M_S = -3/2, -1/2, +1/2, +3/2$ are overlapping (see the vertically aligned three arrows connecting electron-spin levels, M_S, within the same nuclear-spin manifold, M_I, in Fig. 2).

Figure 6 shows the X-band ESR spectrum of ^{14}N@C_{60}/C_{60} in CS_2 at room temperature. The remarkably sharp, three lines are the ESR transitions ($\Delta M_S = +1$ and $\Delta M_I = 0$) for $M_I = +1, 0, -1$ from the low field, respectively. Owing to the high sensitivity for the detection of the ESR transitions with narrow bandwidths, the production of ^{14}N@C_{60} was successfully confirmed. Since the spectrum was measured at the modulation frequency of 200 kHz, further splitting by ∼26 kHz for the outer lines of

$M_I = +1$ and -1 was not resolved.

Towards the coherent control of the molecular spins in N@C$_{60}$, further developments are needed: 1) the prduction efficiency must be improved to obtain a macroscopic amount of the endohedral molecules, 2) the method for purification is to be established for providing a stoichiometrically well-defined material, 3) the instrumentation for the precise control of the molecular spins based on the pulsed-ESR/NMR techniques has to be developped and integrated further, 4) the idea for manipulation of spin-spin interactions, e.g., by a precision control of material structures in nanometer scales or by external fields applied on the material, is to be proposed and implemented, 5) a new protocol for QC/QIP exploiting the advantage of a combination of ESR/NMR transtions is wellcome to be proposed and tested on the novel materials composed of fascinating bulk-phase atomic spins in N@C$_{60}$.

Acknowledgments

In this article, the author, T. W., tried to review the *entangled* story starting from the discovery of C$_{60}$ to the emerging field of manipulation of spins inside the fullerene molecule. One may feel that it is merely a personal record during the survey on this subject, nevertheless, I hope it be a bit of help for some. I would like to express my gratitude for Professor Mikio Nakahara for giving me the opportunity for working on QC/QIP with N@C$_{60}$. Dr. Yukihiro Ota and Dr. Masahito Tada at the Research Center for Quantum Computing are greately acknowledged for their careful reading of the draft. I would like to thank Professor Takayoshi Kuroda for the ESR measurements and discussions. Professor Tatsuhisa Kato (Josai University) and Dr. Takeshi Kodama (Tokyo Metropolitan University) are acknowledged for their valuable comments on the ESR detection of the endohedral fullerenes.

References

1. M. A. Nielsen and I. L. Chuang, *Quantum Computation and Quantum Information* (Cambridge Univ. Press, Cambridge, 2000).
2. M. L. Bellac, *A Short Introduction to Quantum Information and Quantum Computation* (Cambridge Univ. Press, Cambridge, 2005).
3. C. P. Slichter, *Principles of Magnetic Resonance 3rd Ed.*, Section 11.3, (Springer-Verlag, Berlin, 1990).
4. M. S. Dresselhaus, G. Dresselhaus, and P. C. Eklund, *Science of Fullerenes and Carbon Nanotubes* (Academic Press, San Diego, 1996).
5. N. A. Gershenfeld and I. L. Chuang, *Science* **275** (1997) 350.
6. W. S. Warren, N. Gerschenfeld, and I. Chuang, *Science* **277** (1997) 1688.

7. J. A. Jones and M. Mosca, *J. Chem. Phys.* **109** (1998) 1684.
8. I. L. Chuang, N. Gerschenfeld, and M. Kubinec, *Phys. Rev. Lett.* **80** (1998) 3408.
9. I. L. Chuang, L. M. K. Vandersypen, X. Zhou, D. W. Leung, and S. Lloyd, *Nature* **393** (1998) 143.
10. E. Knill, I. Chuang, and R. Laflamme, *Phys. Rev. A* **57** (1998) 3348.
11. D. G. Cory, M. D. Price, and T. F. Havel, *Physica D* **120** (1998) 82.
12. S. L. Braunstein, C. M. Caves, R. Jozsa, N. Linden, S. Popescu, and R. Schack, *Phys. Rev. Lett.* **83** (1999) 1054.
13. L. M. K. Vandersypen, M. Steffen, G. Breyta, C. S. Yannoni, R. Cleve, and I. L. Chuang, *Phys. Rev. Lett.* **85** (2000) 5452.
14. R. Marx, A. F. Fahmy, J. M. Myers, W. Bermel, and S. J. Glaser, *Phys. Rev. A* **62** (2000) 012310.
15. M. Steffen, L. M. K. Vandersypen, and I. L. Chuang, *IEEE* **24** (2001).
16. L. M .K. Vandersypen, M. Steffen, G. Breyta, C. S. Yannoni, M. H. Sherwood, and I. L. Chuang, *Nature* **414** (2001) 883.
17. G. Feher, *Phys. Rev.* **103** (1956) 834.
18. H. W. Kroto, J. R. Heath, S. C. O'Brien, R. F. Curl, and R. E. Smalley, *Nature* **318** (1985) 162.
19. 6. W. Krätschmer, K. Fostiropoulos, L. D. Lamb, and D. R. Huffman, *Nature* **347** (1990) 354.
20. E. Osawa, *Kagaku* **25** (1970) 854 in Japanese.
21. Z. Yoshida and E. Osawa, *Monograph in Chemistry 22: Aromaticity* (Tokyo Kagaku Dojin, 1971) in Japanese.
22. R. D. Johnson, G. Meijer, and D. S. Bethune, *J. Am. Chem. Soc.* **112** (1990) 8983.
23. R. Taylor, J. P. Hare, A. K. Abdul-Sada, and H. W. Kroto, *J. Chem. Soc., Chem. Commun.* (1990) 1423.
24. R. D. Johnson, G. Meijer, J. R. Salem, and D. S. Bethune, *J. Am. Chem. Soc.* **113** (1991) 3619.
25. P. J. Fagan, J. C. Calabrese, and B. Malone, *Science* **252** (1991) 1160.
26. J. M. Hawkins, S. Loren, A. Meyer, and R. Nunlist, *J. Am. Chem. Soc.* **113** (1991) 7770.
27. C. S. Yannoni, R. D. Johnson, G. Meijer, D. S. Bethune, and J. R. Salem, *J. Chem. Phys.* **95** (1991) 9.
28. R. Tycko, R. C. Haddon, G. Dabbagh, S. H. Glarum, D. C. Douglass, and A. M. Mujsce, *J. Phys. Chem.* **95** (1991) 518.
29. R. Tycko, G. Dabbagh, R. M. Fleming, R. C. Haddon, A. V. Makhija, and S. M. Zahurak, *Phys. Rev. Lett.* **67** (1991) 1886.
30. P. A. Heiney, J. E. Fischer, A. R. McGhie, W. J. Romanow, A. M. McCauley, A. B. Smith, III, and D. E. Cox, *Phys. Rev. Lett.* **66** (1991) 2911.
31. T. Matsuo, H. Suga, W. I. F. David, R. M. Ibberson, P. Bernier, A. Zahab, C. Fabre, A. Rassat, and A. Dworkin, *Solid State Commun.* **83** (1992) 711.
32. K. Prassides, *Physica Scripta* **T49**B (1993) 735.
33. R. C. Haddon, A. F. Hebard, M. J. Rosseinsky, D. W. Murphy, S. J. Duclos, K. B. Lyons, B. Miller, J. M. Rosamila, R. M. Fleming, A. R. Kortan, S. H.

Glarum, A. V. Makhija, A. J. Muller, R. H. Eick, S. M. Zahurak, R. Tycko, G. Dabbagh, and F. A. Thiel, *Nature* **350** (1991) 320.
34. K. Tanigaki, S. Kroshima, J. Fujita, and T. W. Ebbesen, *Appl. Phys. Lett.* **63** (1993) 2351.
35. A. Y. Ganin, Y. Takabayashi, Y. Z. Khimyak, S. Margadonna, A. Tamai, M. J. Rosseinsky, and K. Prassides, *Nature Materials* **7** (2008) 367.
36. Y. Kajii, T. Nakagawa, S. Suzuki, Y. Achiba, K. Obi, and K. Shibuya, *Chem. Phys. Lett.* **181** (1991) 100.
37. M. R. Wasielwski, M. P. O'Neil, K. R. Lykke, M. J. Pellin, and D. M. Gruen, *J. Am. Chem. Soc.* **113** (1991) 2774.
38. P. A. Lane, L. S. Swanson, Q.-X. Ni, J. Shinar, J. P. Engel, T. J. Barton, and L. Jones, *Phys. Rev. Lett.* **68** (1992) 887.
39. S. Michaeli, V. Meiklar, M. Schulz, K. Möbius, and H. Levanon, *J. Phys. Chem.* **98** (1994) 7444.
40. G. J. B. Van den Berg, D. J. Van den Heuvel, O. G. Poluektov, I. Holleman, G. Meijer, and E. J. J. Groenen, *J. Magn. Reson.* **131** (1998) 39.
41. A. Sassara, G. Zerza, and M. Chergui, *Chem. Phys. Lett.* **261** (1996) 213.
42. A. Regev, D. Gamliel, V. Meiklyar, S. Michael, and H. Leanon, *J. Phys. Chem.* **97** (1993) 3671.
43. T. Kato, T. Kodama, T. Shida, T. Nakagawa, T. Matsui, S. Suzuki, H. Shiromaru, K. Yamauchi, and Y. Achiba, *Chem. Phys. Lett.* **180** (1991) 446.
44. P.-M. Allemand, G. Srdanov, A. Koch, K. Khemani, and F. Wudl, *J. Am. Chem. Soc.* **113** (1991) 2780.
45. T. Kato, T. Kodama, M. Oyama, S. Okazaki, T. Shida, T. Nakagawa, Y. Matsui, S. Suzuki, H. Shiromaru, K. Yamauchi, and Y. Achiba, *Chem. Phys. Lett.* **186** (1991) 35.
46. T. Kato, T. Kodama, and T. Shida, *Chem. Phys. Lett.* **205** (1993) 405.
47. R. E. Haufler, Y. Chai, L. P. F. Chibante, J. Conceicao, C. Jin, L.-S. Wang, S. Maruyama, and R. E. Smalley, *Mat. Res. Soc. Symp. Proc.* **206** (1991) 627.
48. Y. Chai, T. Guo, C. Jin, R. E. Haufler, and L. P. Chibante, J. Fure, L. Wang, J. M. Alford, and R. E. Smalley, *J. Phys. Chem.* **95** (1991) 7564.
49. F. Diederich, R. Ettl, Y. Rubin, R. L. Whetten, R. Beck, M. Alvarez, S. Anz, D. Sensharma, F. Wudl, K. Khemani, and A. Koch, *Science* **252** (1991) 548.
50. R. Ettl, I. Chao, F. Diederich, and R. L. Whetten, *Nature* **353** (1991) 149.
51. F. Diederich, R. L. Whetten, C. Thilgen, R. Ettl, I. Chao, and M. Alvarez, *Science* **254** (1991) 1768.
52. K. Kikuchi, N. Nakahara, T. Wakabayashi, S. Suzuki, H. Shiromaru, Y. Miyake, K. Saito, I. Ikemoto, M. Kainosho, and Y. Achiba, *Nature* **357** (1992) 142.
53. H. Shinohara, *Rep. Prog. Phys.* **63** (2000) 843.
54. M. Saunders, H. A. Jiménez-Vázquez, R. J. Cross, S. Mroczkowski, D. I. Freedberg, and F. A. L. Anet, *Nature* **367** (1994) 256.
55. A. Khong, R. J. Cross, and M. Saunders, *J. Phys. Chem. A* **104** (2000)

3940.
56. K. Komatsu, M. Murata, and Y. Murata, *Science* **307** (2005) 238.
57. R. F. Kiefl, J. W. Schneider, A. MacFarlane, K. Chow, T. L. Duty, T. L. Estle, B. Hitti, R. L. Licht, E. J. Ansaldo, C. Schwab, P. W. Percival, G. Wei, S. Wlodek, K. Kojima, W. J. Romanow, J. P. McCauley, Jr., N. Coustel, J. E. Fischer, and A. B. Smith, III, *Phys. Rev. Lett.* **68** (1992) 1347.
58. E. J. Ansaldo, J. Boyle, Ch. Neidermayer, G. D. Morris, J. H. Brewer, C. E. Stronach, and R. S. Cary, *Z. Phys. B* **86** (1992) 317.
59. T. Ohtsuki, H. Yuki, M. Muto, J. Kasagi, and K. Ohno, *Phys. Rev. Lett.* **93** (2004) 112501.
60. A. Gromov, W. Krätschmer, N. Krawez, R. Tellgmann, and E. E. B. Campbell, *J. Chem. Soc., Chem. Commun.* (1997) 2003.
61. H. Shinohara, H. Sato, M. Ohkohchi, Y. Ando, T. Kodama, T. Shida, T. Kato, and Y. Saito, *Nature* **357** (1992) 52.
62. K. Kikuchi, S. Suzuki, Y. Nakao, N. Nakahara, T. Wakabayashi, H. Shiromaru, K. Saito, I. Ikemoto, and Y. Achiba, *Chem. Phys. Lett.* **216** (1993) 67.
63. S. Stevenson, G. Rice, T. Glass, K. Harich, F. Cromer, M. R. Jordan, J. Craft, E. Hadju, R. Bible, M. M. Olmstead, K. Maitra, A. J. Fischer, A. L. Balch, and H. C. Dorn, *Nature* **401** (1999) 55.
64. P. Jakes and K.-P. Dinse, *J. Am. Chem. Soc.* **123** (2001) 8854.
65. A. Einstein, B. Podolsky, and N. Rosen, *Phys. Rev.* **47** (1935) 777.
66. D. Bohm, *Phys. Rev.* **85** (1952) 166.
67. D. Bohm and Y. Aharonov, *Phys. Rev.* **108** (1957) 1070.
68. J. S. Bell, *Physics* **1** (1964) 195.
69. J. S. Bell, *Rev. Mod. Phys.* **38** (1966) 447.
70. W. Harneit, *Phys. Rev. A* **65** (2002) 032322.
71. D. Suter and K. Lim, *Phys. Rev. A* **65** (2002) 052309.
72. J. Twamley, *Phys. Rev. A* **67** (2003) 052318.
73. T. Almeida Murphy, Th. Pawlik, A. Weidinger, M. Höhne, R. Alcala, and J.-M. Spaeth, *Phys. Rev. Lett.* **77** (1996) 1075.
74. B. Pietzak, M. Waiblinger, T. Almeida Murphy, A. Weidinger, M. Höhne, E. Dietel, and A. Hirsch, *Chem. Phys. Lett.* **279** (1997) 259.
75. N. Weiden, H. Käss, and K.-P. Dinse, *J. Phys. Chem. B* **103** (1999) 9826.
76. E. Dietel, A. Hirsch, B. Pietzak, M. Waiblinger, K. Lips, A. Weidinger, A. Gruss, and K.-P. Dinse, *J. Am. Chem. Soc.* **121** (1999) 2432.
77. C. Meyer, W. Harneit, K. Lips, A. Weidinger, P. Jakes, and K.-P. Dinse, *Phys. Rev. A* **65** (2002) 061201.
78. T. Suetsuna, N. Dragoe, W. Harneit, A. Weidinger, H. Shimotani, S. Ito, H. Takagi, and K. Kitazawa, *Chem. Eur. J.* **8** (2002) 5080.
79. P. Jakes, K.-P. Dinse, C. Meyer, W. Harneit, and A. Weidinger, *Phys. Chem. Chem. Phys.* **5** (2003) 4080.
80. M. Mehring, J. Mende, and W. Scherer, *Phys. Rev. Lett.* **90** (2003) 153001.
81. M. Mehring, W. Scherer, and A. Weidinger, *Phys. Rev. Lett.* **93** (2004) 206603.
82. W. Scherer and M. Mehling, *J. Chem. Phys.* **128** (2008) 052305.

83. S. C. Benjamin, A. Ardavan, G. A. D. Briggs, D. A. Britz, D. Gunlycke, J. Jefferson, M. A. G. Jones, D. F. Leigh, B. W. Lovett, A. N. Khlobystov, S. A. Lyon, J. J. L. Morton, K. Porfyrakis, M. R. Sambrook, and A. M. Tyryshkin, *J. Phys.: Condens. Matter* **18** (2006) S867.
84. M. A. G. Jones, D. A. Britz, J. J. L. Morton, A. N. Khlobystov, K. Porfyrakis, A. Ardavan, and G. A. D. Briggs, *Phys. Chem. Chem. Phys.* **8** (2006) 2083.
85. E. L. Hahn, *Phys. Rev.* **80** (1950) 580.
86. J. J. L. Morton, A. M. Tyryshkin, A. Ardavan, K. Porfyrakis, S. A. Lyon, and G. A. D. Briggs, *J. Chem. Phys.* **122** (2005) 174504.
87. J. J. L. Morton, A. M. Tyryshkin, A. Ardavan, K. Porfyrakis, S. A. Lyon, and G. A. D. Briggs, *Phys. Rev. Lett.* **95** (2005) 200501.
88. J. J. L. Morton, A. M. Tyryshkin, A. Ardavan, S. C. Benjamin, K. Porfyrakis, S. A. Lyon, and G. A. D. Briggs, *Nat. Phys.* **2** (2006) 40.
89. J. J. L. Morton, A. M. Tyryshkin, A. Ardavan, S. Benjamin, K. Porfyrakis, S. A. Lyon, and G. A. D. Briggs, *Phys. Stat. Sol. B* **243** (2006) 3028.
90. A. Ardavan, J. J. L. Morton, S. C. Benjamin, K. Porfyrakis, G. A. D. Briggs, A. M. Tyryshkin, and S. A. Lyon, *Phys. Stat. Sol. B* **244** (2007) 3874.
91. A. M. Tyryshkin, J. J. L. Morton, A. Ardavan, and S. A. Lyon, *J. Chem. Phys.* **124** (2006) 234508.
92. J. J. L. Morton, N. S. Lees, B. M. Hoffman, and S. Stoll, *J. Magn. Reson.* **191** (2008) 315.
93. J. J. L. Morton, A. M. Tyryshkin, A. Ardavan, K. Porfyrakis, S. A. Lyon, and G. A. D. Briggs, *J. Chem. Phys.* **124** (2006) 014508.
94. J. J. L. Morton, A. M. Tyryshkin, A. Ardavan, K. Porfyrakis, S. A. Lyon, and G. A. D. Briggs, *Phys. Rev. B* **76** (2007) 085418.
95. T. Gaebel, M. Domhan, I. Popa, C. Wittmann, P. Neumann, F. Jelezko, J. R. Rabeau, N. Stavrias, A. D. Greentree, S. Prawer, J. Meijer, J. Twamley, P. R. Hemmer, and J. Warachitrup, *Nature Physics* **2** (2006) 408.
96. J. J. L. Morton, *Nat. Phys.* **2** (2006) 365.
97. G. W. Morley, J. Van Tol, A. Ardavan, K. Porfyrakis, J. Zhang, and G. A. Briggs, *Phys. Rev. Lett.* **98** (2007) 220501.
98. M. Feng, G. J. Dong, and B. Hu, *New. J. Phys.* **8** (2006) 252.
99. B. N. Naydenov, PhD. Thesis, Freie Universität Berlin (2006).
100. B. Naydenov, Ch. Spudat, M. Scheloske, H. I. Suess, J. Hulliger, and W. Harneit, *Phys. Stat. Sol. B* **243** (2006) 2995.
101. W. Harneit, K. Huebener, B. Naydenov, S. Schaefer, and M. Scheloske, *Phys. Stat. Sol. B* **244** (2007) 3879.
102. B. Goedde, M. Waiblinger, P. Jakes, N. Weiden, K.-P. Dinse, and A. Weidinger, *Chem. Phys. Lett.* **334** (2001) 12.
103. C. Meyer, W. Harneit, A. Weidinger, and K. Lips, *Phys. Stat. Sol. B* **233** (2002) 462.
104. L. Franco, S. Ceola, C. Corvaja, S. Bolzonella, W. Harneit, and M. Maggini, *Chem. Phys. Lett.* **422** (2006) 100.
105. B. Corzilius, A. Gembus, N. Weiden, K.-P. Dinse, and K. Hata, *Phys. Stat. Sol. B* **243** (2006) 3273.

106. B. Corzilius, K.-P. Dinse, and K. Hata, *Phys. Chem. Chem. Phys.* **9** (2007) 6063.
107. B. E. Kane, *Nature* **393** (1998) 133.

MOLECULAR MAGNETS FOR QUANTUM COMPUTATION

Takayoshi Kuroda

Department of Chemistry, School of Science and Engineering, Kinki University,
3-4-1 Kowakae, Higashi-Osaka, 577-8502, Japan
E-mail: kuroda@chem.kindai.ac.jp

We review recent progress in molecular magnets especially in the viewpoint of the application for quantum computing. After a brief introduction to single-molecule magnets (SMMs), a method for qubit manipulation by using non-equidistant spin sublevels of a SMM will be introduced. A weakly-coupled dimer of two SMMs is also a candidate for quantum computing, which shows no quantum tunneling of magnetization (QTM) at zero field. In the AF ring Cr_7Ni system, the large tunnel splitting is a great advantage to reduce decoherence during manipulation, which can be a possible candidate to realize quantum computer devices in future.

Keywords: molecular magnets; single-molecule magnets; Mn_{12} complexes; SMM dimer; antiferromagnetic rings

1. Introduction

A term "molecular magnet" has at least two meanings. One is for a molecular compound showing characteristic property of a magnet (ferro- or ferrimagnetic compound which is characterized by spontaneous magnetization together with a hysteresis in the M-H loop) as a result of the cooperative behavior of spins residing in each molecule. There are not a few examples of this category and a term "molecular magnetism" is a name of the field concerning synthesis, structures and magnetic properties of molecule-based compounds having spins. They are definitely different from the usual magnets such as metal oxides or metallic compounds. Another is for a molecule which is a magnet: A molecule itself shows a characteristic behavior of the characteristic for a magnet. So far as known, there is no example of the latter meaning in strictly speaking, but sometimes a "single-molecule magnet (SMM)" is called a "molecular magnet". Although the SMM is not a true magnet, it behaves like a magnet at certain conditions showing hysteresis in a M-H loop together with a quantum tunneling of magnetization (QTM), which is a very interesting phenomenon because a quantum mechanical effect appears in a macroscopic scale. Thus, it can be useful for realizing quantum information processing. In this paper, we review the recent proposals of utilization of molecular magnets for quantum computation. Before starting the review, a brief introduction to SMM is helpful for general readers.

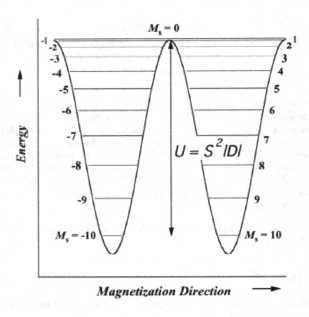

Fig. 1. Double-well potential of the ground state spin sublevels of a $S = 10$ SMM with a magnetic anisotropy constant D.

A molecule with a large spin-value S and a large and negative magnetic anisotropy constant D, is called a SMM. A typical example of the SMM is $[Mn_{12}O_{12}(O_2CMe)_{16}(H_2O)_4]$,[1–3] abbreviated hereafter as Mn_{12}, having $S = 10$ and $D/k_B = -0.72$ K, which have held the record of the highest temperature of the observation of magnetic hysteresis until the appearance of new Mn_6 cluster.[4] The Boltzmann constant is written as k_B. The ground state of the Mn_{12} is split by axial zero-field splitting and the spin Hamiltonian can be expressed in its simplest form as given in Eq. (1):

$$H = g\mu_B H S_z + D\left[S_z^2 - \frac{1}{3}S(S+1)\right], \qquad (1)$$

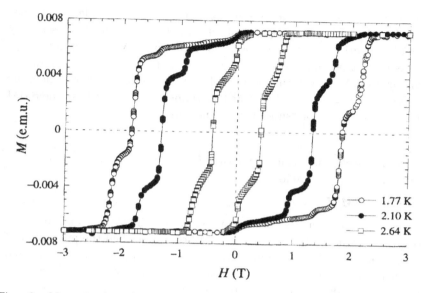

Fig. 2. Magnetization hysteresis loop observed for a single crystal of [Mn$_{12}$O$_{12}$(OAc)$_{16}$(H$_2$O)$_4$] applying the field along the magnetization easy axis. Adapted by permission from Macmillan Publishers Ltd: Nature **383** 145 (1996), copyright (1996).

where μ_B is the Bohr magneton and g is Lande's factor, H is the external field, S is a total spin value and S_z is a spin operator. The first term is the Zeeman term, and the second is the axial zero-field interaction term. The magnetic anisotoropy consntant D gauges the axial zero-field splitting of the ground state. The energy of each M_S sublevel of the ground state is given as $E = M_S g \mu_B H + D\left[M_S^2 - \frac{1}{3}S(S+1)\right]$. Figure 1 shows a double-well potential energy diagram for a $S = 10$ SMM where one molecule reverses its magnetization direction from spin "up" to spin "down" in zero experimental magnetic field. The energy barrier U is estimated to be $S^2|D|$. When temperature T is below the blocking temperature T_b, the "up" spin in the right side of the potential barrier, can not go to the left side by thermal excitation to flip its spin state to "down". This means that no averaging of states with spin "up" and "down" occurs and the molecule still has a certain magnetization at zero-field. This results in the appearance of a hysteresis in the M-H loop, which is the reason why the molecule called single-molecule magnet. Figure 2 shows hysteresis loops observed in a single crystal of Mn$_{12}$. Steps appeared on loops are due to QTM, which is the appearance of quantum effects in macroscopic level. By changing applied field, coincidences of the energies of M_S sublevels on both sides of the barrier in Fig. 1 occur, then magnetization tunneling is induced. This is very hot and attracting topic in physics. One thing we have to mention is the SMM is not a true magnet because this system does not show spontaneous magnetization. If the system was cooled without applying external field, no magnetization was observed even at low temperature. Therefore, the SMM is regarded as a kind of "superparamagnetism".

2. Recent progress in application of molecular magnets to quantum computation

2.1. Mn_{12} SMM

The first report concerning SMM as a possible candidate for quantum computing appears in 2001. Leuenberger and Loss[5] described the possibility of a Mn_{12} SMM to utilize as a qubit for quantum computing. They proposed an implementation of Grover's algorithm[6] by means of the S-matrix of a SMM and time-dependent high-order perturbation theory; the advanced ESR techniques enable to populate coherently and to manipulate many spin states simultaneously by applying one single pulse of an a.c magnetic field containing an appropriate number of matched frequencies. This a.c field creates a nonlinear response of the magnet through multiphoton absorption processes involving particular sequences of σ and π photons which allows the encoding and, similarly, the decoding of states. Finally, the subsequent read-out of the decoded quantum state can be achieved by means of pulsed ESR techniques. These exploit the non-equidistance of energy levels which is a typical of molecular magnets. The mechanism is as follows:

The double-well potential of $S = 10$ system (Fig. 1) has $2S + 1$ spin sublevels from $M_S = -10, -9, \ldots, 9, 10$, with non-equal intervals. The energy interval between M_S and $M_S - 1$ states is calculated to be $\Delta E = |(2M_S - 1)D|$. First, a strong magnetic field in the z direction must be applied to prepare the initial state $|\Psi_0\rangle = |S\rangle$. Then this field is reduced almost to zero (up to the bias δH_z) in such a way that all $|m = M_S\rangle$ states are localized, say, on the left side of the potential barrier. Thus, the magnetic moment pointing along the z axis assumes its maximum value and single spin of a molecular magnet is described by the Hamiltonian (1) where the Zeeman term is treated as perturbation V. To mark specific states, for example, with a certain occupation amplitude, weak oscillating transverse magnetic fields $H_\perp$ was applied to induce multiphoton transitions via virtual states, which can usually be calculated in perturbation

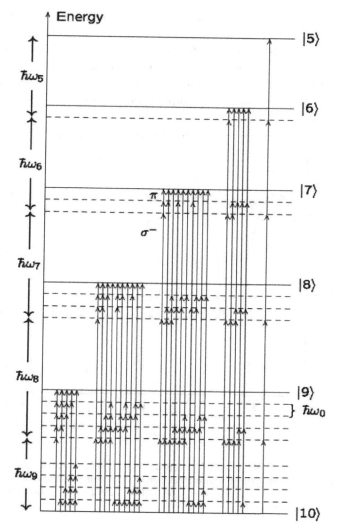

Fig. 3. Energy transition diagrams those contribute to the transition amplitude for $S = 10$ and $m_0 = 5$. Adapted by permission from Macmillan Publishers Ltd: Nature **410**, 789 (2001), copyright (2001).

theory in $H_\perp$. However, the Grover's algorithm requires that all the k-photon transitions, $k = 1, 2, \ldots, S - 1$, have the same amplitudes (and possibly different phases). It works only if the energy levels are not equidistant, which is typically the case in molecular magnets owing to anisotropies.

The nth-order transitions correspond to the nonlinear response of the spin system to strong magnetic fields. Thus, a coherent magnetic pulse of duration T is needed with a discrete frequency spectrum (ω_m), say, for Mn_{12} between 20 and 120 GHz and a single low-frequency ω_0 around 100 MHz. In this regime, the level lifetime in Mn_2 is estimated to be about $\tau_d = 10^{-7}$ s, limited mainly by hyperfine and/or dipolar interactions.

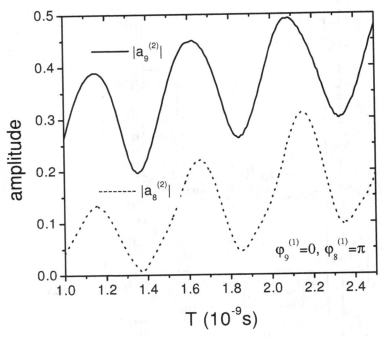

Fig. 4. The amplitudes $|a_9^{(2)}|$ and $|a_8^{(2)}|$ for $m_0 = 8$ oscillate with the pulse duration T in the unit 10^{-9} s. Reprinted figure with permission from B. Zhou, R. Tao, S. Q. Shon, and J. Q. Liang, Physical Review A **66**, 010301 (2002). Copyright (2002) by the American Physical Society.

Figure 3 shows energy transition diagrams those contribute to the transition amplitudes for $S = 10$ and $m_0 = 5$, which describes transitions of 5th order in V in the right well of the spin system (see Fig. 1). The shortest arrows indicate the transitions induced by the low-frequency magnetic fields H_0, while the longer arrows indicate those induced by the high-frequency magnetic fields H_m. Whereas the H_m fields transfer angular momentum to the spin of the molecular magnet by means of σ photons, the H_0 field provides only energy without angular momentum by means of π photons.

In this way all the transition amplitudes are of similar magnitude. For example, the transition from $|10\rangle$ to $|7\rangle$ arises from the absorption of five photons in total, comprising three σ photons and two π photons. As it does not matter in which order the five photons are absorbed, there are 10 different kinds of diagrams. The key point of this technique is the transition probability can be controlled independently and simultaneously by changing the strength of high frequency fields, H_5-H_9, so as to satisfy the requirements of perturbation theory.

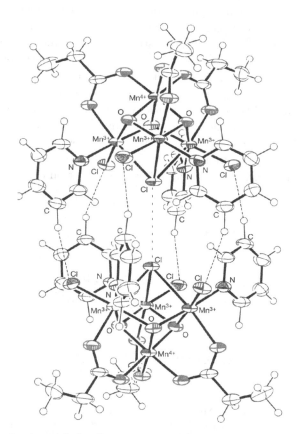

Fig. 5. The structure of the $[Mn_4O_3Cl_4(O_2CEt)_3(py)_3]_2$ dimer, denoted $[Mn_4]_2$. Adapted by permission from Macmillan Publishers Ltd: Nature **416** 406 (2002), copyright (2002).

The feasibility of this system was checked by Zhou et al.[7] by solving the time-dependent Schrösinger equation numerically. A part of their results are shown in Fig. 4, which indicates the amplitudes are sensitive to the time duration of the pulse. They found that accurate time duration of the magnetic pulse as well as the discrete frequency spectrum and the amplitudes are required to design a quantum computing device.

2.2. [Mn$_4$]$_2$ as a dimer of SMM

In 2002, Wernsdorfer and his coworkers[8] reported the magnetic properties of a dimer of a tetranucler manganese complex, [Mn$_4$O$_3$Cl$_4$(O$_2$CEt)$_3$(py)$_3$]$_2$. The monomer Mn$_4$ complex[9] is known as SMM with $S = 9/2$ and $D/k_B = -0.72$ K.[10] The crystal structure of [Mn$_2$]$_4$ is shown in Fig. 5. Each Mn$_4$ complex has a cubane type structure composed of Mn^{4+} ··· O$_3$ ··· Mn$_3^{3+}$ ··· Cl framework surrounded by three propionates bridges, three pyridines, and three chroride anions. Two Mn$_4$ complexes are weakly connected through two kinds of weak interactions, six C-H ··· Cl hydrogen bonds between the pyridine rings on one Mn$_4$ and Cl ions on the other, and a Cl ··· Cl contact, which result in the weak antiferromagnetic coupling ($J = 0.1$ K) between them. This dimer complex shows quite interesting hysteresis behavior as shown in Fig. 6. In addition to the appearance

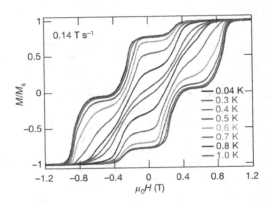

Fig. 6. Magnetization (M) of [Mn$_4$]$_2$ versus applied magnetic field ($\mu_0 H$). The resulting hysteresis loops are shown at different temperatures. Adapted by permission from Macmillan Publishers Ltd: Nature **416** 406 (2002), copyright (2002).

of steps at the interval of ca. 0.6 T, the most characteristic feature of these hysteresis loops is the absence of the magnetization drops at the field zero. This is due to the dimerization of the SMMs and explained by exchange-biased hysteresis. Figure 7 shows energy diagram of spin states of the dimer as a function of applied magnetic field. Each line shows energy variation of each spin states of dimer and the crossing points indicate where magnetization tunneling occurs. Dotted lines labeled 1 to 5 in Fig. 7(b), indicate the strongest tunnel resonances. At zero field, the tunneling from $(-9/2, -9/2)$ to $(9/2, 9/2)$ requires double spin-flip at the same time, which rarely occur due to the low probability.

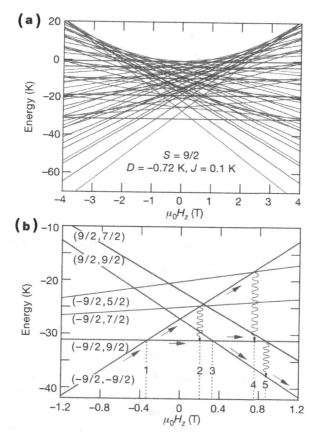

Fig. 7. The spin state energies of $[Mn_4]_2$ as a function of applied magnetic field. (a): Energy versus magnetic field plot for the 100 states of the exchange-coupled dimer of two spin $S = 9/2$ Mn_4 units. (b): Enlargement of (a), showing only levels populated at very low temperature when the field is swept from -1.2 T to $+1.2$ T. A spin state of a dimer is expressed as (M_{S_1}, M_{S_2}). Adapted by permission from Macmillan Publishers Ltd: Nature **416** 406 (2002), copyright (2002).

The absence of a level crossing at zero field makes [Mn$_4$]$_2$ a very interesting candidate as a qubit for quantum computing, because its ground state is the entangled combination of the $(9/2, -9/2)$ and $(-9/2, 9/2)$ states; the coupling of this $S = 0$ system to environmental degrees of freedom should be small, which means decoherence effects should also be small. The absence of tunneling at zero field is also important if SMMs are to be used for information storage, and the option is still retained to switch on tunneling, if and when required, by application of a field.

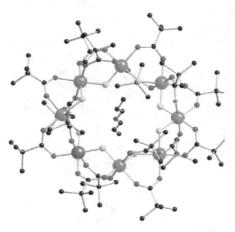

Fig. 8. The molecular structure of [{^{n}Pr$_2$NH$_2$}{Cr$_7$NiF$_8$(O$_2$CCMe$_3$)$_{16}$}] (Cr$_7$Ni) with the hydrogen atoms omitted for clarity. Reprinted figure with permission from M. Affronte, F. Troiani, A. Ghirri, A. Candini, M. Evangelisti1, V. Corradini1, S. Carretta, P. Santini, G. Amoretti, F. Tuna, G. Timco, and R. E. P. Winpenny, Journal of Physics D: Applied Physics **40**, 2999 (2007).

2.3. *Antiferromagnetic rings of* [Cr$_7$Ni]

In ferromagnetic molecular clusters such as Mn$_{12}$ the ground-state tunnel splitting Δ is small compared to the spin decoherence rate $\hbar\Gamma$.[2,3] However, Δ is significantly larger in antiferromagnetic (AF) systems[11,12] which are promising candidates for the observation of macroscopic quantum coherence (MQC). For example, Fe$_{10}$ has known to have a large tunnel splitting Δ/k_B of 2.18 K.[13] More recently, AF rings of Cr$_7$Ni and its family are focused on because their larger Δ/k_B of ca. 13 K.[14] The AF ring with $S = 1/2$ can be a good candidates for quantum bits.[15–17] The molecular structure of [{^{n}Pr$_2$NH$_2$}{Cr$_7$NiF$_8$(O$_2$CCMe$_3$)$_{16}$}][18,19] is shown in Fig. 8, which is representative for Cr$_7$Ni. The ancestor of this family is an octanuclear Cr$_8$ molecule[20] containing eight Cr^{3+} ions, octahedrally coordinated, disposed in an almost perfect octagon with each edge of the octagon bridged by a single F$^-$ anion and two independent carboxylate groups. Cr^{3+} ions have a local spin moment $s = 3/2$ and the preferred arrangement of the eight spin is AF. In order to introduce extra spin to get a desired complex with the $S = 1/2$ ground state, one Cr^{3+} ion is substituted with a divalent transition metal Ni^{2+} ion ($s = 1$) with an ammonium derivative as a cation to compensate the charge balance.

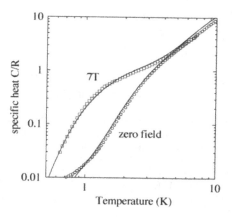

Fig. 9. Temperature dependence of specific heat of Cr$_7$Ni single crystal measured in zero field and at 7 T. Reprinted figure with permission from M. Affronte, F. Troiani, A. Ghirri, A. Candini, M. Evangelisti1, V. Corradini1, S. Carretta, P. Santini, G. Amoretti, F. Tuna, G. Timco, and R. E. P. Winpenny, Journal of Physics D: Applied Physics **40**, 2999 (2007).

The magnetic properties of a single molecular ring of Cr$_7$Ni can be described by the spin Hamiltonian (2):[14]

$$H = \sum_{i=1}^{8} J_i \, \boldsymbol{s}_i \cdot \boldsymbol{s}_{i+1} + \sum_{i=1}^{8} d_i \left[s_{iz}^2 - \frac{1}{3} s_i(s_i + 1) \right]$$

$$+ \sum_{i<j=1}^{8} \boldsymbol{s}_i \cdot \boldsymbol{D}_{ij} \cdot \boldsymbol{s}_j + \mu_B \sum_{i=1}^{8} g_i \boldsymbol{B} \cdot \boldsymbol{s}_i \quad (2)$$

where J_i is an isotropic spin-spin interaction constant between the nearest neighbor spins $\boldsymbol{s}_i$ and $\boldsymbol{s}_{i+1}$, d_i is a single-ion magnetic anisotropy constant at the i-site, $\boldsymbol{D}_{ij}$ is an anisotropic interaction tensor including dipoler interaction.

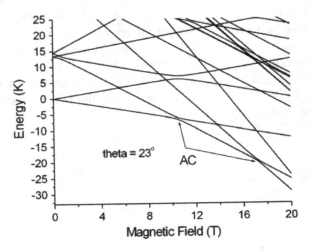

Fig. 10. Zeeman plot of the low-lying energy levels of Cr$_7$Ni. Levels are calculated by numerical diagonalization of the spin Hamiltonian equation (2) with the molecular parameters of Cr$_7$Ni evaluated from experiments shown in the text. An angle $\theta = 23°$ between the applied magnetic field and the axis perpendicular to the plane of the ring was fixed for these calculations. Reprinted figure with permission from M. Affronte, F. Troiani, A. Ghirri, A. Candini, M. Evangelisti1, V. Corradini1, S. Carretta, P. Santini, G. Amoretti, F. Tuna, G. Timco, and R. E. P. Winpenny, Journal of Physics D: Applied Physics **40**, 2999 (2007).

In addition to the magnetic susceptibility measurements, heat capacity measurements under several magnetic fields were conducted as shown in Fig. 9. The data were fitted to equations derived from (2) and the following parameters were obtained: $J_{Cr}/k_B = 17\,K$, $J_{Ni}/J_{Cr} = 0.9 \pm 0.1$, and $|d_i|/k_B = 0.3 \pm 0.15\,K$.[14] The energy diagram shown in Fig. 10 was obtained from these parameters.[20] This is fully consistent with the observation of two peaks at 1.19 and 1.34 meV in the inelastic neutron scattering (INS) spectrum,[21] which correspond to the energy gaps between the ground state and the sublevels of the first excited states. The energy spectrum of the investigated Cr$_7$Ni molecule fully justifies its description in terms of an effective two-level system; besides, the symmetries of the ground-state doublet (S mixing below 1%) suppress the coupling to the higher levels as induced by the transverse magnetic fields which are required for the quantum-gate implementation. In fact, their simulations of the single-qubit gates provide negligible values for the leakage even for gating times of the order of 102 ps, i.e., well below the tens of nanoseconds estimated for the spin decoherence times. While further work is needed for the engineering of the intercluster coupling, these results strongly support the suitability of the Cr$_7$Ni rings for the qubit encoding and manipulation.

3. Summary

Molecular magnets are possible candidates to realizing quantum computation. The utilization of non-equidistant spin sublevels of magnetic clusters for qubit manipulation has proposed but there are some difficulties in realization of the accuracy of pulse duration and pulse manipulation in high frequency. A weakly-coupled dimer of two SMMs shows no QTM at zero field due to the formation of the exchange-biased antiferromagnet, which gives a possibility of utilization of it as memory devices. The large tunnel splitting Δ in the AF ring Cr$_7$Ni is a great advantage to reduce decoherence during manipulation, which can be a possible candidate to realize quantum computer devices in future.

References

1. R. Sessoli, D. Gatteschi, A. Caneschi and M. A. Novak, *Nature* **365** (1993), 141.
2. L. Thomas, F. Lionti, R. Ballou, D. Gatteschi, R. Sessoli and B. Barbara, *Nature* **383** (1996) 145.
3. J. R. Friedman, M. P. Sarachik, J. Tejada and R. Ziolo, *Phys. Rev. Lett.* **76** (1996) 3830.
4. C. J. Milios, A. Vinslava, W. Wernsdorfer, S. Moggach, S. Parsons, S. P. Perlepes, G. Christou and E. K. Brechin, *J. Am. Chem. Soc.* **129** (2007) 2754.
5. M. N. Leuenberger and D. Loss, *Nature* **410** (2001) 789.
6. L. K. Grover, *Phys. Rev. Lett.* **79** (1997), 4709.
7. B. Zhou, R. Tao, S.-Q. Shen and J.-Q. Liang, *Phys. Rev. A* **66** (2002) 010301.
8. W. Wernsdorfer, N. Aliaga-Alcalde, D. N. Hendrickson and G. Christou, *Nature* **416** (2002) 406.
9. D. N. Hendrickson, G. Christou, E. A. Schmitt, E. Libby, J. S. Bashkin, S. Wang, H.-L. Tsai, J. B. Vincent, P. D. W. Boyd, J. C. Huffman, K. Folting, Q. Li and W. E. Streib, *J. Am. Chem. Soc.* **114** (1992) 2455.
10. S. M. J. Aubin, N. R. Dilley, L. Pardi, J. Krzystek, M. W. Wemple, L.-C. Brunel, M. B. Maple, G. Christou and D. N. Hendrickson, *J. Am. Chem. Soc.* **120** (1998) 4991.
11. B. Barbara and E. M. Chudnovsky, *Phys. Lett. A* **145** (1990) 205.
12. I. V. Krive and O. B. Zaslawski, *J. Phys.: Condens. Matter* **2** (1990) 9457.
13. A. Chiolero and D. Loss, *Phys. Rev. Lett.* **80** (1998) 169.

14. F. Troiani, A. Ghirri, M. Affronte, S. Carretta, P. Santini, G. Amoretti, S. Piligkos, G. Timco and R. E. P. Winpenny, *Phys. Rev. Lett.* **94** (2005) 207208.
15. D. Loss and D. P. DiVincenzo, *Phys. Rev. A* **57** (1998) 120.
16. F. Meier and D. Loss, *Phys. Rev. B* **64** (2001) 224411.
17. F. Meier, J. Levy and D. Loss, *Phys. Rev. Lett.* **90** (2003) 047901.
18. F. K. Larsen, E. J. L. McInnes, H. E. Mkami, J. Overgaard, S. Piligkos, G. Rajaraman, E. Rentschler, A. A. Smith, G. M. Smith, V. Boote, M. Jennings, G. A. Timco and R. E. P. Winpenny, *Angew. Chem. Int. Edn.* **42** (2003) 101.
19. M. Affronte, F. Troiani, A. Ghirri, A. Candini, M. Evangelisti, V. Corradini, S. Carretta, P. Santini, G. Amoretti, F. Tuna, G. Timco and R. E. P. Winpenny, *J. Phys. D: Appl. Phys.* **40** (2007) 2999.
20. N. V. Gerbeleu, Y. T. Struchkov, G. A. Timco, A. S. Batsanov, K. M. Indrichan and G. A. Popovich, *Dokl. Akad. Nausk. SSSR.* **313** (1990) 1459.
21. R. Caciuffo, T. Guidi, G. Amoretti, S. Carretta, E. Liviotti, P. Santini, C. Mondelli, G. Timco, C. Muryn and R. E. P. Winpenny, *Phys. Rev. B* **71** (2005) 174407.

ERRORS IN A PLAUSIBLE SCHEME OF QUANTUM GATES IN KANE'S MODEL

Yukihiro Ota

Research Center for Quantum Computing,
Interdisciplinary Graduate School of Science and Engineering, Kinki University,
3-4-1 Kowakae, Higashi-Osaka, 577-8502, Japan
E-mail: yota@alice.math.kindai.ac.jp

We reinvestigate a proposal for universal quantum gates in Kane's model, in which the assumption that electron spin is always downward. We demonstrate a considerable error appears in the spin-flip operations. The transition from the subspace for expressing a qubit to the irrelevant subspace occurs (i.e., a leakage error). We try to eliminate such an error by a composite pulse.

Keywords: silicon-based nuclear spin quantum computer; leakage error

1. Introduction

Many physical systems have been studied for realizing a quantum computer; liquid-state NMR[1–3] [a], solid-state NMR,[3,6] nuclear or electron spins in semiconductors,[7–11] endohedral fullerenes,[12,13] and so on. See Ref. 14 for additional references. Kane[7] proposed a silicon-based nuclear spin quantum computer; the nuclear spin of a dopant atom ^{31}P arrayed in a silicon substrate is a qubit, and the quantum state is measured by detecting the motion of a single electron.

Several schemes for implementing quantum gates in Kane's model have been reported.[15,16] Hill and Goan[16] proposed a nonadiabatic scheme for arbitrary single qubit operations and controlled operations. However, they used the assumption that spin of the donor electrons is always downward.

---

[a]The liquid-state NMR quantum computer is an important model, but may not be a practical one. There is also an objection from a theoretical viewpoint. Braunstein *et al.*[4] pointed out that the effective pure states should be almost separable in the current liquid-state NMR experiments. Their study can be also related to the understanding of the role of entanglement in quantum computing. See Ref. 5 for the additional references on this issue.

In this paper, we show that errors should exist in Hill and Goan's schemes for the gates including spin-flip operations.[17] It is necessary to properly take account of the hyperfine interaction between electron spins and nuclear spins in their composition of the qubits system. The type of the error is a leakage one. This error can occur when the system considerd has three or more energy levels. We review a strategy to suppress it. Unfortunately, this method doesn't work in Kane's model. We will have to reconsider the issue how to represent a qubit and propose more efficient scheme for the leakage elimination in Kane's model.

This paper is organized as follows. We review Kane's model in Sec. 2. In Sec. 3, the Hamiltonian and the computation bases are explained. We show the leakage errors in the spin-flip operations which the authors in Ref. 16 proposed in Sec. 4. We discuss how to suppress such an effect in Sec. 5. Section 6 is devoted to the summary.

2. Kane's model

2.1. *Overview*

The basic devise is a silicon substrate with the phosphorus (^{31}P) as an impurity [Fig. 1]. The impurities are arrayed with equidistant spacing, and the distance between two impurities are assumed to be 20 nm. Each impurity is a donor in the silicon substrate, and it's nuclear spin is $1/2$.[7] Note that one electron can exist around ^{31}P. Namely, a single nucleus-electron system exists in the site where the impurity is put.

The static magnetic field B is applied globally in the direction of z-axis. In order that the electron spin is sufficiently polarized, Kane assumed $B = 2.0$ T and the temperature surrounding the devise is 100 mK. To control spin, the a.c. magnetic field B_{ac} is applied globally on the xy-plain. The impurities are arrayed along x-axis.

Two kinds of gates for applying voltage are set in this devise: an A-gate and a J-gate.[7] The A-gate is set right above each site where the impurity is located. The J-gates are put between the neighboring dopant atoms. First, let us observe the role of the A-gates. We focus on a specific site. The system there is equivalent to a hydrogen atom, neglecting the band structure of Si. The hyperfine interaction (HI) between the nuclear and the electron spins exist. The magnitude of the HI is proportional to the probability density of the electron evaluated at the phosphorus nucleus.[18] When the A-gate is biased positively, the electron wavefunction shifts to the direction of the A-gate [Fig. 2]. This observation is also supported from a quantitative

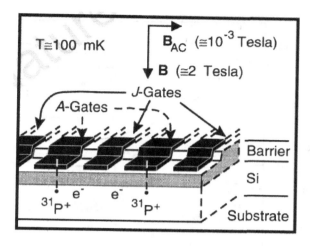

Fig. 1. Illustration of the basic device for Kane's model. Adapted by permission from Macmillan Publishers Ltd: Nature **393**, 133 (1998), copyright (1998).

calculation.[19] The control of the A-gates are performed locally. Then, one can locally control the energy eigenvalues of the system composed of the nuclear and the electron spins.

Next, let us discuss the J-gate. We focus on the neighboring two sites. Two nucleus-electron systems are assumed to interact through the electron exchange interaction (EE).[7] The magnitude of the EE is proportional to the overlap of two electron wavefunctions.[20] Recall that the separation between the impurities is 20 nm. Let us consider the case that the bias voltage of the J-gate is zero. Comparing with the Zeeman energy of the electron spin which is a characteristic energy scale in Kane's model, the value of the EE is very small; it can be regarded as zero.[7] When the J-gate is biased positively, the value of the EE can increase. Note that the behavior of the EE is not a simple decreasing function with respect to the distance between two dopant atoms, due to the band structure of Si.[7,21] In addition, the behavior of the EE is not simple when the J-gate is biased.[22,23] Detailed theoretical studies may be necessary for the accurate control of the EE. The EE is an inevitable element to produce controlled operations. Moreover, the adiabatic control of the EE plays an important role in the measurement scheme. We will briefly review it in the subsequent subsection.

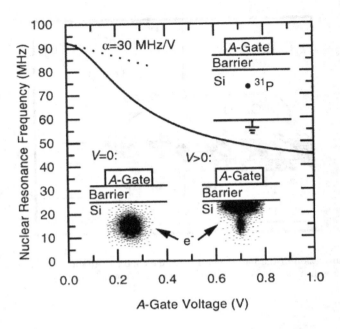

Fig. 2. Behavior of nuclear resonance frequency, applying an electronic field to an A-gate. The decreasing behavior is caused by the reduction of the hyperfine interaction. Adapted by permission from Macmillan Publishers Ltd: Nature **393**, 133 (1998), copyright (1998).

2.2. Readout of a qubit

One of the obstacles in the realization of quantum computers with nuclear spins is the readout of a qubit. In particular, it is very difficult to measure the magnetization of a single nuclear spin. Kane proposed an interesting method; instead of the direct detection of the nuclear spin, one measures the electron spin to which the information of the nuclear spin is transferred.

We briefly explain its main idea.[7,24] Let us consider the system composed of two impurities. Each system is a nucleus-electron system. The nuclear spin on the one site is a qubit for quantum computing. Another is regarded as a detector. If the voltage of the J-gate is zero, the electron spins are almost polarized downward. However, when the voltage is much larger than a threshold,[7] one can find the state of the electron spins depend on the state of the nuclear spins. If two electron spin state is singlet, the electron on the one site cannot move to the other site because of the Pauli exclusion rule. Thus, one may detect the difference between the nuclear spin states by the motion of the single electron. In Ref. 7, the single electron transistor[25] is required to perform the above measurement scheme. Recently, the detection of the single electron spin was reported.[26] The electron charge transport depending on a type of two electron states was also observed.[27]

Direct readout schemes of single nuclear spin have been proposed, as well. In particular, single-spin magnetic resonance force microscopy (MRFM) is promising.[28] In Ref. 11, the application of MRFM to Kane's model was discussed.

2.3. *Requisite techniques for the fabrication*

The devise for Kane's model has not been realized. The precise fabrication technique to array the dopant atoms is necessary for the realization. In particular, the errors caused by the fluctuation of the dopant positions have to be reduced. The strategies to make the devise exist two ways: A top-down approach and a bottom-up approach. Both of them are plausible and have defects.[22]

The single-ion implantation method [29] is suitable for implanting the impurities in the top-down approach. It enables us to make the devise in which the number and the position of them are precisely controlled. Using the focused ion beams, the ions are implanted into a silicon substrate with a high aiming precision; for example, one can array the dopant atoms with the interval of 60 nm [Fig. 3].

3. Hamiltonian and computation bases

3.1. *Hamiltonian*

Kane's model for two qubits is described by the Hamiltonian

$$H(t) = \sum_{i=1}^{2} H^i(t) + J(t)\boldsymbol{\sigma}^{1e} \cdot \boldsymbol{\sigma}^{2e} + \sum_{i=1}^{2} H^i_{ac}(t), \tag{1}$$

$$H^i(t) = -g_n\mu_n B\sigma_z^{in} + \mu_B B\sigma_z^{ie} + A_i(t)\boldsymbol{\sigma}^{ie} \cdot \boldsymbol{\sigma}^{in}, \tag{2}$$

$$H^i_{ac}(t) = B_{ac}\boldsymbol{m}(t) \cdot (-g_n\mu_n\boldsymbol{\sigma}^{in} + \mu_B\boldsymbol{\sigma}^{ie}), \tag{3}$$

where $\boldsymbol{m}(t) = (\cos(\omega_{ac}t), -\sin(\omega_{ac}t), 0)$. The Pauli matrix is written by σ_k ($k = x, y, z$), and the superscript ie (in) in it indicates the electronic (nuclear) spin in the ith site. When $B = 2.0$ T, the value of the Zeeman energies for the electron, $\mu_B B$, and the nucleus, $g_n\mu_n B$ are 0.116 meV and 0.071×10^{-3} meV, respectively. The HI in the ith site between the nuclear spin and the electronic spin is $A_i(t)\boldsymbol{\sigma}^{ie}\cdot\boldsymbol{\sigma}^{in}$. When the voltage in the A-gate

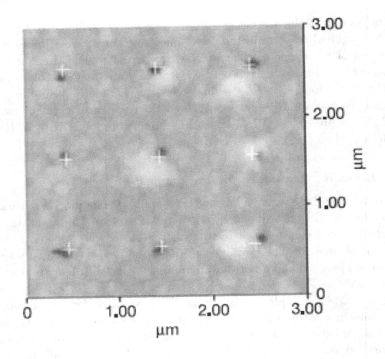

Fig. 3. Experimental results of single ion implantation; typical atomic force microscope image. The white crosses indicate the aimed positions. Adapted by permission from Macmillan Publishers Ltd: Nature **437**, 1128 (2005), copyright (2005).

Table 1. Typical values of parameters.

$\mu_B B/h$ [MHz]	$g_n \mu_n B/h$ [MHz]	A_0/h [MHz]
28.0×10^3	17.2	29.3

is zero, the value of HI is 0.121×10^{-3} meV($\equiv A_0$). The EE between the 1st and the 2nd electronic spins $J(t)\boldsymbol{\sigma}^{1e} \cdot \boldsymbol{\sigma}^{2e}$. The third term in Eq. (1) is the effect of the a.c. magnetic field. We assume $B_{ac} = 2.5$ mT.[16] We summarize the typical values of the parameters in Table 1. In this paper, we discuss only spin-flip operations. Therefore, the value of $J(t)$ is always zero. The energy level and the energy eigenstates for the two-qubit system are discussed in Ref. 17.

3.2. Computation bases

We write the eigenstates for σ_z^{in} as $|0\rangle_i$ and $|1\rangle_i$ ($\sigma_z^{in}|0\rangle_i = |0\rangle_i$ and $\sigma_z^{in}|1\rangle_i = -|1\rangle_i$), and those for σ_z^{ie} as $|\uparrow\rangle_i$ and $|\downarrow\rangle_i$ ($\sigma_z^{ie}|\uparrow\rangle_i = |\uparrow\rangle_i$ and $\sigma_z^{ie}|\downarrow\rangle_i = -|\downarrow\rangle_i$).

Let us diagonalize $H^i(t)$ for a fixed value of $A_i(t)$. We abbreviate the argument t in the equations of its eigenvalues and eigenstates. Introducing $\epsilon = \mu_B B + g_n\mu_n B$, the eigenvalues $E_k^i(A_i)$ ($k = 0, 1, 2, 3$ from the below) are given as follows:[17]

$$E_0^i(A_i) = -A_i - \sqrt{\epsilon^2 + 4A_i^2}, \quad (4)$$

$$E_1^i(A_i) = -\mu_B B + g_n\mu_n B + A_i, \quad (5)$$

$$E_2^i(A_i) = -A_i + \sqrt{\epsilon^2 + 4A_i^2}, \quad (6)$$

$$E_3^i(A_i) = \mu_B B - g_n\mu_n B + A_i. \quad (7)$$

Their corresponding eigenstates $|u_k(A_i)\rangle_i$ are given by[17]

$$|u_0(A_i)\rangle = -\sin\theta_i\, |\uparrow\rangle_i|1\rangle_i + \cos\theta_i\, |\downarrow\rangle_i|0\rangle_i, \quad (8)$$

$$|u_1(A_i)\rangle = |\downarrow\rangle_i|1\rangle_i, \quad (9)$$

$$|u_2(A_i)\rangle = \cos\theta_i\, |\uparrow\rangle_i|1\rangle_i + \sin\theta_i\, |\downarrow\rangle_i|0\rangle_i, \quad (10)$$

$$|u_3(A_i)\rangle = |\uparrow\rangle_i|0\rangle_i, \quad (11)$$

where

$$\cos\theta_i = \frac{\epsilon + \sqrt{\epsilon^2 + 4A_i^2}}{N_i}, \quad (12)$$

$$\sin\theta_i = \frac{2A_i}{N_i}, \quad (13)$$

$$N_i^2 = 2\epsilon\sqrt{\epsilon^2 + 4A_i^2} + 2(\epsilon^2 + 4A_i^2). \quad (14)$$

We choose the computation bases for a single qubit as $|u_0(A_i)\rangle$ and $|u_1(A_i)\rangle$.

3.3. Is a nuclear spin a qubit?

Taking account of $\sin\theta_i \simeq 10^{-3}$ for $A_i = A_0$, the following approximation may be valid: $|u_0(A_i)\rangle \simeq |\downarrow\rangle_i|0\rangle_i$. In addition, $|u_1(A_i)\rangle = |\downarrow\rangle_i|1\rangle_i$. Then, the electron is polarized in the downward direction of the z-axis, and the qubit is represented by up and down of the nuclear spin. Thus, one can find our choice for the computation bases is quite natural, comparing with the original proposal. Note that the single qubit system in Kane's model is a composite one. What is an importnat thing is that one cannot say the qubit is represented by the nuclear spin or the electron spin. In particular, it is nontrivial that the electron state can sustain the downward polarization through the temporal controlling process.

4. Spin-flip operations

4.1. Controlling processes

We keep $A_j(t) = A_0$ ($j \neq i$). By contrast, we adiabatically change the magnitude of $A_i(t)$; we decrease the value of $A_i(t)$ up to a suitable constant value $A(< A_0)$ during $t'_X/2$ (the first step), keep $A_i(t) = A$ during t_X (the second step), and finally increase it from A to A_0 (the third step). In the second step, the a.c. magnetic field is globally applied, while we keep $A_i(t) = A$. An undesired phase difference between $|u_0(A_i)\rangle$ and $|u_1(A_i)\rangle$ is caused by the adiabatically varying processes during $A_i(t)$ (the first and third steps). It can be canceled by carrying out a suitable phase-shift operation.[17]

4.2. Schrödinger equation in the second step

Let us discuss the time evolution in the second step. After instantaneously turning on the a.c. magnetic field, the Schrödinger equation for the ith qubit in a rotating frame is written as follows:[17]

$$i\hbar \frac{d}{dt}|\psi_{\rm rot}(t)\rangle = H^i_{\rm rot}|\psi_{\rm rot}(t)\rangle, \qquad (15)$$

where $H^i_{\rm rot} = \hbar\omega^e \sigma_z^{i\,e}/2 - \hbar\omega^n \sigma_z^{i\,n}/2 + A\boldsymbol{\sigma}^{i\,e} \cdot \boldsymbol{\sigma}^{i\,n} + \mu_B B_{\rm ac} \sigma_x^{i\,e} - g_n\mu_n B_{\rm ac}\sigma_x^{i\,n}$, $\hbar\omega^e = 2\mu_B B + \hbar\omega_{ac}$, and $\hbar\omega^n = 2g_n\mu_n B - \hbar\omega_{ac}$. Let us introduce the matrix representation for an operator M as follows: $M_{ab} \equiv \langle u_a(A)|M|u_b(A)\rangle$ ($a, b = 0, 1, 2, 3$). The Hamiltonian $H^i_{\rm rot}$ is written as the sum of $H^i_{\rm d}$ and $H^i_{\rm mix}$:

$$H^i_{\rm d} = \begin{pmatrix} E^i_0(A) & -\nu_\theta B_{\rm ac} & 0 & 0 \\ -\nu_\theta B_{\rm ac} & E^i_1(A) - \hbar\omega_{ac} & 0 & 0 \\ 0 & 0 & E^i_2(A) & -\nu_{-\theta} B_{\rm ac} \\ 0 & 0 & -\nu_{-\theta} B_{\rm ac} & E^i_3(A) + \hbar\omega_{ac} \end{pmatrix}, \qquad (16)$$

$$H^i_{mix} = \begin{pmatrix} 0 & 0 & 0 & \mu_{-\theta} B_{\rm ac} \\ 0 & 0 & \mu_\theta B_{\rm ac} & 0 \\ 0 & \mu_\theta B_{\rm ac} & 0 & 0 \\ \mu_{-\theta} B_{\rm ac} & 0 & 0 & 0 \end{pmatrix}. \qquad (17)$$

Here, we define $\mu_\theta = \mu_B \cos\theta - g_n\mu_n \sin\theta$ and $\nu_\theta = \mu_B \sin\theta + g_n\mu_n \cos\theta$, and the mixing angle θ is defined for θ_i when $A_i(t) = A$.

4.3. Errors in spin-flip operations

First, we reproduce the results in Ref. 16. We solve Eq. (15), omitting H_{mix}^i. We choose the value of A so as to satisfy the Larmor resonance condition corresponding to the computation bases: $\hbar\omega_{ac} = E_1^i(A) - E_0^i(A)$. The initial state $|\phi\rangle$ is expanded by $|u_0(A)\rangle$ and $|u_1(A)\rangle$. We obtain the resultant state at t_X:

$$|\psi_{\rm rot}(t_X)\rangle = e^{-\frac{i}{\hbar}t_X E_0^i(A)} e^{i\theta_X \sigma_X^i(A)}|\phi\rangle. \qquad (18)$$

We define $\sigma_X^i(A) = |u_0(A)\rangle\langle u_1(A)| + |u_1(A)\rangle\langle u_0(A)|$ and $\theta_X = \nu_\theta B_{ac} t_X/\hbar$. Under the Larmor resonance condition for the ith qubit, the time-evolution of the remaining j th qubit ($j \neq i$) is approximately expressed by a gate producing the rotation around z-axis.[17]

Table 2. Values of $F(|\phi\rangle)$ for $\theta_X = \pi/4$. Reprinted table with permission from Y. Ota, S Mikami, and I. Ohba, Physical Review A **73**, 032321 (2006). Copyright (2006) by the American Physical Society.

		$F(|\phi\rangle)$
A/A_0	$|\phi\rangle = |0\rangle_i$	$|\phi\rangle_i = (|0\rangle_i + |1\rangle_i)/\sqrt{2}$
0.75	0.72514	0.70458
0.5	0.72458	0.70454
0.25	0.73390	0.70526

Now, we demonstrate the existence of a considerable error in the above scheme. We solve Eq. (15) without omitting $H_{\rm mix}^i$.[17] Then, we compare the exact solution with the approximate solution (18) by calculating the following quantity: $F(|\phi\rangle) = |\langle\phi|e^{\frac{i}{\hbar}H_{\rm rot}^i t_X}e^{-\frac{i}{\hbar}H_d^i t_X}|\phi\rangle|$. If $F(|\phi\rangle)$ is adequately close to one, the dynamics generated only by H_d^i is a good approximation for the genuine one. The exact solution is shown in Ref. 17. Choosing $\theta_X = \pi/4$, we calculate $F(|\phi\rangle)$ for three values of the HI ($A/A_0 = 0.75, 0.5, 0.25$) and different two initial states. We show the results in Table 2. They imply that the values of $F(|\phi\rangle)$ are almost on the order of 10^{-1}, regardless of the initial states and the magnitude of the HI.

5. Suppression of leakage errors

We have shown there exist errors in a plausible scheme of quantum gates in Kane's model. The type of the error is a leakage error; there exists the non-zero transition from the subspace which is used to represent qubits to the other subspace. When experimental systems have three or more energy levels, such an error is always possible to occur. Several strategies have been proposed to eliminate. The methods based on the quantum Zeno effect and the successive application of unitary kicks may be hopeful.[30] In addition, a kind of composite pulses can be useful.[31-33]

Tian and Lloyd[31] showed a general theory to eliminate leakage errors. Unfortunately, we will find that it is difficult to cancel the leakage error in the current case, even if we use it. Nevertheless, it is worth while to understand its key idea, due to develop more efficient proposals. Then, we review their strategy and show the limitation in Kane's model.

5.1. *Model*

Let us introduce a general model for discussing leakage elimination. We use the unit system in which $\hbar = 1$. We assume the system concerned is a d-level one and its Hamiltonian is $H_0 = \sum_{i=1}^{d} \varepsilon_i |i\rangle\langle i|$, where $\varepsilon_i \in \mathbb{R}$ and $\varepsilon_i > \varepsilon_j$ if $i > j$. We write the differences between two energy eigenvalues as $\omega_{ij} = \varepsilon_i - \varepsilon_j \, (> 0) \, (i > j)$. We assume $\omega_{ij} \neq \omega_{kl}$ if $(i, j) \neq (k, l)$. Namely, the differences between any two energy levels are not equal to each other. The Hilbert space for the system is written by $\mathcal{H} \simeq \mathbb{C}^d$. The coding space $\mathcal{C}$ is a subspace of $\mathcal{H}$. The lowest and the second lowest states are used to represent a qubit; $\mathcal{C} = \text{Span}\{|1\rangle, |2\rangle\}$. The control Hamiltonian is given by $H_c = \sum_{k>l} (\gamma_{kl} |k\rangle\langle l| + \text{h.c.})$, where $\gamma_{kl} \in \mathbb{C}$. A standard way to make a quantum gate is to use $H_1(t) \equiv H_0 + H_c \cos(\omega_{21} t + \varphi)$ from $t = 0$ to $t = T_{\text{op}}$. The second term corresponds to the a.c. magnetic field in Kane's model. Notice that the leakage errors are possible to exist.

To cancel the transitions to the higher energy levels, the operation time T_{op} is divided into several small intervals, and the parameters in cosine pulse are tuned in each time interval. The total number of the time intervals is N, and the value of the rth time interval is $\tau_r = t_r - t_{r-1} \, (r = 1, 2, \ldots, N)$, where $t_0 = 0$ and $t_N = T_{\text{op}}$. The value of N is

$$N = 1 + 2(d - 2). \tag{19}$$

Note that the undesired transitions from $\mathcal{C}$ exist $2(d-2)$ ways. Thus, we uses the additional $2(d-2)$ pulses to cancel the transition to the higher energy levels.

The operator P is the projection for the coding space $\mathcal{C}$: $P = |2\rangle\langle 2| + |1\rangle\langle 1|$. The projection for the irrelevant subspace is $I - P \equiv Q$, where I is the identity operator. The set of the pairs of integers indicated undesired transitions is written as S_L; for example, $S_L = \{(3, 1), (3, 2), (4, 1), (4, 2)\}$

when $d = 4$, and the element $(3, 1)$ means the transition between $|1\rangle$ and $|3\rangle$. We assume that $k > l$ if $(k, l) \in S_L$. In particular, $P|k\rangle = 0$ if $(k, l) \in S_L$.
The total Hamiltonian for $t \in [t_{r-1}, t_r]$ is

$$H_N(t) = H_0 + H_c \alpha_{(r)} \cos[\omega_{(r)}(t - t_{r-1}) + \varphi_{(r)}]. \tag{20}$$

When $r = 1$, we set $\omega_{(1)} = \omega_{21}$, $\alpha_{(1)} = 1 (\equiv \alpha_{21})$, and $\varphi_{(1)} = \varphi (\equiv \varphi_{21})$. For $r \geq 2$, $\omega_{(r)}$s are chosen by satisfying the following conditions: $\omega_{(r)} \neq \omega_{(s)}$ if $r \neq s$, and $\omega_{(r)} = \omega_{kl}$, where $(k, l) \in S_L$. When $d = 4$, $\omega_{(2)} = \omega_{31}$, $\omega_{(3)} = \omega_{41}$, $\omega_{(4)} = \omega_{32}$, and $\omega_{(5)} = \omega_{42}$, for example. Hereafter, instead of $\alpha_{(r)}$ and $\varphi_{(r)}$, we write α_{kl} and φ_{kl}.

We write the time evolution operators generated by H_0 and $H_N(t)$ as $U_0(t)$ and $U_N(t)$, respectively. The initial state can be prepared in $\mathcal{C}$. Therefore, we can analyze $U_N(T_{\text{op}})P$. Obviously,

$$U_N(T_{\text{op}})P = PU_N(T_{\text{op}})P + QU_N(T_{\text{op}})P. \tag{21}$$

5.2. Determination of parameters for leakage elimination

Let us consider the case of the equidistant time interval: $\tau_r = T_{\text{op}}/N (\equiv \tau)$. Thus, the free parameters are α_{kl} and φ_{kl} ($(k, l) \in S_L$). The aim is to vanish or reduce the error term $QU_N(T_{\text{op}})P$ in Eq. (21). The authors in Ref. 31 showed a recipe to do it, based on the time-dependent perturbation theory. Using the first Born approximation,[34] we obtain $U_N(T_{\text{op}}) = U_0(T_{\text{op}})V_N(T_{\text{op}})$, where

$$V_N(T_{\text{op}}) = I + V_{N,21}^{(1)}(\tau) + \sum_{(i,j) \in S_L} V_{N,ij}^{(1)}(\tau), \tag{22}$$

$$V_{N,ij}^{(1)}(\tau) = -i\alpha_{ij} \sum_{k>l} \frac{1}{2} \left\{ \gamma_{kl} \tau \left[e^{i\varphi_{ij}} C_+(\tau; kl|ij) \right. \right.$$
$$\left. \left. + e^{-i\varphi_{ij}} C_-(\tau; kl|ij) \right] |k\rangle\langle l| + \text{h.c.} \right\}. \tag{23}$$

The coefficients in Eq. (23) are given by

$$C_+(\tau; kl|ij) = \frac{e^{i(\omega_{kl}+\omega_{ij})\tau} - 1}{i(\omega_{kl}+\omega_{ij})\tau}, \tag{24}$$

$$C_-(\tau; kl|ij) = \frac{e^{i(\omega_{kl}-\omega_{ij})\tau} - 1}{i(\omega_{kl}-\omega_{ij})\tau} \quad ((i, j) \neq (k, l)), \tag{25}$$

$$C_-(\tau; ij|ij) = 1. \tag{26}$$

Note that $\tau = T_{op}/N$. Since $[U_0(T_{op}), P] = [U_0(T_{op}), Q] = 0$,

$$PU_N(T_{op})P = U_0(T_{op})PV_N(T_{op})P, \quad (27)$$

$$QU_N(T_{op})P = U_0(T_{op})QV_N(T_{op})P. \quad (28)$$

As far as the first Born approximation is valid, we find the leakage error is caused by $QV_{N,ij}^{(1)}(\tau)P$.

To clear the order of the leakage error, we introduce a small constant $\delta\,(>0)$, which satisfies

$$1 > \delta > \max\left\{|\gamma_{ij}|\tau, \frac{1}{\omega_{ij}\tau}\right\}. \quad (29)$$

The order of $C_\pm(\tau; kl|ij)$ can be estimated as follows: $|C_\pm(\tau; kl|ij)| \sim O(\delta)$, except for $C_-(\tau; ij|ij)$.

Now, let us explain how to choose the proper parameters α_{kl} and φ_{kl} to reduce the leakage errors. First, we focus on $QV_{N,21}^{(1)}(\tau)P$. We find that $\langle 2|QV_{N,21}^{(1)}(\tau)P|1\rangle = \langle 1|QV_{N,21}^{(1)}(\tau)P|2\rangle = 0$ because $Q|2\rangle = Q|1\rangle = 0$. In addition, $\langle l|QV_{N,21}^{(1)}(\tau)P|k\rangle = 0$ when $(k,l) \in S_L$. Thus, the remaining matrix element is $|\langle k|QV_{N,21}^{(1)}(\tau)P|l\rangle| \sim O(\delta^2)$, where $(k,l) \in S_L$. To eliminate this term, we use the contributions from $QV_{N,kl}^{(1)}(\tau)P$. In general, we find for $(k,l) \in S_L$

$$QV_{N,kl}^{(1)}(\tau)P = -i\alpha_{kl}e^{-i\varphi_{kl}}\frac{1}{2}\gamma_{kl}\tau|k\rangle\langle l| + QW_{N,kl}^{(1)}(\tau)P, \quad (30)$$

$$QW_{N,kl}^{(1)}(\tau)P = -i\alpha_{kl}e^{i\varphi_{kl}}\sum_{(m,n)\in S_L}\frac{\gamma_{mn}\tau}{2}C_+(\tau; mn|kl)|m\rangle\langle n|$$

$$-i\alpha_{kl}e^{-i\varphi_{kl}}\sum_{(m,n)\in \overline{S}_{kl}}\frac{\gamma_{mn}\tau}{2}C_-(\tau; mn|kl)|m\rangle\langle n|, \quad (31)$$

where $\overline{S}_{kl} = S_L - \{(k,l)\}$. Then, we impose

$$-\langle k|QV_{N,21}^{(1)}(\tau)P|l\rangle = \langle k|Q(V_{N,kl}^{(1)}(\tau) - W_{N,kl}^{(1)}(\tau))P|l\rangle, \quad (32)$$

for $(k,l) \in S_L$. Therefore, we obtain the conditions which the parameters α_{kl} and φ_{kl} have to satisfy:

$$\alpha_{kl}e^{-i\varphi_{kl}} = -e^{i\varphi}C_+(\tau; kl|21) - e^{-i\varphi}C_-(\tau; kl|21). \quad (33)$$

Note that $|\alpha_{kl}e^{-i\varphi_{kl}}| \sim O(\delta)$. Thus, $|\langle k|QW_{N,kl}^{(1)}(\tau)P|l\rangle| \sim O(\delta^3)$. Consequently, if we impose Eq. (32), we find the arbitrary matrix elemetns in the error term is given by $|\langle i|QU_N(T_{op})P|j\rangle| \sim O(\delta^3)$. Note that $|\langle i|QU_1(T_{op})P|j\rangle| \sim O(\delta^2)$. Accordingly, if there exist $T_{op}(=N\tau)$ such that δ is very small, $QU_N(T_{op})P$ can be neglected.

In the above discussion, we used the first Born approximation to evaluate $V_N(T_{op})$. One may take account of the higher-order corrections in a similar manner.[31]

5.3. Leakage free quantum gates

The resultant time evolution operator is given as follows:

$$U_N(T_{op})P = U_0(T_{op})PV^{(1)}_{N,21}(\tau)P + O(\delta^3). \quad (34)$$

When the second term in Eq. (34) can be omitted, the time evolution operator is leakage-free. The time evolution operator $U_0(T_{op})$ corresponds to a phase-shift gate. The spin-flip operation is expressed by

$$PV^{(1)}_{N,21}(\tau)P = P - i\frac{\tau}{2}(\gamma_{21}e^{i\varphi}|2\rangle\langle 1| + \text{h.c.})$$
$$= e^{-i\tau(\gamma_{21}e^{i\varphi}|2\rangle\langle 1|+\text{h.c.})/2}P + O(|\gamma_{21}|^2\tau^2). \quad (35)$$

Note that the projection P corresponds to the identity operator in $\mathcal{C}$.

Equation (34) is a basic unit for creating the quantum gates which we want. Namely, it is necessary to choose α and φ, use it repeatedly, and design an approximate gate for the target one. If the number of the repetition of $U_N(T_{op})P$ is K, we need KN pulses for creating a leakage-free quantum gate. The number of the pulses required can reduce in typical experimental systems.[32]

5.4. Leakage elimination of spin-flip operations in Kane's model

Let us the case of Kane's model. The greatest value of γ_{ij} is characterized by $\mu_B B_{ac} \sim 10^{-4}$ meV. The least value of ω_{ij} is characterized by $2A \sim A_0 \sim 10^{-4}$ meV. Note that $\delta > \gamma_{ij}\tau$ and $\omega_{ij}\tau > \delta^{-1}$ by definition of δ. Thus, it may be difficult to find a suitable δ to eliminate the leakage error.

6. Summary

We have reinvestigated the spin-flip operations for Kane's model. Our choice of a qubit is different from those in Ref. 16, in which the electron spin state is assumed to be always down. The considerable error for the scheme in Ref. 16 exists. The physical origin of such an error is the mixing effect between the computational bases and the irrelevant states by H^i_{mix}, which is neglected in Ref. 16. We have tried to eliminate it based on the scheme in Ref. 31, but it is not easy to suppress it. It is necessary to revisit the issue how to represent a qubit, as well as to develop efficient scheme for the leakage elimination, in Kane's model.

Acknowledgments

The author acknowledges I. Ohba for valuable discussions. This work is supported by "Open Research Center" Project for Private Universities: matching fund subsidy from MEXT (Ministry of Education, Culture, Sports, Science and Technology).

References

1. L. M. K. Vandersypen and I. L. Chuang, *Rev. Mod. Phys.* **76** (2004) 1037.
2. Y. Kondo, M. Nakahara and S. Tanimura, in *Physical Realizations of Quantum Computing: Are the DiVincenzo Criteria Fulfilled in 2004 ?*, edited by M. Nakahara, S. Kanemitsu, M. M. Salomaa and S. Takagi (World Scientific, Singapore, 2006).
3. I. S. Oliverira, T. J. Bonagamba, R. S. Sarthour, J. C. C. Freitas and E. deAzevedo, *NMR Quantum Information Processing* (Elsevier, Amsterdam, 2007).
4. S. L. Braunstein, C. M. Caves, R. Jozsa, N. Linden, S. Popescu and R. Schack, *Phys. Rev. Lett.* **83** (1999) 1054.
5. R. Schack and C. M. Caves, *Phys. Rev. A* **60** (1999) 4354; N. C. Menicucci and C. M. Caves, *Phys. Rev. Lett.* **88** (2002) 167901; L. Gurvits and H. Barnum, *Phys. Rev. A* **68** (2003) 042312; E. Biham, G. Brassard, D. Kenigsberg and T. T. Mor, *Theor. Comp. Science* **320** (2004) 15; Y. Shimoni, D. Shapira and O. Biham, *Phys. Rev. A* **69** (2004) 062303; Y. Shimoni, D. Shapira and O. Biham, *Phys. Rev. A* **72** (2005) 062308; T. M. Yu, K. R. Brown and I. L. Chuang, *Phys. Rev. A* **71** (2005) 032341; Y. Ota, S. Mikami, M. Yoshida and I. Ohba, *J. Phys. A: Math. Theor.* **40** (2007) 14263.
6. H. Kampermann and W. S. Veeman, *Quantum Inf. Process.* **1** (2002) 327; G. M. Leskowitz, N. Ghaderi, R. A. Olsen and L. J. Mueller, *J. Chem. Phys.* **119** (2003) 1643; S. Watanabe and S. Sasaki, *Jpn. J. Appl. Phys.* **42** (2003) L1350; J. Baugh, O. Moussa, C. A. Ryan, R. Laflamme, C. Ramanathan, T. F. Havel and D. G. Cory, *Phys. Rev. A* **73** (2006) 022305.
7. B. E. Kane, *Nature*, **393** (1998) 133; *Fortschr. Phys.* **48** (2000) 1023.
8. R. Vrijen, E. Yablonovitch, K. Wang, H. W. Jiang, A. Baladin, V. Roychowdhury, T. Mor and D. DiVincenzo, *Phys. Rev. A* **62** (2000) 012306.
9. T. D. Ladd, J. R. Goldman, F. Yamaguchi, Y. Yamamoto, E. Abe and K. M. Itoh, *Phys. Rev. Lett.* **89** (2002) 017901.
10. A. J. Skinner, M. E. Davenport and B. E. Kane, *Phys. Rev. Lett.* **90** (2003) 087901.
11. C. D. Hill, L. C. L. Hollenberg, A. G. Fowler, C. J. Wellard, A. D. Greentree and H.-S. Goan, *Phys. Rev. B* **72** (2005) 045350.
12. W. Harneit, *Phys. Rev. A* **65** (2002) 032322.

13. D. Suter and K. Lim, *Phys. Rev. A* **65** (2002) 052309.
14. M. Nakahara and T. Ohmi, *Quantum Computing: From Linear Algebra To Physical Realizations* (CRC Press, New York, 2008).
15. C. Wellard, L. C. L. Hollenberg and H. C. Pauli, *Phys. Rev. A* **65** (2002) 032303; A. G. Fowler, C. J. Wellard and L. C. L. Hollenberg, *Phys. Rev. A* **67** (2003) 012301.
16. C. D. Hill and H.-S. Goan, *Phys. Rev. A* **68** (2003) 012321; *Phys. Rev. A* **70** (2004) 022310.
17. Y. Ota, S. Mikami and I. Ohba, *Phys. Rev. A* **73** (2006) 032321.
18. J. D. Bjorken and S. D. Drell, *Relativistic Quantum Mechanics* (McGraw-Hill, New York, 1964).
19. C. I. Pakes, C. J. Wellard, D. N. Jamieson, L. C. L. Hollenberg, S. Prawer, A. S. Dzurak, A. R. Hamilton and R. G. Clark, *Microelectron. J.* **33** (2002) 1053.
20. C. Herring, *Rev. Mod. Phys.* **34** (1962) 631.
21. B. Koiller, X. Hu and S. D. Sarma, *Phys. Rev. Lett.* **88** (2002) 027903.
22. C. J. Wellard, L. C. Hollenberg, E. Parisoli, L. M. Kettle, H.-S. Goan, J. A. McIntosh and D. N. Jamieson, *Phys. Rev. B* **68** (2003) 195209.
23. L. M. Kettle, H.-S. Goan, S. C. Smith, L. C. L. Hollenberg and C. J. Wellard, *J. Phys.: Condens. Matter* **16** (2004) 1011.
24. G. P. Berman. D. K. Campbell, G. D. Doolen and K. E. Nagaev, *J. Phys.: Condens. Matter* **12** (2000) 2945.
25. O. Astafiev, Y. A. Pashkin, T. Yamamoto, Y. Nakamura and J. S. Tsai, *Phys. Rev. B* **69** (2004) 180507(R).
26. J. M. Elzerman, R. Hanson, L. H. W. van Beveren, B. Witkamp, L. M. K. Vandersypen and L. P. Kouwenhoven, *Nature* **430** (2004) 431; M. Xiao, I. Martin, E. Yablonovitch and H. W. Jiang, *Nature* **430** (2004) 435.
27. C. Boehme and K. Lips, *Phys. Rev. Lett.* **91** (2003) 246603; A. R. Stegner, C. Boehme, H. Huebl, M. Stutzmann, K. Lips and M. S. Brandt, *Nature Phys.* **2** (2006) 835; H. Huebl, F. Hoehne, B. Grolik, A. R. Stegner, M. Stutzmann and M. S. Brandt, *Phys. Rev. Lett.* **100** (2008) 177602.
28. D. Rugar, B. C. Stipe, H. J. Mamin, C. S. Yannoni, T. D. Stowe, K. Y. Yasumura and T. W. Kenny, *Appl. Phys. A* **72** [Suppl.] (2001) S3; G. P. Berman, G. W. Brown, M. E. Hawley and V. I. Tsifrinovich, *Phys. Rev. Lett.* **87** (2001) 097902; G. P. Berman, F. Borgonovi, G. Chapline, S. A. Gurvitz, P. C. Hammel, D. V. Pelekhov, A. Suter and V. I. Tsifrinovich, *J. Phys. A:Math. Gen.* **36** (2003) 4417.
29. T. Shinada, A. Ishikawa, C. Hinoshita, M. Koh and I. Ohdomari, *Jpn. J. Appl. Phys.* **39** (2000) L265; T. Shinada, H. Koyama, C. Hinoshita, K. Imamura and I. Ohdomari, *Jpn. J. Appl. Phys.* **41** (2002) L287; T. Shinada, S. Okamoto, T. Kobayashi and I. Ohdomari, *Nature* **437** (2005) 1128.
30. L.-A. Wu, M. S. Byrd and D. A. Lidar, *Phys. Rev. Lett.* **89** (2002) 127901; P. Facchi, D. A. Lidar and S. Pascazio, *Phys. Rev. A* **69** (2004) 032314; D. Dhar, L. K. Grover and S. M. Roy, *Phys. Rev. Lett.* **96** (2006) 100405.
31. L. Tian and S. Lloyd, *Phys. Rev. A* **62** (2000) 050301(R).
32. Z. Zhou, S.-I. Chu and S. Han, *Phys. Rev. B* **66** (2002) 054527.

33. M. Steffen, J. M. Martinis and I. L. Chuang, *Phys. Rev. B* **68** (2003) 224518.
34. A. Messiah, *Quantum Mechanics: Two Volumes Bound as One* (Dover, New York, 1999).

YET ANOTHER FRAMEWORK FOR QUANTUM SIMULTANEOUS NONCOOPERATIVE BIMATRIX GAMES

AKIRA SAITOH[1], ROBABEH RAHIMI[1], MIKIO NAKAHARA[2]

[1] *Interdisciplinary Graduate School of Science and Engineering, Kinki University, 3-4-1 Kowakae, Higashi Osaka, Osaka 577-8502, Japan*
[2] *Department of Physics, Kinki University, 3-4-1 Kowakae, Higashi Osaka, Osaka 577-8502, Japan*

We introduce a conceptually reasonable framework of quantum simultaneous noncooperative bimatrix games with naturally-behaving game referees. Although an entangling operation arranged by a game referee is so far a commonly used tool to obtain an advantage over classical counterparts in the conventional formulation of quantum games, we should recall that the classical metagame theory introduced in 1960s resolves many dilemmas without introducing a special correlated system prepared by a game referee. We introduce a quantum metagame framework with which we can eliminate dilemmas under the metalevel one for certain games, without introducing an entangling operation.

Keywords: Noncooperative bimatrix game; Quantumness

1. Introduction

Noncooperative bimatrix games[1] are miniature of behavioral and/or economical competition involving multiple opponents. Many of these games have been gathered in an arena of quantum information science to exhibit how quantumness are useful to resolve dilemmas that appear in classical settings. Formulations of quantum game theory so far followed the original proposals by Meyer[2] and that of Eisert et al.[3] except for a number of different formulations.[5] Since the particular example of Penny flipping in Meyer's work utilized a single-qubit quantumness, it was claimed by van Enk[6] that its mechanism can be implemented with a classical superposition-like state. In contrast, Eisert et al's formulation has become a standard way to extend a classical game to a quantum counterpart despite an unnatural use of entangling operations to be discussed later here. The standard formulations of quantum games appeared so far allow a referee to prepare a correlated system to which players apply local operations as their strategies. It has been shown that quantumness of the correlated system is the resource to resolve a dilemma that cannot be resolved in a classical counterpart in several noncooperative games.[3,4]

A drawback of the standard formulations is that there are unnatural settings of games assumed in quantum noncooperative games. One unnatural setting is that a referee of a quantum game seems to support players to achieve better payoffs by preparing a correlated system. To impose such a task to a referee cannot be naturally justified. The other unnatural setting is that observables are represented by diagonal Hermitian matrices in most quantum games; they can carry off-diagonal elements in principle. In addition, discussions on quantum games are sometimes unclear due to heavily extended mathematical formulations. For example, introducing a joint probability of players' choices only for quantum strategies[4] does not seem to be a fair formulation. Indeed joint probabilities may appear in quantum games as a result of entangling operations, but classical joint probabilities may also appear in case we consider joint strategy extensions of classical games. As for noncooperative simultaneous games, we need to impose independence of strategies. Nevertheless players behave as if they may use joint strategies in quantum extensions. Does this fact suggest that advantages found in conventional quantum extensions are due to a player-friendly referee who sets up a certain system to let players play with joint strategies instead of independent strategies? Despite these unnatural settings, quantum noncooperative games have been studied extensively to seek for the use of a quantum nature.

Although noncooperative games have been most successful applications of a quantum extension, in a general game theory, weaving quantumness into a game is not always beneficial for players. It was mathematically shown by Kargin[7] that there always exists a classical coordination game, similar to a given quantum coordination game, that provides a better payoff for a Bayesian Nash equilibrium if the utility function of the given quantum coordination game is dependent only on the first player's type. Thus it is indispensable to restrict the way of extending classical games into quantum counterparts if the extension is made for the purpose of a certain benefit for players.

As the quantum game theory has developed in the community of quantum physicists, a certain attention has been paid to naturalness of incorporating quantum nature into games with a careful solicitude. It was discussed by Özdemir et al.[8] that a natural quantum extension of a game should reproduce a classical game. A quantum game is usually said natural when it is reduced to a classical counterpart for some set of quantum strategies given by local unitary operations available for players. Nevertheless, it is not a discussion on the prepared initially correlated quantum system supplied by a game referee for the benefit of game players.

Much attention should have been paid to the fact that dilemmas in some classical simultaneous noncooperative games, such as that in Prisoner's Dilemma (PD), were resolved by (classical) metagame extensions in old literatures (*e.g.*, Refs. 9,10). The metagame extension is not to change the rule of a game but to allow players to use predictions on the other players' strategies to choose cleverer strategies.[14,15] For example, Alice may consider what is a proper strategy for Bob if she chooses a certain strategy. Then Alice can choose her optimal strategy based on her prediction. It was shown that the dilemma in PD is resolved in a level-two (or full-full) metagame extension but not in a level-one (or full) metagame extension.[9]

In this contribution, we propose a way to extend two-player classical simultaneous noncooperative bimatrix games using a quantum nature without using a correlated quantum system supplied to players. It is a variant of metagame extensions: we introduce clever players who locally prepare quantum states based on their predictions on the other player's strategies. Such a game is a faithful extension of a classical metagame and we call it a *quantum metagame*.

The paper starts with a brief review of conventional classical and quantum simultaneous noncooperative game theories in Sec. 2. A discussion on a certain unnaturalness of conventional quantum games is made in Sec. 3. It is followed by an overview of the classical metagame theory in Sec. 4. In Sec. 5, we extend a classical metagame theory to a quantum metagame theory and show an example to clarify the advantage of a quantum metagame in comparison to its classical counterpart. Then we discuss on the obtained results in Sec. 6. The paper is concluded in Sec. 7.

2. Conventional Classical and Quantum Simultaneous Noncooperative Bimatrix Games

2.1. *Classical simultaneous noncooperative game*

Let us begin with a formal definition of a classical simultaneous noncooperative game. The game consists of the following constituents. (i) There are m players $1, ..., m$. (ii) Player l can choose a strategy $s_l \in \{1_l, 2_l, ..., q_l\}$. (iii) A *strategy profile* $\mathbf{s} = <s_1, ..., s_m>$, which is a set of the instantaneous strategies of all individual players, is a single instance input to the referee. (iv) A referee assigns a payoff $p_l(\mathbf{s})$ to each player l. Payoffs for a single player l constitute a local *payoff function* $p_l(\mathbf{s})$ where $\mathbf{s} \in \{<1_1,...,1_m>, <1_1,...,1_{m-1}, 2_m>, \cdots, <q_1,...,q_l>\}$. Thus a game is characterized by the group of local payoff functions, $\{p_l(\mathbf{s})\}_l$. (v) It is assumed that a player will choose a *rational strategy* (or, *rational outcome*) so as to maximize her/his income.

To find rational outcomes, we should follow a formalism as below. Let us introduce the following function for player l.

$$M_l(\mathbf{r}) = \{\mathbf{s} | p_l(\mathbf{s}) \leq p_l(\mathbf{r})\}.$$

This is the set of less or equally preferable strategy profiles for player l in comparison to a particular strategy profile $\mathbf{r}$. Then the set of rational choices for player l for a set of fixed strategies of the other players is given by

$$C_l(\bar{s}_1,...,\bar{s}_{l-1},\bar{s}_{l+1},...,\bar{s}_m)$$
$$= \{\mathbf{r} = <\bar{s}_1,...,\bar{s}_l,...,\bar{s}_m> | \forall s_l :<\bar{s}_1,...,\bar{s}_{l-1}, s_l, \bar{s}_{l+1},...,\bar{s}_m> \in M_l(\mathbf{r})\}.$$

This is the set of strategy profiles $\mathbf{r}$ such that player l cannot achieve a larger payoff by changing her/his strategy for a set of fixed strategies $\bar{s}_1,...,\bar{s}_{l-1},\bar{s}_{l+1},...,\bar{s}_m$ of the other players. Finally, we can define the set of rational outcomes for player l as

$$R_l = \cup_{\bar{s}_1,...,\bar{s}_{l-1},\bar{s}_{l+1},...,\bar{s}_m} C_l(\bar{s}_1,...,\bar{s}_{l-1},\bar{s}_{l+1},...,\bar{s}_m).$$

A game may have certain equilibrium strategy profiles (or, simply equilibria) from which no player wants to change one's strategy. The set of equilibria is given by

$$E = \cap_l R_l.$$

When a game is given, its equilibria are important properties to investigate. One curious phenomenon is that sometimes equilibria are not optimum strategy profiles, *i.e.*, sometimes there is a non-equilibrium strategy profile with which players can obtain better payoffs. Appearance of such a *dilemma* is an indisputably insightful property of the nature of a game. Let us recall a formal definition of an optimum strategy profile here. In general, a strategy profile $\bar{\mathbf{s}} = <\bar{s}_1,...,\bar{s}_m>$ is *Pareto optimum* if and only if there exists no other strategy profile $\mathbf{s}$ such that $p_l(\mathbf{s}) \geq p_l(\bar{\mathbf{s}})$ for $\forall l$ with strict inequality for at least one l. Resolving a dilemma to achieve a Pareto-optimum equilibrium is a usual motivation to introduce nonstandard extension of a game.

One of the most famous examples of noncooperative simultaneous games possessing a dilemma is the Prisoner's Dilemma game (PD). The game is played by two players, Alice (player a) and Bob (player b), and it is a *bimatrix game* since the payoff function becomes a bimatrix when $m = 2$. PD is characterized by the payoff matrix shown in Table 1 each of whose elements is $(p_a(s_a, s_b), p_b(s_a, s_b))$ with $s_a, s_b \in \{C, D\}$. Symbols C and D stand for Cooperate and Defect, respectively.

Table 1. Classical payoff matrix of the PD game. ♠: Alice's rational outcome, ♣: Bob's rational outcome. The set of payoffs for the equilibrium is denoted in bold face.

		Bob	
		C	D
Alice	C	(3, 3)	(0, 4)♣
	D	♠(4, 0)	♠**(2, 2)**♣

Let us find the set R_a of Alice's rational outcomes and the set R_b of Bob's rational outcomes. If we fix the strategy of Bob to C (D), Alice's choice is D (D) to maximize her income. Thus $R_a = \{<D, C>, <D, D>\}$. If we fix the strategy of Alice to C (D), Bob's choice is D (D) to maximize his income. Thus $R_b = \{<C, D>, <D, D>\}$. Therefore the equilibrium is $R_a \cap R_b = \{<D, D>\}$ giving the payoff $(2, 2)$, which is not Pareto optimum.

2.2. *Quantum simultaneous noncooperative game*

Researches on simultaneous noncooperative games encountered a conceptual leapfrog when the dilemma of PD was shown to be resolved by introducing a certain entangled quantum state and local unitary operations as strategies.[3] In a quantum simultaneous noncooperative game, a referee prepares a certain entangled state $|\psi_{\text{ent}}\rangle = U_{\text{ent}}|\psi_{\text{in}}\rangle$ of an m-partite system, where $|\psi_{\text{in}}\rangle$ is an initial state and U_{ent} is a certain entangling operation. Each player chooses a strategy $s_l \in \{1_l, ..., q_l\} \subseteq U(d_l)$ where d_l is the dimension of the Hilbert space of the lth component of the m-partite system. Then a referee applies $U_{\text{ent}}^\dagger$ to obtain $|\psi_{\text{out}}\rangle = U_{\text{ent}}^\dagger \bigotimes_l s_l U_{\text{ent}}|\psi_{\text{in}}\rangle$. The payoff function $p_l(\mathbf{s}) = \langle\psi_{\text{out}}|W_l|\psi_{\text{out}}\rangle$ is used to determine the payoff for each player l where W_l is a certain observable.

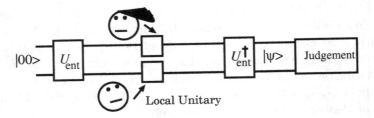

Fig. 1. Quantum circuit of PD introduced by Eisert et al.

The original quantum PD by Eisert et al.[3] is characterized by the following settings. The entangling operation is $U_{\text{ent}} = \frac{1}{\sqrt{2}} \begin{pmatrix} 1 & 0 & 0 & i \\ 0 & 1 & -i & 0 \\ 0 & -i & 1 & 0 \\ i & 0 & 0 & 1 \end{pmatrix}$. Players Alice ($a$) and Bob ($b$) can take strategies $s_x(\theta_x, \phi_x, \psi_x) = \begin{pmatrix} e^{i\phi_x} \cos\theta_x/2 & e^{i\psi_x} \sin\theta_x/2 \\ -e^{-i\psi_x} \sin\theta_x/2 & e^{-i\phi_x} \cos\theta_x/2 \end{pmatrix}$ with $0 \leq \theta_x \leq \pi$ and $0 \leq \phi_x, \psi_x \leq \pi/2$ (here, $x = a$ or b). Observables used by a referee are $W_a = \text{diag}(3, 0, 4, 2)$ and $W_b = \text{diag}(3, 4, 0, 2)$. The entire process of the game is illustrated as Fig. 1. It was shown[3] that $s_a = s_b(\theta_b = 0, \phi_b = \pi/2, \psi_b = 0)$ yields a Pareto-optimum equilibrium with the payoff $(3, 3)$. It was also claimed[3] that this game reproduces the classical PD by imposing the constraint $s_a, s_b \in \{I, iY\}$ with the standard Pauli I and Y matrices. Operators I and iY correspond to classical strategies C and D, respectively; this is because any element of $\{I, iY\}^{\otimes 2}$ commutes with U_{ent}.

It was later shown by Benjamin and Hayden[11] that there would be no equilibrium if s_x were a general SU(2) operator. It is, however, a highly conceptual matter as to which kind of constraints should be imposed to s_x and U_{ent}. One way of thinking is that we can impose any condition as far as the game can reproduce its classical counter part naturally. Then it is possible to simplify the quantum PD game in the following way.

In a simplified quantum PD game,[12] we employ $s_x \in \{I, iY, Z\}$ with the other settings kept unchanged. The payoff matrix for this game is shown in Table 2. The sets of rational outcomes are $R_a = \{<iY, I>, <Z, iY>, <Z, Z>\}$ and $R_b = \{<I, iY>, <iY, Z>, <Z, Z>\}$. Therefore the equilibrium is $<Z, Z>$ and the outcome is $(3, 3)$; this is Pareto optimum.

Table 2. Payoff matrix of the simplified quantum PD game.[12] ♠: Alice's rational outcome, ♣: Bob's rational outcome. The set of payoffs for the equilibrium is denoted in bold face.

		Bob		
		I	iY	Z
Alice	I	(3, 3)	(0, 4)♣	(2, 2)
	iY	♠(4, 0)	(2, 2)	(0, 4)♣
	Z	(2, 2)	♠(4, 0)	♠(**3, 3**)♣

3. Is entanglement or nonclassical correlation an important resource?

A pure entangled state prepared by a referee is a resource to resolve a dilemma as we have seen above. Nevertheless, entanglement is not a necessary resource to resolve a dilemma. We will prove this fact for a general noncooperative bimatrix game.

3.1. *Strategically equivalent games without using entanglement*

Here we mathematically prove that entanglement is not a necessary resource to resolve a dilemma in a two-player noncooperative bimatrix game. First, we recall the well-known definition of *strategically equivalent* games.[1]

Definition 3.1. Two noncooperative bimatrix games G and G' with the individual payoff matrices in the same dimension are *strategically equivalent* if and only if the set of equilibrium points in G is identical to the set of equilibrium points in G'.

It is easy to prove the following lemma:

Lemma 3.1. *Consider a general pure-strategy two-player noncooperative bimatrix game with the strategy sets $\{a_i\}$ and $\{b_j\}$ of Alice and Bob, respectively, whose payoff matrix has (i,j) elements $< p_A(a_i, b_j), p_B(a_i, b_j) >_{i,j}$. The game can be played simultaneously, or firstly by a leader and later by a follower. Then, its strategically equivalent game is a pure-strategy two-player noncooperative bimatrix game with the same strategy sets and payoff matrix elements $< c_1[p_A(a_i, b_j) + c_2], c_1[p_B(a_i, b_j) + c_3] >_{i,j}$ where $c_1 > 0$ and $c_2, c_3 \in \mathbb{R}$ are constants common to $\forall i, j$.*

Proof. It is easy to show that payoff matrices with elements $< c_1[p_A(a_i, b_j) + c_2], c_1[p_B(a_i, b_j) + c_3] >_{i,j}$ lead to the same behaviors of players as those of the original game. First, the offsets c_2 for Alice and c_3 for Bob do no effect on the rational strategies of players. In addition, it is clear that payoff matrices linearly proportional to each other result in the same rational strategies. $\square$

Second, we show that, for any pure-strategy two-player quantum noncooperative bimatrix game, there exists a strategically equivalent game that uses no entangled state during the game process.

Theorem 3.1. *Suppose that the time evolution during a game process is governed by unitary transformations except for the measurement stage for the final judgement of a referee. Consider a pure-strategy two-player quantum noncooperative bimatrix game with the strategy set $\{a_i\}$ of Alice and the strategy set $\{b_j\}$ of Bob ($a_i \in \mathrm{SU}(d_a), b_i \in \mathrm{SU}(d_b)$ with d_a and d_b the dimensions of the Hilbert spaces of systems sent to Alice and Bob, respectively); the output state to measure, $|\psi_{\mathrm{out}}(a_i, b_j)\rangle \in \mathcal{H}_{d_a} \otimes \mathcal{H}_{d_b}$, dependent on a_i and b_j; the payoff $< \langle \psi_{\mathrm{out}}(a_i, b_j)|W_A|\psi_{\mathrm{out}}(a_i, b_j)\rangle, \langle \psi_{\mathrm{out}}(a_i, b_j)|W_B|\psi_{\mathrm{out}}(a_i, b_j)\rangle >$ with W_A and W_B the $d_a d_b \times d_a d_b$ Hermitian matrices. Then, there exists a strategically equivalent game that involves no entangled state during the process.*

Proof. In the game, the final state can be written as $|\psi_{\mathrm{out}}\rangle = U^\dagger(a_i \otimes b_j)U|\psi_{\mathrm{in}}\rangle$ where $|\psi_{\mathrm{in}}\rangle$ is the initial state, U and $U^\dagger$ are the operations applied by a referee before and after players' local operations. Now we consider another game with the input state represented by a density matrix of a pseudo-pure state, $p|\psi_{\mathrm{in}}\rangle\langle\psi_{\mathrm{in}}| + (1-p)I/(d_a d_b)$ and referee's observables $W'_A = W_A/p$ and $W'_B = W_B/p$ where $0 < p \leq 1$. Then the output state to measure is $\rho_{\mathrm{out}} = p|\psi_{\mathrm{out}}(a_i, b_j)\rangle\langle\psi_{\mathrm{out}}(a_i, b_j)| + (1-p)I/(d_a d_b)$. This leads to the payoff matrix with matrix elements

$$< \mathrm{Tr}\rho_{\mathrm{out}} W'_A, \mathrm{Tr}\rho_{\mathrm{out}} W'_B > = < \langle \psi_{\mathrm{out}}(a_i, b_j)|W_A|\psi_{\mathrm{out}}(a_i, b_j)\rangle + \frac{1-p}{pd_a d_b}\mathrm{Tr}W_A,$$

$$\langle \psi_{\mathrm{out}}(a_i, b_j)|W_B|\psi_{\mathrm{out}}(a_i, b_j)\rangle + \frac{1-p}{pd_a d_b}\mathrm{Tr}W_B > .$$

Thus this game is strategically equivalent to the original game. It is known[13] that, for any pure state $|\phi\rangle$ that may be entangled, $p|\phi\rangle\langle\phi| + (1-p)I/d$ is separable for a sufficiently small $p > 0$ where d is the dimension of the Hilbert space. Therefore, the strategically equivalent game set up with a sufficiently small p involves no entanglement throughout the game process. □

As we have seen, entanglement in a pure-strategy quantum noncooperative bimatrix game is not a necessary resource to resolve dilemmas of its classical counterpart. It is always possible to find a strategically equivalent game using no entanglement according to the above theorem. It seems, however, still beneficial to utilize a certain quantum-correlated state prepared by a referee because the proof of the theorem relies on a game using a pseudo-pure state that may be pseudo-entangled.

Indeed, it is a convenient story to enjoy the benefit of quantumness in a game, but we should not escape from the criticism that it is unnatural to introduce an initially correlated system for the benefit of players. Preparing a quantum-correlated system might be useful but not easily justifiable because there is no reason for a referee to prepare such a device for the benefit of players.

Furthermore, it is questionable if a certain correlation between local systems is a requirement to achieve an advantage of a quantum game. We will show a fancy PD game in the next subsection to make it clear that a referee even does not have to prepare a correlated system to put a certain advantage to a quantum version of PD.

3.2. *Solving the PD dilemma with fancy observables*

Now our interest is if a referee can contribute to an advantage of a quantum game without preparing correlated subsystems for players.

Let us, tentatively, assume that a game referee willingly sets up systems so that noncooperative selfish players can achieve rational outcomes resulting in a Pareto-optimum equilibrium. In such a case, the referee does not have to introduce an entangled system for players. As a simple case, we introduce a fancy quantum PD game. Let us restrict players' strategies to $s_a, s_b \in \{I, X, H\}$ acting on the local (independent) state $|0\rangle$ where X is a standard Pauli X operator and $H = \frac{1}{\sqrt{2}} \begin{pmatrix} 1 & 1 \\ 1 & -1 \end{pmatrix}$ is the Hadamard gate. Let the referee to use fancy observables to determine payoffs. To determine the payoff for Alice, the observable $W_a = \begin{pmatrix} 3 & 0 & 0 & 3/2 \\ 0 & 0 & 0 & 0 \\ 0 & 0 & 4 & 0 \\ 3/2 & 0 & 0 & 2 \end{pmatrix}$ is used, and to determine the payoff for Bob, the observable

$$W_b = \begin{pmatrix} 3 & 0 & 0 & 3/2 \\ 0 & 4 & 0 & 0 \\ 0 & 0 & 0 & 0 \\ 3/2 & 0 & 0 & 2 \end{pmatrix}$$ is used. Payoffs are calculated for individual strategy profiles. For example, when the strategy profile is $< H, X >$, the payoff is $(\text{Tr}\rho_{HX} W_A, \text{Tr}\rho_{HX} W_B) = (1, 3)$ where $\rho_{HX} = H \otimes X |00\rangle\langle 00| H^\dagger \otimes X^\dagger$. The payoff matrix of the game is given by Table 3. The rational outcomes

Table 3. Payoff matrix of the fancy quantum PD game. ♠: Alice's rational outcome, ♣: Bob's rational outcome. The sets of payoffs for the equilibria are denoted in bold face.

		Bob		
		I	X	H
Alice	I	(3, 3)	(0, 4)♣	(3/2, 7/2)
	X	♠(4, 0)	♠(2, 2)♣	♠(3, 1)
	H	(7/2, 3/2)	(1, 3)♣	♠(**3, 3**)♣

for Alice and Bob are $R_a = \{< X, I >, < X, X >, < X, H >, < H, H >\}$ and $R_b = \{< I, X >, < X, X >, < H, X >, < H, H >\}$. Thus the equilibria are $R_a \cap R_b = \{< X, X >, < H, H >\}$. Thus players can achieve the payoff $(3, 3)$ for one of the two equilibria, $< H, H >$. It is clear that a referee friendly to players can prepare the fancy observables that we have seen.

In a brief summary, we have discussed, in this section, that entanglement is not a necessary resource to resolve a dilemma in a two-player noncooperative bimatrix game because its strategically equivalent games without utilizing entanglement always exist. We have further shown that preparation of a correlated system seems not a necessary resource to resolve a dilemma. In a quantum PD game, a referee can easily prepare fancy observables so that selfish players can achieve a Pareto optimum equilibrium without introducing correlated local systems.

These facts suggest that we should be more pedantic to extend a classical game to a quantum version. A referee in a realistic world should be strict to keep the rule of a game. More natural extension of a game is to introduce clever players who can behave on the basis of predictions of the other player's strategy without introducing a referee friendly to players. Such an extension is called a metagame extension.[9] We point out in the next section that some dilemmas of bimatrix games are known to be resolved by using a metagame extension without using a special system prepared by a referee, since 1960s.

4. Overview of the Metagame Theory

It was in 1960s and 1970s that a natural extension of a game was developed so that players can behave in a cleverer manner on the basis of predictions of other players' strategies.[9] Such an extension is thought to be realistic because players in a real world must choose strategies based on predictions or estimations of other players' behaviors.[9,14] The reason why it is a natural extension is that the rule of a game is kept unchanged, including the behavior of a referee.

How deeply a player thinks over other players' behaviors is called the *metalevel*. This word is not exactly defined, but it can be interpreted in the following way. A metalevel-one metagame extension is to allow a player to choose one's strategy based on predictions on other players' strategies. A metalevel-two metagame extension is to allow a player to choose one's strategy based on which strategies other players are likely to choose if the player herself/himself chooses a certain strategy.

A typical example of metagame extensions is the meta-PD games. First we will see the metalevel-one PD game with a "clever" Bob. In this game, Bob predicts the strategy of Alice and chooses his meta-strategy. The meta-strategy x/y indicates that Bob chooses $x \in \{C, D\}$ if Alice's strategy is predicted to be C; he chooses $y \in \{C, D\}$ if Alice's strategy is predicted to be D. The payoff matrix is then extended to one described as Table 4. The sets of rational outcomes are $R_a = \{< D, C/C >, < D, D/D >,$

Table 4. The extended payoff matrix for the metalevel-one PD game with a clever Bob. ♠: Alice's rational outcome, ♣: Bob's rational outcome. The set of payoffs for the equilibrium is denoted in bold face.

		Bob			
		C/C	D/D	C/D	D/C
Alice	C	(3,3)	(0,4)♣	♠(3,3)	(0,4)♣
	D	♠(4,0)	♠**(2,2)**♣	(2,2)♣	♠(4,0)

$< D, D/C >, < C, C/D >\}$ and $R_b = \{< D, D/D >, < D, C/D >, < C, C/D >, < C, D/C >\}$. Thus the equilibrium is $E = R_A \cap R_B = \{< D, D/D >\}$, which is not Pareto optimum. The dilemma remains unsolved in this metalevel-one game.

Table 5. The extended payoff matrix for the metalevel-two PD game with an even cleverer Alice.[9,10] ♠: Alice's rational choice, ♣: Bob's rational choice. The equilibria (with the resultant payoffs denoted in bold face) are those marked with both of ♠ and ♣.

Alice \ Bob	C/C	D/D	C/D	D/C
C/C/C/C	(3,3)	(0,4)♣	(3,3)	(0,4)♣
D/D/D/D	♠(4,0)	♠**(2,2)**♣	(2,2)♣	♠(4,0)
D/D/D/C	♠(4,0)	♠(2,2)	(2,2)	(0,4)♣
D/D/C/D	♠(4,0)	♠(2,2)	♠**(3,3)**♣	♠(4,0)
D/D/C/C	♠(4,0)	♠(2,2)	♠(3,3)	(0,4)♣
D/C/D/D	♠(4,0)	(0,4)♣	(2,2)	♠(4,0)
D/C/D/C	♠(4,0)	(0,4)♣	(2,2)	(0,4)♣
D/C/C/D	♠(4,0)	(0,4)♣	♠(3,3)	♠(4,0)
D/C/C/C	♠(4,0)	(0,4)♣	♠(3,3)	(0,4)♣
C/D/D/D	(3,3)♣	♠(2,2)	(2,2)	♠(4,0)
C/D/D/C	(3,3)	♠(2,2)	(2,2)	(0,4)♣
C/D/C/D	(3,3)♣	♠(2,2)	♠**(3,3)**♣	♠(4,0)
C/D/C/C	(3,3)	♠(2,2)	♠(3,3)	(0,4)♣
C/C/D/D	(3,3)	(0,4)♣	(2,2)	♠(4,0)
C/C/D/C	(3,3)	(0,4)♣	(2,2)	(0,4)♣
C/C/C/D	(3,3)	(0,4)♣	♠(3,3)	♠(4,0)

Second we will see the metalevel-two PD game with an "even cleverer" Alice. She predicts the metalevel-one strategy x/y of Bob and chooses her metalevel-two strategy $e/f/g/h$ which indicates that she chooses (i) e if she predicts Bob chooses C/C, (ii) f if she predicts Bob chooses D/D, (iii) g if she predicts Bob chooses C/D, (iv) h if she predicts Bob chooses D/C. In this game, the payoff matrix becomes the one given by Table 5. We can now find equilibria from rational outcomes of players marked in the table. The equilibria are $< D/D/D/D, D/D >$, $< D/D/C/D, C/D >$, and $< C/D/C/D, C/D >$ with corresponding payoffs $(2,2)$, $(3,3)$, and $(3,3)$, respectively. Therefore, one finds that the metalevel-two extension of PD resolves the dilemma of the original PD.

We have briefly visited the historical milestone of the metagame theory in this section. Its concept is substantially different from that of a conventional quantum game theory in the sense that the metagame theory extends players' behaviors and keeps the behavior of a referee unchanged while the quantum game theory extends the behavior of a referee and players' behaviors.

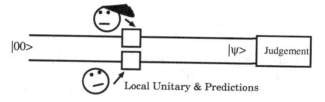

Fig. 2. The quantum circuit of a two-player noncooperative bimatrix game without U_{ent}. We may allow a certain prediction for a player depending on a metagame rule.

5. Quantum Metagame

As we have seen, a level-two metagame extension is a strong tool for players to resolve a dilemma without asking a referee for a special setup in a certain game. One may be curious if there is still any room to introduce a quantum nature for the benefit of players. For a fair comparison to classical counterparts, we consider a quantum game setting in which only local operations and predictions are allowed for individual players as illustrated in Fig. 2. Here, we show that a metalevel-one quantum PD game eliminates the dilemma of PD (under a certain rule imposed for the metagame). In addition, we show that a dilemma of Samaritan's Dilemma (SD), another well-known noncooperative 2×2 bimatrix game, is resolved by introducing a quantum metalevel-one SD. Thus quantumness is useful to eliminate dilemmas in a shallower metalevel in certain noncooperative bimatrix games.

5.1. *From quantum PD to quantum meta-PD*

Let us begin with a quantum PD game without entangling operation although one might expect that such a game does not achieve any improvement over the classical counterpart.

We assume that $|0\rangle$ and $|1\rangle$ correspond to C and D, respectively. Alice's strategy is set to $|s_a\rangle = \alpha_0|0\rangle + \alpha_1|1\rangle$ with $|\alpha_0|^2 + |\alpha_1|^2 = 1$ and Bob's strategy is set to $|s_b\rangle = \beta_0|0\rangle + \beta_1|1\rangle$ with $|\beta_0|^2 + |\beta_1|^2 = 1$. A payoff for player $x \in \{a, b\}$ is calculated by $\text{Tr}(W_x|s_a\rangle\langle s_a| \otimes |s_b\rangle\langle s_b|)$ with $W_a = \text{diag}(3, 0, 4, 2)$ and $W_b = \text{diag}(3, 4, 0, 2)$. Let us introduce parameters $A = |\alpha_0|^2$ and $B = |\beta_0|^2$. With these parameters, the payoff for Alice is $p_a = -2A + AB + 2B + 2$ and that for Bob is $p_b = -2B + AB + 2A + 2$. Now it is obvious that this game is equivalent to the classical mixed-strategy PD game. Alice's rational outcome is $A = 0$ for a fixed B. Bob's rational outcome is $B = 0$ for a fixed A. Thus the equilibrium is $< A = 0, B = 0 >$, leading to the payoff $(2, 2)$. There is no improvement of the payoff as was expected.

For the next step, we extend this game to produce a metalevel-one quantum PD game. There is an infinite number of ways for the extension because payoffs are functions of continuous variables. Here we employ the following extension with a "clever" Alice. (i) Regard A as a function f of a predicted value B' of B. (ii) Set Alice's possible meta-strategies f to $f_1(B') = B'$ and $f_2(B') = 1 - B'$. A payoff matrix to consider in this setting is the one whose elements are expected payoffs with the row indices f_1 and f_2 for Alice and the continuous column index B' for virtual Bob (in her mind), as shown in Table 6. The set of rational outcomes of Alice

Table 6. Payoff matrix of the metalevel-one quantum PD game with a clever Alice. ♠: Alice's rational outcome, ♣: Bob's rational outcome (specific values of B' indicated with #'s are shown as rational outcomes).

Alice \ Bob	B' (continuous)
f_1	#♠:$B' \leq 1/2$# $(B'^2 + 2, B'^2 + 2)$ #♣:$B' = 1$#
f_2	#♠:$1/2 \leq B'$# $(-(B' - \frac{5}{2})^2 + \frac{25}{4}, -(B' + \frac{3}{2})^2 + \frac{25}{4})$ #♣:$B' = 0$#

is $R_a = \{< f_1, B' > |_{B'=0}^{1/2}, < f_2, B' > |_{B'=1/2}^{1}\}$ and that of virtual Bob is $R_b = \{< f_1, B' = 1 >, < f_2, B' = 0 >\}$. There is no equilibrium because $R_a \cap R_b = \emptyset$. The transition of strategy $< A, B' >$ occurs in the following manner:

$$< 0, 0 > \to < 0, 1 > \to < 1, 1 > \to < 1, 0 > \to < 0, 0 > \to \cdots.$$

The average payoff is then $(9/4, 9/4)$, which is slightly better than $(2, 2)$. [a]

A similar result is obtained when we exchange the roles of Alice and Bob. (i) Regard B as a function g of a predicted value A' of A. (ii) Set Bob's possible meta-strategies g to $g_1(A') = A'$ and $g_2(A') = 1 - A'$.
In this case, the equilibrium becomes an empty set again.

In a brief summary, the metalevel-one PD game has no equilibrium and thus the original dilemma does not exist.

---

[a]One may find this result unnatural because Alice may consider the security level (the least possible payoff she can achieve). The security level for the strategy f_1 is 2 and that for f_2 is 0. Therefore, Alice should prefer f_1 although we cannot actually determine the equilibrium due to such a common sense when we follow the formal way to find equilibria.

5.2. From quantum Samaritan's Dilemma to quantum meta-Samaritan's Dilemma

Let us consider another well-known noncooperative 2 × 2 bimatrix game, the Samaritan's Dilemma (SD), among others. It is characterized by the payoff matrix of Table 7. As is clear from the marks in the table to indicate

Table 7. Payoff matrix of classical SD. ♠: Alice's rational outcome, ♣: Bob's rational outcome.

		Bob	
		W	L
Alice	A	♠(3, 2)	(−1, 3)♣
	NA	(−1, 1)♣	♠(0, 0)

rational outcomes of individual players, there is no equilibrium point in this classical SD.

First we consider a quantum SD involving only local operations before measurements. Let the strategies of Alice and Bob be $|s_a\rangle$ and $|s_b\rangle$, respectively, represented by the same equations as those used in the quantum PD, with $A = |\alpha_0|^2$ and $B = |\beta_0|^2$. The observables used to determine the payoffs are $W_a = \mathrm{diag}(3, -1, -1, 0)$ for deciding Alice's payoff $p_a = \langle s_a|\langle s_b|W_a|s_a\rangle|s_b\rangle$ and $W_b = \mathrm{diag}(2, 3, 1, 0)$ for deciding Bob's payoff $p_b = \langle s_a|\langle s_b|W_b|s_a\rangle|s_b\rangle$. We have $p_a = A(5B - 1) - B$ and $p_b = B(1 - 2A) + 3A$. The set of rational outcomes of Alice and Bob are $R_a = \{< A = 0, B \leq 1/5 >, < A = 1, B \geq 1/5 >\}$ and $R_b = \{< A \leq 1/2, B = 1 >, < A \geq 1/2, B = 0 >\}$, respectively. Thus $R_a \cap R_b = \emptyset$ and there is no equilibrium. The dilemma remains unsolved.

Second we consider a metagame extension of the above quantum SD. Alice's strategies are certain possible ways to set the value of A. One possible setting is to choose a strategy $A \in \{f_1, f_2\}$ with the functions $f_1 = B'$ and $f_2 = 1 - B'$ where B' is the predicted value of B. The payoff matrix is then represented as Table 8. The set of rational outcomes of Alice and Bob are $R_a = \{< f_1, B' > |_{B' \leq 1/5, 1/2 \leq B'}, < f_2, B' > |_{1/5 \leq B' \leq 1/2}\}$ and $R_b = \{< f_1, B' = 1 >, < f_2, B' = 0 >\}$, respectively. Thus the equilibrium is $R_a \cap R_b = < f_1, B' = 1 >$. This results in the payoff $(3, 2)$. In this way, a Pareto-optimum equilibrium is achieved.

This result suggests that Alice can control the game to attain the payoff $(3, 2)$ by choosing her strategy from f_1 and f_2 based on the predicted value of B. The prediction is realistic if the game is played many times.

Table 8. Payoff matrix of the metalevel-one quantum SD game with a clever Alice. ♠: Alice's rational outcome, ♣: Bob's rational outcome (specific values of B' indicated with #'s are shown as rational outcomes).

Alice \ Bob	B' (continuous)
f_1	#♠:$B' \leq 1/5, 1/2 \leq B'$# $(5B'^2 - 2B', -2B'^2 + 4B')$ #♣:$B' = 1$#
f_2	#♠:$1/5 \leq B' \leq 1/2$# $(-5B'^2 + 5B' - 1, 2B'^2 - 4B' + 3)$ #♣:$B' = 0$#

6. Discussions

Quantumness introduced by a game referee was thought to be useful for solving a dilemma.[3] Nevertheless, it is questionable if a real referee wants to prepare an initially correlated system for the benefit of players when we extend a classical game to a quantum counterpart by allowing players to handle quantum states. A natural setting of a quantum noncooperative bimatrix game is to allow players to use local operations and predictions (LO&P) only and a referee to use diagonal Hermitian matrices as observables to calculate payoffs. This is because a game referee should try to keep the rule of a noncooperative game. Nevertheless, the majority of works in quantum games have their basis on a preparation of a certain quantum correlation before players touch local systems.

In a setting in which a game referee can prepare an entangled state for the benefit of players, several dilemmas of noncooperative games have been resolved in literature.[3,16] It became well-known that correlations without entanglement can be used to eliminate dilemmas in several games.[16] As we have seen in Sec. 3.1, entanglement is not a necessary resource in conventional noncooperative bimatrix quantum games because it is always possible to find a strategically equivalent game that does not involve entanglement. We further discussed that there is another way to resolve a dilemma of PD by introducing certain off-diagonal elements in observables used to determine payoffs, in Sec. 3.2.

It should be noted that there is a game whose dilemma can be resolved with only local superpositions and diagonal observables. It is the Chicken Game (CG) which is one of the popular noncooperative bimatrix games. CG is characterized by the payoff matrix represented as Table 9. Suppose that Alice uses a state $\alpha_0|0\rangle + \alpha_1|1\rangle$ and Bob uses a state $\beta_0|0\rangle + \beta_1|1\rangle$ with $A = |\alpha_0|^2$ and $B = |\beta_0|^2$, as strategies. Let us set the observables to determine the payoff of Alice and that of Bob as $W_a = \text{diag}(3, 1, 4, 0)$ and $W_b = \text{diag}(3, 4, 1, 0)$, respectively. Then the payoff of Alice is $p_a = A(1 - B) + 3B$ and that of Bob is $p_b = B(1 - A) + 3A$. The set of rational

Table 9. Payoff matrix of classical CG. ♠: Alice's rational outcome, ♣: Bob's rational outcome. There are two equilibria (with the resultant payoffs denoted in bold face) that are not best for players.

		Bob	
		S	C
Alice	S	(3,3)	♠(**1**, **4**)♣
	C	♠(**4**, **1**)♣	(0,0)

outcomes of Alice is $R_a = \{< A = 0, B = 1 >, < A = 1, 0 \leq B \leq 1 >\}$ and that of Bob is $R_b = \{< 0 \leq A \leq 1, B = 1 >, < A = 1, B = 0 >\}$. Thus the set of equilibria is $R_a \cap R_b = \{< A = 1, B = 1 >< A = 1, B = 0 >\}$. The equilibrium $< A = 1, B = 1 >$ leads to the payoff (3,3); the dilemma is resolved.

As for CG, a classical mixed-strategy extension is also known to resolve the dilemma. This is essentially equivalent to the above extension in the sense that we have only to regard A (B) as the probability that Alice (Bob) chooses S and $1 - A$ ($1 - B$) as the probability that Alice (Bob) chooses C, in order to regard the above quantum CG as a classical mixed-strategy CG.

Thus the benefit of quantum correlation is not obvious; it is highly dependent on the rule or settings of a game. This paper has mainly pursued the possibility to eliminate dilemmas in noncooperative bimatrix games using LO&P without a player-friendly referee. We utilized Howard's metagame theory to extend quantum games (without referee-prepared entanglement) to quantum metagames. It has been shown that dilemmas in PD and SD can be eliminated by level-one quantum metagame extensions. This suggests that a local superposition and predictions of strategies of the other player are useful resources for players to achieve a better outcome without an aid from a game referee. We can also show that a game strategically equivalent to a conventional quantum bimatrix game that uses entanglement can be played in principle using only LO&P as we will see in Appendix A.

Quantum games may enjoy their heyday for the time being. It is, however, not predictable if these games are of practical use in future. As was claimed in the original paper by Eisert et al.,[3] quantum games can be implemented in a simulation using classical resources. This is true, of course, as long as we play a game with a small number of players with individual local Hilbert spaces small enough. We should discuss a very large-scale game to claim the advantage of quantum games because such a game cannot be simulated by classical computers in a polynomial time in general.

7. Conclusion

We considered quantum simultaneous noncooperative bimatrix games without a referee-prepared correlated system. We revisited the metagame theory and applied it to extending such quantum games to quantum metagames. It has been shown that a level-one quantum metagame extension eliminates original dilemmas in the games of Prisoner's Dilemma and Samaritan's Dilemma.

Appendix A. Playing quantum-PD equivalent game using local operations and predictions

In a conventional quantum PD (CQPD), a game referee prepares $|\psi_{\text{ent}}\rangle = (|00\rangle + i|11\rangle)/\sqrt{2}$ for the benefit of players so that they can achieve the payoff $(3,3)$. We have already shown in Theorem 3.1 that strategically equivalent games always exist for such a noncooperative bimatrix game, so that we do not have to utilize a prepared entanglement as a resource. A strategically equivalent game for the CQPD is the one utilizing $\rho_{\text{pe}} = p|\psi_{\text{ent}}\rangle\langle\psi_{\text{ent}}| + (1-p)I/4$ with $0 < p \leq 1$, which is a separable state for $p \leq 1/3$. One might be curious about the possibility to play a game strategically equivalent to CQPD without asking a game referee to prepare a state like $|\psi_{\text{ent}}\rangle$ or ρ_{pe}. This is not an easy task because a usual metagame extension involves an extension of a payoff matrix.

Players of PD should accomplish the three tasks for this purpose: (i) Generation of ρ_{pe}, (ii) Local unitary operations, (iii) Applying an operation effectively equivalent to $U_{\text{ent}}^{\dagger}$.

Task (i) is possible if the players, Alice and Bob, can generate ρ_{pe} by themselves on the basis of the metagame extension. In a metagame, players can use local operations and predictions (LO&P) before the last measurement made by a referee. A single prediction of Alice involves predicting every probability w_β that a pure state $|\beta\rangle = U_B(\beta)|0\rangle$ is chosen by Bob with $U_B \in \text{U}(2)$. Then Alice can choose a mixed strategy against the probability distribution $(w_{\beta'}, |\beta'\rangle)_{\beta'}$. Such a strategy is given by $(w_\alpha, |\alpha[(w_{\beta'}, |\beta'\rangle)_{\beta'}]\rangle)_\alpha$ with α a function of the predicted probability distribution $(w_{\beta'}, |\beta'\rangle)_{\beta'}$ of Bob's mixed strategy. The general representation of the ensemble state that can be generated in average by Alice under LO&P is

$$\rho_A = \sum_{\alpha[(w_{\beta'},|\beta'\rangle)_{\beta'}]} w_\alpha |\alpha\rangle\langle\alpha| \otimes \sum_{\beta'} w_{\beta'} |\beta'\rangle\langle\beta'|.$$

Thus any separable state can be generated by Alice and virtual Bob in Alice's mind. Note that this state is not a real state but a state made by Alice and virtual Bob. An actual probability distribution of Bob's mixed strategy may be different. The state describes the system consisting of Alice's real quantum system and virtual Bob's quantum system in her mind. The real-virtual state ρ_A becomes a real-real state when Bob behaves in the exact way as predicted in Alice's mind. Thus it is, in principle, possible to generate ρ_{pe}.

Task (ii) is easily achieved.

Task (iii) seems impossible at a glance, but it is possible. Let $\rho_{\text{pe}} = \Lambda(|00\rangle\langle 00|)$. Each time of a (real) play, the players generate a component $|\alpha\rangle|\beta\rangle = U_A \otimes U_B |00\rangle$ of ρ_{pe} by applying local unitary operations U_A and U_B to $|00\rangle$. Thus the inverse of Λ can be effectively generated by each time applying $U_A^\dagger \otimes U_B^\dagger$ after local operations in Task(ii).

Although it is questionable to consider such unrealistically clever players, it has been shown that a game strategically equivalent to CQPD can be played by players using LO&P only.

References

1. J. F. Nash, *Ann. of Math.* **54**, 286 (1951).
2. D. A. Meyer, *Phys. Rev. Lett.* **82**, 1052 (1999).
3. J. Eisert, M. Wilkens, and M. Lewenstein, *Phys. Rev. Lett.* **83**, 3077 (1999).
4. T. Cheon and A. Iqbal, *J. Phys. Soc. Japan* **77**, 024801 (2008).
5. J. Wu, quant-ph/0405032.
6. S. J. van Enk, *Phys. Rev. Lett.* **84**, 789 (2000).
7. V. Kargin, *Int. J. Game Theory*, "online-first" issue, (http://dx.doi.org/10.1007/s00182-007-0106-1).
8. S. K. Özdemir, J. Shimamura, and N. Imoto, *New J. Phys.* **9**, 43 (2007).
9. N. Howard, *General Systems* **11**, 167 (1966); ibid, 187 (1966).
10. N. Howard, *Paradoxes of Rationality: Theory of Metagames and Political Behavior* (MIT Press, Cambridge, 1971).
11. S. C. Benjamin and P. M. Hayden, *Phys. Rev. Lett.* **87**, 069801 (2001).
12. S. J. van Enk and R. Pike, *Phys. Rev. A* **66**, 024306 (2002).
13. S. L. Braunstein, C. M. Caves, R. Jozsa, N. Linden, S. Popescu, and R. Schack, *Phys. Rev. Lett.* **83**, 1054 (1999).
14. E. Akiyama, in *Frontiers of game theories* (Saiensu-sha, Tokyo, 2005, in Japanese), edited by T. Ikegami and H. Matsuda, pp. 66-75; G. Masumoto, ibid, pp. 76-83; M. Nakayama, ibid, pp.84-91; M. Yabuuchi, ibid, pp.92-100.

15. G. Masumoto and T. Ikegami, *BioSystems* **80**, 219 (2005).
16. J. Shimamura, *Playing games in quantum realm*, PhD Thesis, Osaka University (2004).

CONTINUOUS–VARIABLE TELEPORTATION OF SINGLE–PHOTON STATES AND AN ACCIDENTAL CLONING OF A PHOTONIC QUBIT IN TWO–CHANNEL TELEPORTATION

Toshiki Ide

Okayama Institute for Quantum Physics,
1-9-1 Kyoyama, Okayama City, Okayama, 700-0015, Japan
E-mail: toshiki_ide@pref.okayama.jp

The properties of continuous–variable teleportation of single–photon states are investigated. The output state is different from the input state due to the non–maximal entanglement in the EPR beams. The photon statistics of the teleportation output are determined and the correlation between the field information β obtained in the teleportation process and the change in photon number is discussed. The results of the output photon statistics are applied to the transmission of a qubit encoded in the polarization of a single photon. The information encoded in the polarization of a single photon can be transferred to a remote location by two–channel continuous variable quantum teleportation. However, the finite entanglement used in the teleportation causes random changes in photon number. If more than one photon appears in the output, the continuous variable teleportation accidentally produces clones of the original input photon. In this paper, it derives the polarization statistics of the N–photon output components and shows that they can be decomposed into an optimal cloning term and completely unpolarized noise. It is found that the accidental cloning of the input photon is nearly optimal at experimentally feasible squeezing levels, indicating that the loss of polarization information is partially compensated by the availability of clones.

Keywords: Quantum teleportation; Quantum cloning; Entanglement; Continuous variables

1. Introduction

Quantum teleportation is a method for Alice (sender) to transmit an unknown quantum input state to Bob (receiver) at a distant place by sending only classical information using a shared entangled state as a resource. Originally quantum teleportation was proposed for discrete variables in two–dimensional Hilbert spaces.[1] Later it was applied to continuous variables (two components of the electromagnetic field) in infinite–dimensional Hilbert spaces.[2] However, continuous–variable (CV) teleportation ideally requires a maximally entangled state which has infinite energy as a resource. Nevertheless it has been shown theoretically that quantum teleportation realized by using non–maximally entangled state can still transfer non–classical features of quantum states.[3] Experimentally, such a CV teleportation has been realized by Furusawa et al.[4] In [3], the physics of CV teleportation was described in terms of Wigner function. Reference [5] described it in terms of discrete basis states. Reference [6] formulates the whole process of the quantum teleportation by a transfer operator which is acting on arbitrary input states.

In the experiment of [4], a coherent state was used as an input state. But any quantum state can be teleported by this method. Therefore, the transfer of non–classical states is of interest. In the following, the transfer operator formalism derived in [6] is used for analyzing the photon statistics of the output state of a one photon state teleportation. It is shown that the change in photon number is strongly dependent on the field measurement result obtained in the process of teleportation. This result is then applied to the two mode teleportation of a polarized photon, illustrating the possibility of using CV teleportation for the transfer of single–photon qubits.

One of the most fascinating aspects of quantum optics is the insight it offers into the relation between the continuous field variables and photon numbers. In many cases, quantum protocols can be implemented by using either a photon or a continuous variable approach. In particular, this is true for teleportation and cloning, where both approaches have been realized experimentally.[4,7-11] Recently, there have also been efforts to combine both approaches, e.g. by applying homodyne detection to photon number states,[12,13] or by adding and subtracting photons from squeezed and coherent light.[14-16] In the light of these technological advances, it is interesting to take a closer look at some of the possibilities inherent in the application of continuous–variable protocols to photon number states.

Since CV teleportation works for any input state, it is in principle straightforward to apply it to photon number inputs.[17-20] However, the transmission process does not preserve photon number, so it is necessary to evaluate the effects of photon loss and photon addition. Specifically, a qubit encoded in the polarization state of a single photon can be either lost or multiplied in the CV teleportation process. If the photon is multiplied, the quantum information carried by its polarization is distributed to all output photons, resulting in an accidental cloning of the initial qubit. Photon multiplication errors should therefore be evaluated in terms of their cloning fidelity, which is of course limited by the fact that ideal cloning of quantum states is impossible.[21,22]

In the present paper, we see how the whole process of the CV teleportation is described and formulated by the so-called "transfer operator" and then the photon number statistics is analyzed for the output of a single-photon teleportation. We can transfer the quantum information that is encoded in the polarization of a photon by expanding this scheme to a two-channel one. The transfer operator also can be expanded to the two-mode one and it analyzes the photon statistics of the output of a photonic qubit teleportation. And it is shown that the output density matrix can be decomposed into an optimal cloning term and completely unpolarized white noise. The N-photon outputs are identified and the cloning fidelities is derived. It shows that they are close to the optimal cloning fidelities at experimentally feasible squeezing levels. This result indicates that the transfer of quantum information is mostly limited by the availability of clones in the output. Interestingly, photons are cloned even though the transmitted field signal is not amplified. As the discussion shows, the cloning effects can be quantified in terms of the Gaussian field noise added in the teleportation process. It can be therefore conjectured that accidental cloning is a general effect of Gaussian field noise on photonic qubits or unknown polarized light. Finally, it is summarized for the results and their possible relevance.

2. Transfer operator

Figure 1 shows the schematic sets of the quantum teleportation according to [4]. Alice transmits an unknown quantum state $\mid \psi \rangle_A$ to Bob. Alice and Bob share EPR beams in advance. The quantum state of the EPR beams reads[5,6]

$$\mid q \rangle_{R,B} = \sqrt{1-q^2} \sum_{n=0}^{\infty} q^n \mid n \rangle_R \mid n \rangle_B, \qquad (1)$$

where R, B are the modes for reference and Bob each. q is a parameter which stands for the degree of entanglement. It varies from 0 to 1 with 1 being maximal entanglement and 0 being no entanglement. The degree of entanglement depends on the squeezing achieved in the parametric amplification. In the experiment of [4], 3dB ($q = 0.33$) squeezed light was used. Squeezing of up to 10dB ($q = 0.82$) should be possible with available technology.

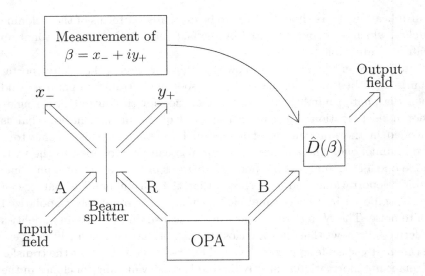

Fig. 1. Schematic representation of the quantum teleportation setup.

Alice mixes her input state with the reference EPR beam by a 50% beam–splitter and performs an entanglement measurement of the complex field value $\beta = x_- + iy_+$, where

$$\hat{x}_- = \hat{x}_A - \hat{x}_R,$$
$$\hat{y}_+ = \hat{y}_A + \hat{y}_R. \tag{2}$$

This measurement projects A and R onto the eigenstate

$$|\beta\rangle_{A,R} = \frac{1}{\sqrt{\pi}} \sum_{n=0}^{\infty} \hat{D}_A(\beta) \, |n\rangle_A \, |n\rangle_R, \tag{3}$$

where $\hat{D}_A(\beta)$ is a displacement operator acting on the mode A with a displacement amplitude of β. The output state $|\psi(\beta)\rangle_B$ conditioned by the measurement process can be written as a projection of the initial product state onto the eigenstate $|\beta\rangle_{A,R}$,

$$|\psi(\beta)\rangle_B = {}_{A,R}\langle\beta \, | \, \psi\rangle_A \, |q\rangle_{R,B}$$
$$= \sqrt{\frac{1-q^2}{\pi}} \sum_{n=0}^{\infty} q^n \, |n\rangle_{BA}\langle n | \, \hat{D}_A(-\beta) \, |\psi\rangle_A. \tag{4}$$

$|\psi(\beta)\rangle_B$ is not normalized since the probability of obtaining the field measurement value β is given by $P_q(\beta) = \langle\psi_{\text{out}}(\beta) \, | \, \psi_{\text{out}}(\beta)\rangle$.

After Bob gets the information of the field measurement value β from Alice, Bob applies a displacement to the output state by mixing the coherent field of a local oscillator with the output EPR beam B. The output state $|\psi_{\text{out}}(\beta)\rangle_B = \hat{D}_B(\beta) |\psi(\beta)\rangle_B$ may also be written as

$$|\psi_{\text{out}}(\beta)\rangle_B = \hat{T}_q(\beta) |\psi\rangle_A, \tag{5}$$

where $\hat{T}_q(\beta)$ is a transfer operator which represents all processes of the quantum teleportation.[6] In its diagonalized form, it reads

$$\hat{T}_q(\beta) = \sqrt{\frac{1-q^2}{\pi}} \sum_{n=0}^{\infty} q^n \hat{D}(\beta) |n\rangle_{BA}\langle n| \hat{D}(-\beta). \tag{6}$$

When $q \to 1$, $\hat{T}_q(\beta)$ becomes similar to the equivalence operator $\hat{1}$ except for a change of the mode from A to B which indicates a perfect teleportation with no modification to the input state. For general q, the transfer operator at $\beta = 0$ is diagonal in the photon number states.

3. Single–photon state teleportation

In the experiment of [4], a coherent state was transfered. In this case, the output state is modified but is still a coherent state.[6] In case of a one photon input state, the output state is

$$\begin{aligned}
\hat{T}_q(\beta) |1\rangle &= \sqrt{\frac{1-q^2}{\pi}} \sum_{n=0}^{\infty} q^n \hat{D}(\beta) |n\rangle\langle n| \hat{D}(-\beta) |1\rangle \\
&= \sqrt{\frac{1-q^2}{\pi}} \sum_{n=0}^{\infty} q^n \hat{D}(\beta) |n\rangle\langle n| \hat{D}(-\beta)\hat{a}^\dagger |0\rangle \\
&= \sqrt{\frac{1-q^2}{\pi}} \sum_{n=0}^{\infty} q^n \hat{D}(\beta) |n\rangle\langle n| (\hat{a}^\dagger + \beta^*)\hat{D}(-\beta) |0\rangle \\
&= \sqrt{\frac{1-q^2}{\pi}} e^{-(1-q^2)\frac{|\beta|^2}{2}} \hat{D}((1-q)\beta) \left((1-q^2)\beta^* |0\rangle + q|1\rangle\right).
\end{aligned}\tag{7}$$

For simplicity, the suffixes for the modes are not explicitly given. In general, this output state is quite different from the original one photon input state.

β represents a coherent field measurement performed on the input state. The probability distribution over field measurement values β for one photon input states is given by

$$\begin{aligned} P_q(\beta) &= \langle \psi_{\text{out}}(\beta) \mid \psi_{\text{out}}(\beta) \rangle \\ &= \langle 1 \mid \hat{T}_q^\dagger(\beta)\hat{T}_q(\beta) \mid 1 \rangle \\ &= \frac{1-q^2}{\pi} e^{-(1-q^2)|\beta|^2} \left((1-q^2)^2|\beta|^2 + q^2\right). \end{aligned} \quad (8)$$

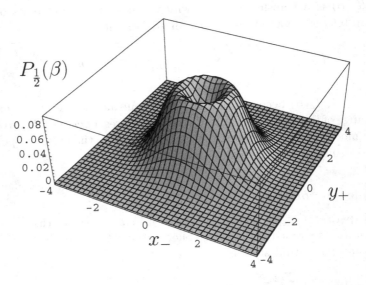

Fig. 2. The probability distribution of $P_q(\beta)$ for the field measurement value $\beta = x_- + iy_+$ for $q = \frac{1}{2}$.

Figure 2 shows the probability distribution $P_q(\beta)$ for $q = \frac{1}{2}$. The circular symmetric distribution with a dip in the middle and a maximum for amplitude of $|\beta| = 1$ is characteristic of the one photon input state.

The properties of the output state can be investigated experimentally by several kinds of detection setups. If Bob has a photon counting setup which can discriminate photon numbers, he can obtain the photon statistics of the output state. The overall photon statistics of the output are obtained by integrating over β,

$$\begin{aligned} P_q(n) &= \int d^2\beta |\langle n \mid \hat{T}_q(\beta) \mid 1 \rangle|^2 \\ &= \frac{1+q}{2} \left(\frac{1-q}{2}\right)^{n+1} \left(1 + \left(\frac{1+q}{1-q}\right)^2 n\right). \end{aligned} \quad (9)$$

$P_q(n)$ is the probability of counting n-photons after the teleportation. Figure 3 shows the probability distribution over n in the case of $q = \frac{1}{2}$. The probability of detecting one photon is maximal and the probabilities of higher photon numbers get less likely as photon number increases.

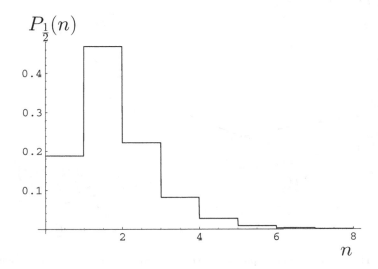

Fig. 3. The probability of n-photon counting $P_q(n)$ for one photon state teleportation in the case of $q = \frac{1}{2}$.

The changes in photon number may be summarized in terms of photon loss $n = 0$, successful photon transfer $n = 1$, and photon gain $n \geq 2$. The q dependence of these probabilities is given by

$$P_q(0) = \frac{1}{4}(1 - q^2), \tag{10}$$

$$P_q(1) = \frac{1}{4}(1 + q + q^2 + q^3), \tag{11}$$

$$P_q(n \geq 2) = \frac{1}{4}(2 - q - q^3). \tag{12}$$

Since the input state is a one photon state, Eq. (11) gives the fidelity of the teleportation. Figure 4 shows the q dependence of $P_q(0), P_q(1), P_q(n \geq 2)$. In the case of a maximally entangled state ($q \rightarrow 1$), Bob can receive nothing but a one photon state ($P_q(1) \rightarrow 1$) and all other probabilities vanish ($P_q(0) \rightarrow 0, P_q(n \geq 2) \rightarrow 0$), which indicates perfect teleportation. In the case of non-maximally entangled states ($q < 1$), the probabilities of zero–photon counting and more than one–photon counting appear. Note that the probability of photon gain $n \geq 2$ is always greater than that of photon loss $n = 0$.

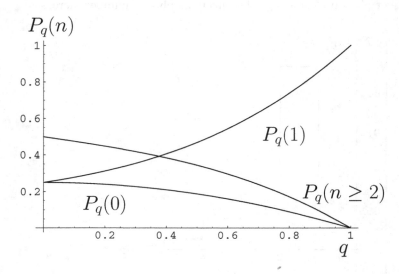

Fig. 4. The probability of zero photon, one photon and more than one photon counting for one photon state teleportation.

In order to investigate the change in the photon number given by Eq. (9) in more detail, now the conditional probability distributions over β for the cases of zero–photon, one–photon, and more than one–photon counting is derived. These probabilities can be obtained from $P_q(n, \beta) = |\langle n | \hat{T}_q(\beta) | 1 \rangle|^2$. The results read

$$P_q(0, \beta) = \frac{1-q^2}{\pi} e^{-2(1-q)|\beta|^2} (1-q)^2 |\beta|^2, \qquad (13)$$

$$P_q(1, \beta) = \frac{1-q^2}{\pi} e^{-2(1-q)|\beta|^2} \left((1-q^2)^2 |\beta|^2 + q^2\right), \qquad (14)$$

$$P_q(n \geq 2, \beta) = P_q(\beta) - P_q(0, \beta) - P_q(1, \beta). \qquad (15)$$

Figure 5 shows these different contributions to the total probability distribution over $|\beta|$ in case of $q=0.5$. For high values of $|\beta|$, every probability vanishes. In the $\beta = 0$ case, we see that only the probability of obtaining a one photon output is non–zero. This means that the output of the EPR beam B is already a one photon state without any additional displacement. In general, the photon number states are the eigenstates of $\hat{T}_q(\beta = 0)$, as can be seen from Eq. (6). A measurement value of $\beta = 0$ thus indicates that the output beam photon number is equal to the input beam photon number. Note also that a low field amplitude provides little information about the phase of the teleported field. Therefore, the lack of phase information in the input photon number state is preserved in the teleportation process.

In the region of $|\beta| \ll 1$, the probability of obtaining a one photon output is nearly constant and that of zero–photon and more than one–photon are increasing with the increase of $|\beta|$. The increase of these two probabilities gives rise to the peak of the total probability. The probability of photon loss ($n = 0$) and photon gain ($n \geq 2$) are crossing at around $|\beta| = 1$. Below $|\beta| = 1$ the probability of photon loss ($n = 0$) is greater than that of photon gain ($n \geq 2$). Above $|\beta| = 1$, the probability of photon gain ($n \geq 2$) is dominant. For high field measurement results $|\beta| \gg 1$, the teleportation process generates more photons in the output. Since β can be considered as a measurement of the coherent input field amplitude, this result is similar to the correlation of field measurement and photon number discussed in [23].

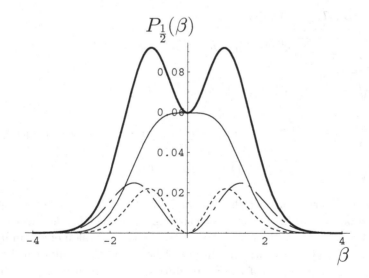

Fig. 5. The probability of n–photon counting $P_q(n)$ for one photon state teleportation in the case of $q = \frac{1}{2}$. The thick solid line shows $P_{\frac{1}{2}}(\beta)$, the thin solid line shows $P_{\frac{1}{2}}(1, \beta)$, the dotted line shows $P_{\frac{1}{2}}(0, \beta)$ and the dot–bar line shows $P_{\frac{1}{2}}(n \geq 2, \beta)$.

4. Application to a single–photon polarization

Since the polarization of the light field can be used to encode information using single photons, the case of a polarization sensitive CV teleportation is of considerable interest. The results obtained for the single–mode case can be applied to the teleportation of two polarization modes, H and V, by applying a separate transfer operator to each mode. Note that this does not imply that the teleportation H and V has to be conducted separately. For example, the two dimensional measurement amplitude (β_H, β_V) could also be obtained by measuring the circular polarization components $(\beta_H \pm i\beta_V)$. The CV teleportation of polarized photons therefore does not require any previous knowledge of the input polarization and the results obtained for successful transfers and for polarization flips can be applied directly to the teleportation of an unknown photon polarization.

The output state of a one photon state with polarization H can be written as a product of the one photon teleportation given in equation (7) and a vacuum teleportation, which is a special case of the conventional coherent state teleportation discussed in [6]. The result reads

$$\hat{T}_{Hq}(\beta_H)\hat{T}_{Vq}(\beta_V) \mid 1\rangle_H \mid 0\rangle_V$$
$$= \sqrt{\frac{1-q^2}{\pi}} \sum_{n=0}^{\infty} q^n \hat{D}_H(\beta_H) \mid n\rangle_{HH}\langle n \mid \hat{D}_H(-\beta_H) \mid 1\rangle_H$$
$$\otimes \sqrt{\frac{1-q^2}{\pi}} \sum_{m=0}^{\infty} q^m \hat{D}_V(\beta_V) \mid m\rangle_{VV}\langle m \mid \hat{D}_V(-\beta_V) \mid 0\rangle_V$$
$$= \sqrt{\frac{1-q^2}{\pi}} e^{-(1-q^2)\frac{|\beta_H|^2}{2}} \hat{D}((1-q)\beta_H)\left((1-q^2)\beta_H^* \mid 0\rangle_H + q \mid 1\rangle_H\right)$$
$$\otimes \sqrt{\frac{1-q^2}{\pi}} e^{-(1-q^2)\frac{|\beta_V|^2}{2}} \hat{D}((1-q)\beta_V) \mid 0\rangle_V. \tag{16}$$

This result describes all details of a single-photon teleportation, including the information obtained from the measurement result (β_H, β_V). In the following, however, it will be concentrated on the transmission errors induced by the teleportation. As will be discussed below, these results can be expressed entirely in terms of the probabilities $P_q(n)$ given by equation (9) and the well known coherent state fidelity of $(1 + q)/2$.

The fidelity of the teleportation is the total chance of successfully transmitting a photon with the correct polarization P_{trans}. It is equal to the product of the probabilities for successfully teleporting a single photon, $P_q(1)$, and the probability of successfully teleporting the vacuum, $(1+q)/2$, as given by

$$P_{\text{trans}}(q) = \underbrace{\int d^2\beta_H |_H\langle 1 | \hat{T}_{Hq}(\beta_H) | 1\rangle_H|^2}_{= P_q(1)} \underbrace{\int d^2\beta_V |_V\langle 0 | \hat{T}_{Vq}(\beta_V) | 0\rangle_V|^2}_{= \frac{1+q}{2}}$$

$$= \left(\frac{1+q}{2}\right)^2 \frac{1+q^2}{2}. \tag{17}$$

The total chance of a polarization flip while preserving the total photon number of one, P_{flip}, is equal to the product of the probabilities for a vacuum output in a one photon teleportation, $P_q(0)$, and the reverse situation where the vacuum input produces a one photon output. Since $\hat{T}_q(\beta)$ is hermitian, however, these two probabilities are equal and the result reads

$$P_{\text{flip}}(q) = \underbrace{\int d^2\beta_H |_H\langle 0 | \hat{T}_{Hq}(\beta_H) | 1\rangle_H|^2}_{= P_q(0)} \underbrace{\int d^2\beta_V |_V\langle 1 | \hat{T}_{Vq}(\beta_V) | 0\rangle_V|^2}_{= P_q(0)}$$

$$= \left(\frac{1+q}{2}\right)^2 \left(\frac{1-q}{2}\right)^2. \tag{18}$$

Finally, there are also probabilities for changes in the total photon number. The chance of obtaining no photon, P_{zero}, is equal to the product of the probabilities for a vacuum output in a one photon teleportation, $P_q(0)$, and for the successful teleportation of the vacuum given by the coherent state fidelity $(1+q)/2$,

$$P_{\text{zero}}(q) = \underbrace{\int d^2\beta_H |_H\langle 0 | \hat{T}_{Hq}(\beta_H) | 1\rangle_H|^2}_{= P_q(0)} \underbrace{\int d^2\beta_V |_V\langle 0 | \hat{T}_{Vq}(\beta_V) | 0\rangle_V|^2}_{= \frac{1+q}{2}}$$

$$= \left(\frac{1+q}{2}\right)^2 \frac{1-q}{2}. \tag{19}$$

The total chance of obtaining more than one photon in the output can then be obtained by

$$P_{n\geq 2}(q) = 1 - P_{\text{flip}}(q) - P_{\text{trans}}(q) - P_{\text{zero}}(q)$$

$$= 1 - \left(\frac{1+q}{2}\right)^2 \frac{5 - 4q + 3q^2}{4}. \tag{20}$$

Figure 6 shows the q dependence of the above probabilities. The fidelity P_{trans}, increases with increasing entanglement q, while the probabilities of the various error sources decrease. P_{trans} exceeds $1/2$ around $q \sim 0.7$ and $2/3$ around $q \sim 0.8$, illustrating that the entanglement requirements for high fidelity single photon transfers could be fulfilled using the best squeezing sources presently available. The dominant source of error is the chance of generating additional photons, $P_{n\geq 2}$, while the probability of flipping a polarization, P_{flip}, is always significantly lower than all the other probabilities. Therefore, the photon loss and gain processes are a more serious problem than the flip of a polarization for the transmission of the qubit. It is interesting to compare this new type of error with the post–selection problems inherent in the previously realized teleportation by entangled photon pairs[7] using coincidence counting as a trigger.

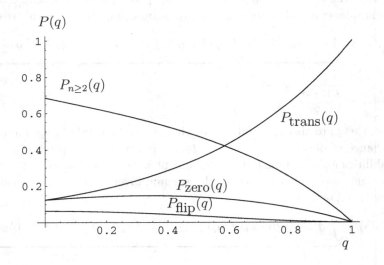

Fig. 6. The q dependence of the probabilities of the polarized photon teleportation.

In particular, the fidelity of CV teleportation can also be increased by post–selecting the one photon outputs, eliminating the multi photon and photon loss errors. The conditional fidelity is now given by $F_{\text{cond}} = P_{\text{trans}}/(P_{\text{trans}} + P_{\text{flip}})$. This fidelity is already 2/3 at $q = 0$ and reaches 10/11 at $q = 1/2$.

While such post selection may be difficult to realize due to the limited quantum efficiency and saturation characteristics of convensional photon-detectors, it illustrates now well the polarization property itself is preserved in the CV teleportation.

5. Transfer of polarization by two–mode CV teleportation

Conventional CV teleportation transfers only a single mode with a well defined polarization. In order to transfer the polarization of a single photon, it is therefore necessary to teleport two modes in parallel. Since CV teleportation preserves the coherence of the modes, it is not important which pair of orthogonal polarization modes is selected, as long as the four mode entangled state used in the teleportation is unpolarized.[20,24]

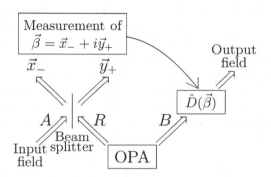

Fig. 7. Schematic representation of the two–mode quantum teleportation setup. The entangled state is generated by four–mode squeezing in an optical parametric amplifier (OPA). Four separate homodyne detection measurements are used to obtain the polarization components of the complex displacement amplitudes $\vec{\beta} = (\beta_H, \beta_V)$, and the corresponding complex two mode displacement amplitude is added to the output field B.

Figure 7 illustrates the extension of CV teleportation to the two mode system using the Jones vectors of the fields to express the polarization. The teleportation setup uses the four mode squeezed state generated by an optical parametric amplifier (OPA) to generate entanglement between two polarization modes in beam R and two polarization modes in beam B. In the photon number basis of the horizontally (H) and vertically (V) polarized modes, this four–mode squeezed entangled state can be written as

$$\mid \text{EPR}(q)\rangle_{R,B} = (1-q^2) \sum_{N=0}^{\infty} q^N \sum_{n=0}^{N} \mid n; N-n\rangle_R \mid n; N-n\rangle_B. \quad (21)$$

The amount of squeezing is given by q ($0 \leq q < 1$), which is related to the logarithmic attenuation of the field r by $q = \tanh r$. Since the amount of squeezing is the same for all polarizations, the entangled state is completely unpolarized and the entanglement between the polarizations of R and B in each N–photon subspace is maximal.

The teleportation is then performed by mixing the two mode input in beam A with the reference R at a 50/50 beam splitter and measuring the quadrature components $\vec{x}_-$ of the difference and $\vec{y}_+$ of the sum. In practice, this requires four separate homodyne detection measurements, to obtain the real and imaginary parts of the two dimensional Jones vector, $\vec{\beta} = \vec{x}_- + i\vec{y}_+$. However, the choice of polarizations for the Jones vector measurement has no effect on the teleportation itself since the Bell measurement performed on R and A simply projects the two beams onto a maximally entangled state displaced by a coherent amplitude of $\vec{\beta}$.[6] Due to the entanglement between R and B, the field value $\vec{\beta}$ corresponds to a field difference between the unknown input A and the output beam B. This difference can be corrected by a two mode field displacement $\hat{D}_{\text{pol}}(\vec{\beta})$. The complete teleportation process can then be summarized by a two mode transfer operator, which is a straightforward extension of the single–mode case derived in [6]. This operator reads

$$\hat{T}_{\text{pol}}(\vec{\beta}) = \frac{1-q^2}{\pi} \sum_{N=0}^{\infty} \sum_{n=0}^{N} q^N \hat{D}_{\text{pol}}(\vec{\beta}) \hat{\Pi}_N \hat{D}_{\text{pol}}(-\vec{\beta}), \qquad (22)$$

where the operator $\hat{\Pi}_N$ is the projection operator of the N photon subspace,

$$\hat{\Pi}_N = \sum_{n=0}^{N} |n, N-n\rangle\langle n, N-n|. \qquad (23)$$

It should be noted that the choice of polarization modes in the preparation of the squeezed state and in the homodyne measurements have no effect on the transfer process itself. The transfer operator is completely defined by the Jones vector $\vec{\beta}$ obtained in the homodyne detections.

As explained in [6], the transfer operator determines both the probabilities of measuring $\vec{\beta}$ and the output state of the teleportation. The output density matrix is a mixture over all possible measurement outcomes given by

$$\hat{\rho}_{\text{out}} = \int d^4\vec{\beta}\, \hat{T}_{\text{pol}}(\vec{\beta}) |\psi_{\text{in}}\rangle\langle\psi_{\text{in}}| \hat{T}_{\text{pol}}^\dagger(\vec{\beta}). \qquad (24)$$

In this integral, the Jones vector $\vec{\beta}$ is averaged over all possible polarizations, so the polarization of the output depends only on the input polarization.

Since it will be relevant in the following analysis, it may be useful to consider the case of a vacuum input. In this case, the teleportation simply adds Gaussian noise to the field, resulting in a thermal state output given by

$$\hat{R}_{\text{vac}} = \int d^4\vec{\beta}\, \hat{T}_{\text{pol}}(\vec{\beta})\, |\,0;0\rangle\langle 0;0\,|\, \hat{T}^\dagger_{\text{pol}}(\vec{\beta})$$
$$= \left(\frac{1+q}{2}\right)^2 \sum_N \left(\frac{1-q}{2}\right)^N \hat{\Pi}_N. \tag{25}$$

In the wave picture, the teleportation error can be interpreted as Gaussian field noise with a variance of V_q equal to the average photon number added to each mode.[3] According to the thermal state given by (25), this error is related to the squeezing parameter q by

$$V_q = \frac{1-q}{1+q}. \tag{26}$$

The value of the teleportation error V_q lies between zero for error free teleportation with infinitely squeezed light, and one for the limit of classical teleportation using a pair of vacuum states instead of entanglement. In the particle picture, V_q is the average number of photons added per mode, so two–mode teleportation adds a total average of $2V_q$ photons in the output field.

Since the vacuum state input is completely unpolarized, the output state has no polarization either. However, the teleportation process will transfer the polarization of the input state to the output. In the next section, it will show how the transfer operator can be used to describe this transfer of polarization in the case of a single–photon polarization qubit.

6. CV teleportation of a single–photon polarization qubit

It is now considered for the case of a single–photon input of unknown polarization. Such input states can be described by a superposition of two basis states, $|\,H\rangle = \hat{a}^\dagger_H\,|\,0;0\rangle$ and $|\,V\rangle = \hat{a}^\dagger_V\,|\,0;0\rangle$. Alternatively, the unknown quantum information can be described by a creation operator $\hat{a}^\dagger_{\text{in}}$, so that the input state is given by

$$|\,\psi_{\text{in}}\rangle = \hat{a}^\dagger_{\text{in}}\,|\,0;0\rangle$$
$$\text{where} \quad \hat{a}_{\text{in}} = c^*_H \hat{a}_H + c^*_V \hat{a}_V. \tag{27}$$

By using the basis independent properties of the operator $\hat{a}_{\text{in}}$, it can keep track of the quantum information in the single–photon input as it is transferred to the multi photon output components.

According to the transfer operator formalism, the output density matrix for the single-photon qubit teleportation is given by

$$\hat{\rho}_{\text{out}} = \int d^4\vec{\beta}\, \hat{T}_{\text{pol}}(\vec{\beta})\, \hat{a}_{\text{in}}^\dagger \mid 0;0\rangle\langle 0;0 \mid \hat{a}_{\text{in}} \hat{T}_{\text{pol}}^\dagger(\vec{\beta}). \tag{28}$$

To solve this integral, it needs to consider the effects of the transfer operator on the unknown operator $\hat{a}_{\text{in}}$, which defines the polarization of the input qubit. For this purpose, it is convenient to define the component β_{in} of the Jones vector $\vec{\beta}$ with the same polarization as the unknown input,

$$\beta_{\text{in}} = c_H^* \beta_H + c_V^* \beta_V. \tag{29}$$

It is then possible to commute the transfer operators $\hat{T}_{\text{pol}}(\vec{\beta})$ and the operators of the unknown input polarization, $\hat{a}_{\text{in}}$ and $\hat{a}_{\text{in}}^\dagger$, using the relations

$$\hat{T}_{\text{pol}}(\vec{\beta})\hat{a}_{\text{in}}^\dagger = (q\hat{a}_{\text{in}}^\dagger + (1-q)\beta_{\text{in}}^*)\hat{T}_{\text{pol}}(\vec{\beta})$$
$$\hat{a}_{\text{in}}\hat{T}_{\text{pol}}(\vec{\beta}) = \hat{T}_{\text{pol}}(\vec{\beta})(q\hat{a}_{\text{in}} + (1-q)\beta_{\text{in}}). \tag{30}$$

The output density matrix can then be written as

$$\hat{\rho}_{\text{out}} = \int d^4\vec{\beta}(q^2 \hat{a}_{\text{in}}^\dagger \hat{r}_{\text{vac}} \hat{a}_{\text{in}} + (1-q)q(\hat{a}_{\text{in}}^\dagger \beta_{\text{in}} \hat{r}_{\text{vac}} + \hat{r}_{\text{vac}} \beta_{\text{in}}^* \hat{a}_{\text{in}}) $$
$$+ (1-q)^2 \beta_{\text{in}} \hat{r}_{\text{vac}} \beta_{\text{in}}^*), \tag{31}$$

where $\hat{r}_{\text{vac}}$ is an abbreviation for the operator obtained by applying the transfer operator to the vacuum density matrix,

$$\hat{r}_{\text{vac}}(\vec{\beta}) = \hat{T}_{\text{pol}}(\vec{\beta}) \mid 0;0\rangle\langle 0;0 \mid \hat{T}_{\text{pol}}^\dagger(\vec{\beta}). \tag{32}$$

Since β_{in} is unknown, the integral in Eq. (31) still depends on the coefficients c_H and c_V defining the direction of the unknown input polarization. To obtain an integral that is independent of the unknown polarization, it needs to convert these values back into operators independent of the measurement outcome $\vec{\beta}$. This transformation can be accomplished by making use of the fact that the vacuum teleportation $\hat{T}_{\text{pol}}(\vec{\beta}) \mid 0;0\rangle$ results in a coherent state with an amplitude of $(1-q)\vec{\beta}$.[6] Therefore, this state is a right eigenstate of $\hat{a}_{\text{in}}$, and it can transform β_{in} into $\hat{a}_{\text{in}}$ using

$$\hat{a}_{\text{in}} \hat{r}_{\text{vac}} = (1-q)\beta_{\text{in}} \hat{r}_{\text{vac}}$$
$$\hat{r}_{\text{vac}} \hat{a}_{\text{in}}^\dagger = (1-q)\beta_{\text{in}}^* \hat{r}_{\text{vac}}. \tag{33}$$

It is thus possible to convert all factors of β_{in} in Eq. (31), resulting in an integral where only $\hat{r}_{\text{vac}}$ depends on $\vec{\beta}$,

$$\hat{\rho}_{out} = \int d^4\vec{\beta} \left(q^2 \hat{a}_{\text{in}}^\dagger \hat{r}_{\text{vac}} \hat{a}_{\text{in}} + q \left(\hat{a}_{\text{in}}^\dagger \hat{a}_{\text{in}} \hat{r}_{\text{vac}} + \hat{r}_{\text{vac}} \hat{a}_{\text{in}}^\dagger \hat{a}_{\text{in}} \right) \right.$$
$$\left. + \hat{a}_{\text{in}} \hat{r}_{\text{vac}} \hat{a}_{\text{in}}^\dagger \right) \tag{34}$$

It can now perform the integration of $\hat{r}_{\text{vac}}$, the result of which is equal to the output state of vacuum teleportation $\hat{R}_{\text{vac}}$ given in Eq. (25). The output of single–photon teleportation can therefore be expressed in terms of applications of the input operator $\hat{a}_{\text{in}}$ to the unpolarized thermal state $\hat{R}_{\text{vac}}$,

$$\hat{\rho}_{out} = q^2 \hat{a}_{\text{in}}^\dagger \hat{R}_{\text{vac}} \hat{a}_{\text{in}} + q \left(\hat{R}_{\text{vac}} \hat{a}_{\text{in}}^\dagger \hat{a}_{\text{in}} + \hat{a}_{\text{in}}^\dagger \hat{a}_{\text{in}} \hat{R}_{\text{vac}} \right) + \hat{a}_{\text{in}} \hat{R}_{\text{vac}} \hat{a}_{\text{in}}^\dagger. \tag{35}$$

The result can be simplified by considering the commutation relations of the thermal state and arbitrary creation and annihilation operators, namely

$$\hat{a}_{\text{in}} \hat{R}_{\text{vac}} = \frac{1-q}{2} \hat{R}_{\text{vac}} \hat{a}_{\text{in}}$$
$$\hat{R}_{\text{vac}} \hat{a}_{\text{in}}^\dagger = \frac{1-q}{2} \hat{a}_{\text{in}}^\dagger \hat{R}_{\text{vac}}. \tag{36}$$

It is then possible to rearrange the operator ordering in Eq. (35) so that the operator $\hat{a}_{\text{in}}$ acts only as a creation operator on the thermal output $\hat{R}_{\text{vac}}$. In this simplified form, the output reads

$$\hat{\rho}_{out} = \left(\frac{1+q}{2}\right)^2 \underbrace{\hat{a}_{\text{in}}^\dagger \hat{R}_{\text{vac}} \hat{a}_{\text{in}}}_{\text{photon added state}} + \left(\frac{1-q}{2}\right) \underbrace{\hat{R}_{\text{vac}}}_{\text{white noise}}. \tag{37}$$

As indicated, it is now possible to interpret the output as a mixture of a photon added state polarized by the application of the creation operator $\hat{a}_{\text{in}}^\dagger$ to $\hat{R}_{\text{vac}}$, and a completely unpolarized white noise component represented by the thermal state $\hat{R}_{\text{vac}}$ itself. It may be interesting to note that the photon added state is a two mode version of the single–photon–added thermal state investigated in recent experiments[25,26] because of its non–classical features such as the negativity of the Wigner function.[27] The photon added term in Eq. (37) thus describes the teleportation of non–classical features in the field quadrature statistics of the single–photon state. For the purpose of photon cloning however, only the photon number distributions are relevant. In that context, it is significant that the application of a creation operator also describes the effects of photon bunching and of stimulated emission used in previous photon cloning experiments.[8-10] As it shall show in the following, the photon added state is indeed equivalent to a mixture of optimally cloned N-photon outputs. It is therefore possible to interpret the photon added state as optimal cloning and the thermal white noise background as a measure of the non–optimal nature of accidental cloning.

More specifically, Eq. (37) provides a simple quantification of the unpolarized white noise background added in the teleportation in terms of the statistical weights of the two components. Since $\hat{R}_{\text{vac}}$ is normalized, the statistical weights are $(1-q)/2$ for the white noise and $(1+q)/2$ for the photon added state representing optimal cloning. The ratio of the two components is thus exactly equal to the Gaussian field error V_q derived in sec. 5. Since V_q is a very intuitive measure of the teleportation error in terms of field noise, it may be interesting to see how this continuous variable noise measure is related to the cloning errors in the discrete photon number statistics.

7. Cloning fidelity of the N-photon output

In the previous section, it has been derived for the complete output state of the two mode teleportation of a single–photon polarization qubit. The total output photon number of this state is random, so it is not possible to predict how many clones (if any) are generated. Since the optimal cloning fidelity is a function of the number of clones produced, it is necessary to separate the output density matrix of Eq. (37) into its N-photon output components, representing the accidental occurrences of $1 \to N$ cloning events,

$$\hat{\rho}_{\text{out}} = \sum_{N=0}^{\infty} P(N) \hat{\rho}_N. \qquad (38)$$

The decomposition of the unpolarized thermal states $\hat{R}_{\text{vac}}$ in Eq. (37) is given by Eq. (25). In the photon added component that represents optimal cloning, the number of photons is raised by one due to the application of the creation operator $\hat{a}_{\text{in}}^\dagger$. Therefore, the decomposition of the output density matrix into N-photon subspaces includes a white noise term given by $\hat{\Pi}_N$ and an optimal cloning term given by $\hat{a}_{\text{in}}^\dagger \hat{\Pi}_{N-1} \hat{a}_{\text{in}}$. Since the optimal cloning part has no zero–photon component, $\hat{\Pi}_{-1}$ should be defined as zero. The decomposition of the output density matrix into N-photon subspaces then reads

$$\hat{\rho}_{\text{out}} = \frac{(1+q)^2}{2(1-q)} \sum_{N=0}^{\infty} \left(\frac{1-q}{2}\right)^N \left(\left(\frac{1+q}{2}\right)^2 \hat{a}_{\text{in}}^\dagger \hat{\Pi}_{N-1} \hat{a}_{\text{in}} + \left(\frac{1-q}{2}\right)^2 \hat{\Pi}_N\right)$$

$$= \frac{1}{V_q(1+V_q)^3} \sum_{N=0}^{\infty} \left(\frac{V_q}{1+V_q}\right)^N \left(\hat{a}_{\text{in}}^\dagger \hat{\Pi}_{N-1} \hat{a}_{\text{in}} + V_q^2 \hat{\Pi}_N\right). \quad (39)$$

Here, the entanglement parameter q has been converted into the more intuitive measure of Gaussian field error V_q. It is then possible to relate the photon number distribution and the cloning errors directly to the Gaussian noise error of the CV teleportation.

The statistical weights of the contributions in Eq. (39) are determined by the traces of the operators. Specifically, the trace of the white noise term $\hat{\Pi}_N$ is $N+1$ and the trace of the optimal cloning term $\hat{a}_{\text{in}}^\dagger \hat{\Pi}_{N-1} \hat{a}_{\text{in}}$ is $(N+1)N/2$. The probability $P(N)$ of obtaining an N–photon output is given by the product trace of the total density matrix in the N–photon subspace. In terms of the Gaussian field noise V_q, this photon number distribution reads

$$P(N) = \text{Tr}\{\hat{\Pi}_N \hat{\rho}_{\text{out}}\}$$

$$= \frac{(N+1)(N+2V_q^2)}{2V_q(1+V_q)^3} \left(\frac{V_q}{1+V_q}\right)^N. \quad (40)$$

$P(0)$ is the probability of losing the photon, $P(1)$ is the probability of single–photon teleportation, and $P(N \geq 2)$ are the probabilities of accidental $1 \to N$ cloning. Since the generation of additional photons is itself a kind of teleportation error, the probability of generating accidental clones increases with V_q. Specifically, the average photon number in the output is $1 + 2V_q$ (the original photon plus twice the average photon number added per mode). Even at $V_q = 1$, the average photon number is only three and the probability of obtaining a high number of output photons drops with $(1/2)^N$. Accidental cloning probabilities are therefore generally low for high numbers of clones. However, the probabilities of obtaining two, three or four clones can be quite significant, as shown in Fig. 8. Specifically, the probabilities of obtaining N clones at an experimentally feasible error of $V_q = 0.25$ requiring about 6 dB squeezing are $P(2) = 26.1\%$, $P(3) = 10.2\%$, and $P(4) = 3.4\%$. The accidental generation of clones should therefore be a very common occurrence if qubits are teleported using squeezed state entanglement at presently available levels of squeezing.

It can now evaluate the quality of accidental $1 \to N$ cloning by separating the N–photon output $\hat{\rho}_N$ into its optimal cloning component $\hat{C}_N$ and

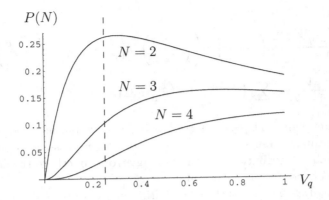

Fig. 8. Cloning probabilities $P(N)$ for different output photon numbers N plotted to V_q. From top to bottom, the curves show $P(2), P(3), P(4)$. The dashed vertical line indicates $V_q = 0.25$.

its white noise component $\hat{W}_N$,

$$\hat{\rho}_N = \eta_N \hat{C}_N + (1 - \eta_N)\hat{W}_N. \qquad (41)$$

The density matrices $\hat{C}_N$ and $\hat{W}_N$ are independent of the teleportation errors V_q and can be generated from the projection operators $\hat{\Pi}_N$ into the N–photon subspaces by

$$\hat{C}_N = \frac{2}{N(N+1)} \hat{a}_{\text{in}}^\dagger \hat{\Pi}_{N-1} \hat{a}_{\text{in}}$$

$$\hat{W}_N = \frac{1}{N+1} \hat{\Pi}_N. \qquad (42)$$

Therefore, the N–photon output is fully characterized by the single parameter η_N. Since this parameter defines the fraction of optimally cloned N–photon outputs, it will be called as the cloning efficiency of accidental $1 \to N$ cloning. The cloning efficiency η_N is a function of teleportation error V_q and photon number N. Using Eq. (37), it finds that this ratio is

$$\eta_N = \frac{1}{1 + 2V_q^2/N}. \qquad (43)$$

Figure 9 illustrates this dependence of cloning efficiency η_N on the teleportation error V_q for several output photon numbers N. Not surprisingly, the teleportation error V_q reduces the cloning efficiency. However, it is interesting to note that only the square of V_q enters into the relation, indicating that the cloning efficiency rapidly approaches one for low teleportation errors. For example, an experimentally feasible error of $V_q = 0.25$ requiring about 6 dB of squeezing already gives a two photon cloning efficiency of $\eta_2 = 16/17$, or about 94 %. It can be therefore conclude that the accidental cloning observed in CV teleportation at presently available levels of squeezed state entanglement will be nearly optimal. Another significant feature of accidental cloning is that the minimal cloning efficiencies obtained at the classical teleportation limit ($V_q = 1$) have a photon number dependent value of $N/(N+2)$. Thus, the cloning efficiencies for high N are always close to one, indicating that the generation of a large number of clones is quite robust against teleportation errors.

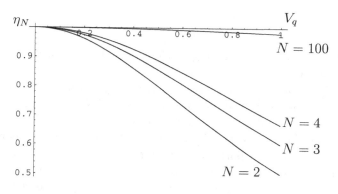

Fig. 9. Cloning efficiency η_N for different output photon numbers N plotted to V_q. From bottom to top, the curves show η_2, η_3, η_4 and η_{100}.

Up to now, the discussion was based only on the formal decomposition of the output into an optimal cloning component and a white noise term. Experimentally, this decomposition is not directly observable. Instead, cloning is characterized by the cloning fidelity F_N, defined as the fraction of output photons with the same polarization as the input photon, $\langle \hat{a}_{in}^\dagger \hat{a}_{in} \rangle / N$. This fidelity is usually measured by splitting up the output light into a sufficiently large number of channels, so that the polarization of each photon can be detected separately.[8,10] It is then possible to evaluate cloning by conventional photon counting.

To determine the fidelity F_N of a $1 \to N$ cloning process with cloning efficiency η_N, it can be used for the fidelities of the optimal cloning component $\hat{C}_N$ and the white noise components $\hat{W}_N$. Since the white noise component is completely unpolarized, exactly half of the photons will have the same polarization as the input, corresponding to a fidelity of 1/2. For the optimal cloning term, the cloning fidelity is given by

$$F_{\text{opt.}} = \frac{1}{N}\text{Tr}\left\{\hat{a}^{\dagger}_{\text{in}}\hat{a}_{\text{in}}\hat{C}_N\right\} = \frac{2N+1}{3N}. \tag{44}$$

This result is equal to the $1 \to N$ cloning fidelity of an optimal cloning machine,[22] proving the conjecture that $\hat{C}_N$ represents optimal cloning. The accidental cloning fidelity F_N of the mixture of $\hat{C}_N$ and $\hat{W}_N$ defined by the cloning efficiency η_N can now be obtained by taking the weighted average of the white noise fidelity of 1/2 and the optimal cloning fidelity given by Eq. (44),

$$\begin{aligned}F_N &= \eta_N \frac{2N+1}{3N} + (1-\eta_N)\frac{1}{2} \\ &= \frac{2}{3} + \frac{1-V_q^2}{3(N+2V_q^2)}.\end{aligned} \tag{45}$$

As this relation shows, the cloning efficiencies at $V_q = 1$ all correspond to a cloning fidelity of 2/3, which is the optimal cloning fidelity for $N \to \infty$. Moreover, 2/3 is also the optimal cloning fidelity obtained if the cloning is performed using only local measurements and classical communications instead of a direct quantum mechanical interaction between the original and the clones. At $V_q = 1$, this is exactly what happens, since there is no entanglement and the teleportation is performed by a local measurement of the input photon and the generation of a coherent state of the appropriate amplitude in the output. It is therefore obvious that $V_q = 1$ is still close to the optimal limit for high N. On the other hand, cloning fidelities above 2/3 are only possible because of the entanglement used in the teleportation.

Figure 10 shows the cloning fidelity F_N for two, three, and five output photons. In addition, the high N limit is indicated by the result for $N = 100$, which is basically indistinguishable from a flat line at $F_\infty = 2/3$. The dotted line indicates a realistic teleportation error of $V_q = 0.25$, achievable by about 6 dB of squeezing. The cloning fidelities at this noise level are already quite close to the optimal fidelities. Specifically, the fidelities at $V_q = 0.25$ ($V_q = 0$) are $F_2 = 0.814$ ($F_2 = 0.833$), $F_3 = 0.767$ ($F_3 = 0.778$), $F_4 = 0.742$ ($F_4 = 0.75$). It should therefore be possible to observe nearly optimal accidental quantum cloning in the CV teleportation of single–photon qubits under presently realizable experimental conditions.

In order to put the above results into a wider context, it may be interesting to recall that the cloning effect has been obtained without any field amplification. In fact, a minimal noise amplification can produce optimal clones, and this situation can be realized by adjusting the gain of CV

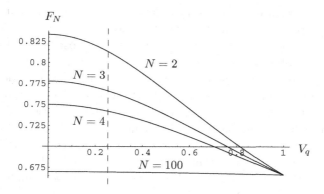

Fig. 10. Cloning fidelities F_N for different output photon numbers N plotted to V_q. From top to bottom, the curves show F_2, F_3, F_4 and F_{100}. The dashed vertical line indicates $V_q = 0.25$.

teleportation, as it has been already pointed out elsewhere.[28] However, accidental cloning occurs at a gain of one, due to the Gaussian field noise added in the teleportation. Since all linear optics processes should be equivalent, it can conjecture that the results describe the accidental cloning effects of any kind of Gaussian field noise added to the field of a single–photon qubit state. In particular, V_q can be modified to include a variety of Gaussian errors in addition to the limitation of squeezed state entanglement. In this case, V_q could even exceed one, indicating fidelities below 2/3. However, all successful CV teleportation experiments should reduce V_q well below one, corresponding to cloning fidelities close to optimal cloning.

8. Conclusions

The properties of CV teleportation of single–photon states have been investigated. The difference between the input state and the output state is due to the non–maximal entanglement in the EPR beams shared by Alice and Bob. The field measurement conditions the output state. Nearly zero values of the field measurement tend to preserve the initial one photon input state because the eigenstates of the field measurement are close to a photon number states . In the intermediate range of field measurement values, photon loss and photon gain processes occur during teleportation. At high values of the field measurement, the probability of photon gain is dominant, corresponding to the high amplitude observed. An application of this analysis to the teleportation of a polarized photon shows that the photon loss and gain processes are a more serious problem than the polarization flips. The results imply that unconditional CV teleportation of single–photon polarization could be considered an alternative to the post–selected scheme[7] using entangled photon pairs.

The results show that the accidental multiplication of a single–photon polarization qubit transferred by CV teleportation can be interpreted as a nearly optimal quantum cloning process. A CV teleportation system operating at feasible squeezing levels thus tends to act both as a fax and a copy machine on the teleported qubits. Even though the cloning happens as a consequence of the noise introduced by non–maximal entanglement, the effect is similar to intentional telecloning, where a special multi–photon entangled state needs to be prepared beforehand.[29] The analysis indicates that telecloning is a natural feature of the CV teleportation process when it is applied to single–photon qubits. This observation might be useful in the implementation of multi party protocols, where the distribution of quantum information to several parties is desirable.

It is also interesting to note that the clones are generated without field amplification, simply by the random addition of Gaussian field noise. It can therefore conjecture that accidental cloning is a general effect of Gaussian field noise on photonic qubits. Even though this kind of cloning without amplification can never be optimal, the additional error can be described by mixing the optimal cloning output with a completely unpolarized component of the density matrix. It has been thus successfully converted a Gaussian field error into an N–photon cloning error, with the cloning efficiency η_N describing the exact fraction of the optimal cloning state in the output. In the context of recent investigations into the quantum mechanics of the photon–field dualism,[12–16] this result may shed some light on the fundamental relations between photons and field noise.

In conclusion, the accidental cloning of photonic qubits in CV teleportation is a phenomenon that may be both useful in the development of new technologies for quantum information networks and in the exploration of the fundamental physics behind the dualism of photons and fields. Hopefully, the analysis presented above will be a fruitful contribution to both.

References

1. C.H. Bennett, G. Brassard, C. Crepeau, R.Jozsa, A. Peres and W.K. Wootters, *Phys. Rev. Lett.* **70** (1993) 1895.
2. L. Vaidman, *Phys. Rev. A* **49** (1994) 1473.
3. S. L. Braunstein and H. J. Kimble, *Phys. Rev. Lett.* **80** (1998) 869.

4. A. Furusawa, J. L. Sørensen, S. L. Braunstein, C. A. Fuchs, H. J. Kimble and E. J. Polzik, *Science* **282** (1998) 706.
5. S. J. van Enk, *Phys. Rev. A* **60** (1999) 5095.
6. H. F. Hofmann, T. Ide, T. Kobayashi and A. Furusawa, *Phys. Rev. A* **62** (2004) 062304.
7. D. Bouwmeester, J. W. Pan, K. Mattle, M. Eibl, H. Weinfurter and A. Zeilinger, *Nature* **390** (1997) 575.
8. A. Lamas–Linares, C. Simon, J. C. Howell and D. Bouwmeester, *Science* **296** (2002) 712.
9. F. De Martini, V. Bužek, F. Bovino and C. Sias, *Nature* **419** (2002) 815.
10. W.T.M. Irvine, A.L. Linares, M.J.A. de Dood and D. Bouwmeester, *Phys. Rev. Lett.* **92** (2004) 047902.
11. U. L. Andersen, V. Josse and G. Leuchs *Phys. Rev. Lett.* **94** (2005) 240503.
12. A. I. Lvovsky, H. Hansen, T. Aichele, O. Benson, J. Mlynek and S. Schiller, *Phys. Rev. Lett.* **87** (2001) 050402.
13. A. I. Lvovsky and J. Mlynek, *Phys. Rev. Lett.* **88** (2002) 250401.
14. T. Opatrny, G. Kurizki and D.–G. Welsch, *Phys. Rev. A* **61** (2000) 032302.
15. A. Zavatta, S. Viciani and M. Bellini, *Science* **306** (2004) 660.
16. A. Ourjoumtsev, R. Tualle–Brouri and P. Grangier, *Phys. Rev. Lett.* **96** (2006) 213601.
17. R. E. S. Polkinghorne and T. C. Ralph, *Phys. Rev. Lett.* **83** (1999) 2095.
18. T.C. Ralph, *Phys. Rev. A* **61** (2001) 044301.
19. T.C. Ralph, *Phys. Rev. A* **65** (2002) 012319.
20. T. Ide, H. F. Hofmann, T. Kobayashi and A. Furusawa, *Phys. Rev. A* **65** (2002) 012313.
21. W. K. Wooters and W. H. Zurek, *Nature* **299** (1996) 802.
22. N. Gisin and S. Massar, *Phys. Rev. Lett.* **79** (1997) 2153.
23. H. F. Hofmann, T. Kobayashi and A. Furusawa, *Phys. Rev. A* **62** (2000) 013806.
24. A. Dolinska, B.C. Buchler, W.P. Bowen, T.C. Ralph and P.K. Lam, *Phys. Rev. A* **68** (2003) 052308.
25. V. Parigi, A. Zavatta and M. Bellini, *Proc. of SPIE* **6305** (2006) 63050Z.
26. A. Zavatta, V. Parigi and M. Bellini, e–print arXiv:0704.0179 (2007).
27. G.S. Agarwal and K. Tara, *Phys. Rev. A* **46** (1992) 458.
28. H. F. Hofmann and T. Ide, *New J. Phys.* **8** (2006) 130.
29. M. Murao, D. Jonathan, M. B. Plenio and V. Vedral, *Phys. Rev. A* **59** (1999) 156.